AF829182

How to Be a Design Academic

How to Be a Design Academic

From Learning to Leading

Edited by

Alethea Blackler and Evonne Miller

CRC Press
Taylor & Francis Group
Boca Raton London New York

CRC Press is an imprint of the
Taylor & Francis Group, an **informa** business

First edition published 2021
by CRC Press
6000 Broken Sound Parkway NW, Suite 300, Boca Raton, FL 33487-2742

and by CRC Press
2 Park Square, Milton Park, Abingdon, Oxon, OX14 4RN

© 2021 Taylor & Francis Group, LLC

CRC Press is an imprint of Taylor & Francis Group, LLC

The right of Alethea Blackler and Evonne Miller to be identified as the authors of the editorial material, and of the authors for their individual chapters, has been asserted in accordance with sections 77 and 78 of the Copyright, Designs and Patents Act 1988.

Reasonable efforts have been made to publish reliable data and information, but the author and publisher cannot assume responsibility for the validity of all materials or the consequences of their use. The authors and publishers have attempted to trace the copyright holders of all material reproduced in this publication and apologize to copyright holders if permission to publish in this form has not been obtained. If any copyright material has not been acknowledged please write and let us know so we may rectify in any future reprint.
Except as permitted under U.S. Copyright Law, no part of this book may be reprinted, reproduced, transmitted, or utilized in any form by any electronic, mechanical, or other means, now known or hereafter invented, including photocopying, microfilming, and recording, or in any information storage or retrieval system, without written permission from the publishers.

For permission to photocopy or use material electronically from this work, access www.copyright.com or contact the Copyright Clearance Center, Inc. (CCC), 222 Rosewood Drive, Danvers, MA 01923, 978-750-8400. For works that are not available on CCC please contact mpkbookspermissions@tandf.co.uk

Trademark notice: Product or corporate names may be trademarks or registered trademarks and are used only for identification and explanation without intent to infringe.

Library of Congress Cataloging-in-Publication Data

Names: Blackler, Alethea, editor. | Miller, Evonne, editor.
Title: How to be a design academic : from learning to leading / edited by Alethea Blackler and Evonne Miller.
Description: Boca Raton : CRC Press, 2021. | Includes bibliographical references and index.
Identifiers: LCCN 2020044206 (print) | LCCN 2020044207 (ebook) | ISBN 9780367362904 (hardback) | ISBN 9780429351693 (ebook)
Subjects: LCSH: Design--Study and teaching (Higher) | College teachers--Vocational guidance.
Classification: LCC NK1170 .H69 2021 (print) | LCC NK1170 (ebook) | DDC 745.4023--dc23
LC record available at https://lccn.loc.gov/2020044206
LC ebook record available at https://lccn.loc.gov/2020044207

ISBN: 978-0-367-36290-4 (hbk)
ISBN: 978-0-429-35169-3 (ebk)

Typeset in Times
by Deanta Global Publishing Services, Chennai, India

This book was conceived and planned at the wonderful Binna Burra Mountain Lodge in Lamington National Park, Queensland, in February 2019. In September 2019, the Lodge and parts of the surrounding National Park were destroyed at the start of Australia's "Black Summer" of catastrophic bushfires. As we write this, the Lodge is beginning to re-open and we hope it will continue to refresh and inspire others in future.

The book is dedicated to the hundreds of people and billions of animals who lost their lives in the Black Summer, and the thousands who lost homes and livelihoods.

Contents

SECTION 3 Being an Effective Design Educator

PART 2 Becoming a Leader

SECTION 4 Leading Yourself

SECTION 5 Leading Others

Preface

This book caters to the growing number of new and experienced design academics, and aims to help them manage their careers and have a positive impact on the world through their work. Design is a growing area in academia, and is now being taken seriously as a research approach that can have impact both within its own research field and as a framework for interdisciplinary research. Academics who work in this space have a wider variety of backgrounds than many others, and this book is tailored to their unique needs in establishing and growing their careers.

From career planning to teaching and research leadership, this book provides insights into the day-to-day activities of contemporary design academics and the secrets of their success. The book reflects on how to build a career as a design academic and offers advice on how to manage the challenges, requirements and processes that come with the everyday and extraordinary parts of an academic role in design. This first book dedicated to being an academic in the design fields explores a range of relevant issues:

- Crafting a unique career path, whether you are from design practice or various other academic disciplines, wanting to build industry engagement or seeking to move into leadership
- Being a design academic, including planning and executing daily tasks, maintaining work–life balance and life-long learning
- Being a successful design educator, including positioning design education within the institution, designing curricula and building student engagement
- Being a successful design researcher, including finishing that PhD and then building your research field or group

Edited by two professors in design from very different backgrounds, and with contributions from a range of authors with different disciplinary approaches, priorities and career pathways, this book provides a comprehensive guide to being a successful design academic in today's challenging environment. This book is designed to equip the reader with a wealth of knowledge in this important area, along with an understanding of how it can be applied to their own careers.

Acknowledgements

This edited collection has been a collaborative effort between our colleagues, primarily from the School of Design at Queensland University of Technology in Brisbane, Australia. It arose from the QUT Design Lab "Book Project," which brought design academics together to collaborate on writing books in their areas of expertise. We are grateful to the QUT Design Lab and School of Design for their financial contributions and support. We want to particularly acknowledge our contributors and their unique insights, and their dedication to completing this project despite the challenges of the COVID-19 pandemic. We would also like to thank our peer reviewers who helped us to shape and improve the individual chapters. Finally, we also thank CRC Press Executive Editor, Cindy Carelli, and Senior Editorial Assistant, Erin Harris, for their invaluable advice and support throughout.

Editors

Alethea (Thea) Blackler (PhD) is Discipline Leader for Industrial Design, Interaction Design, Visual Communication and Fashion Design within the School of Design at QUT. Her principle area of research interest is intuitive interaction, in which she is one of the world leaders. She pioneered the first empirical work in the field, leading a federally funded Australian Research Council Discovery project on Facilitating Intuitive Interaction for Older People, as well as editing a journal issue and a book on intuitive interaction. She is also working in the areas of older people and technology, design for dementia and designing active play for young children. She has attracted and worked with external partners in government, community and industry on various projects and has had three Australian Research Council grants. She is a very experienced HDR supervisor, with 8 current PhD students and 15 PhD completions. She has published more than 80 peer-reviewed papers, been invited to give presentations internationally and is the recipient of several awards, including a 2018 Ig-Nobel.

Evonne Miller is Professor of Design Psychology and Director of the QUT Design Lab in the School of Design, at Queensland University of Technology. Evonne has secured over $3,500,000 in competitive grant funding and over 100 peer-reviewed publications in the fields of urban design, population ageing, health and wellbeing, climate change and sustainability, disaster and resilience. Her research focuses on how to design environments – built, technical, socio-cultural and natural – that better engage and support all users, especially older people in residential aged care. Her 2020 book, *Creating Great Places: Evidence-based urban design for health and wellbeing* (Routledge, with A/Prof Deb Cushing) outlines how designers can create great places that are inclusive, sustainable and salutogenic. Evonne has supervised over 20 PhD students, and has a special interest in mentoring early career researchers.

List of Contributors

Shayne Beaver
QUT School of Design
Queensland University of Technology
Brisbane, Australia

Alethea Blackler
QUT Design Lab
Queensland University of Technology
Brisbane, Australia

Dean Brough
QUT Design Lab
Queensland University of Technology
Brisbane, Australia

Glenda Caldwell
QUT Design Lab
Queensland University of Technology
Brisbane, Australia

Marianella Chamorro-Koc
QUT Design Lab
Queensland University of Technology
Brisbane, Australia

Richard Evans
QUT Design Lab
Queensland University of Technology
Brisbane, Australia

Marcus Foth
QUT Design Lab
Queensland University of Technology
Brisbane, Australia

Brett Fyfield
QUT School of Design
Queensland University of Technology
Brisbane, Australia

Rafael Gomez
QUT Design Lab
Queensland University of Technology
Brisbane, Australia

Nick Kelly
QUT Design Lab
Queensland University of Technology
Brisbane, Australia

Jeremy Kerr
QUT Design Lab
Queensland University of Technology
Brisbane, Australia

Tim Lindgren
QUT Design Lab
Queensland University of Technology
Brisbane, Australia

Heather McKinnon
QUT Design Lab
Queensland University of Technology
Brisbane, Australia

Deanna Meth
Queensland University of Technology
Brisbane, Australia

Evonne Miller
QUT Design Lab
Queensland University of Technology
Brisbane, Australia

Gjoko Muratovski
Myron E. Ullman Jr. School of Design
University of Cincinnati
Cincinnati, Ohio

Markus Rittenbruch
QUT Design Lab
Queensland University of Technology
Brisbane, Australia

Leo Rezayan
QUT School of Design
Queensland University of Technology
Brisbane, Australia

Shannon Satherley
QUT School of Design
Queensland University of Technology
Brisbane, Australia

Lisa Scharoun
QUT Design Lab
Queensland University of Technology
Brisbane, Australia

Andrew Scott
QUT School of Design
Queensland University of Technology
Brisbane, Australia

Levi Swann
QUT Design Lab
Queensland University of Technology
Brisbane, Australia

Jennifer Seevinck
QUT Design Lab
Queensland University of Technology
Brisbane, Australia

Sheona Thomson
QUT School of Design
Queensland University of Technology
Brisbane, Australia

Jane Turner
QUT School of Design
Queensland University of Technology
Brisbane, Australia

Anastasia Tyurina
QUT Design Lab
Queensland University of Technology
Brisbane, Australia

Tim Williams
QUT School of Design
Queensland University of Technology
Brisbane, Australia

Oksana Zelenko
QUT Design Lab
Queensland University of Technology
Brisbane, Australia

1 Being a Design Academic
From Beginning to Winning

Alethea Blackler and Evonne Miller

CONTENTS

1.1 WHY DO YOU NEED THIS BOOK?

This edited book offers an in-depth look into how design academics, from early career academics to full professors, navigate the changing and often turbulent waters of academia with confidence and impact. Being an academic, we believe, is one of the greatest jobs there is: as well as educating, influencing and mentoring undergraduate and postgraduate students through our teaching, we get to conduct innovative research that may have a positive impact on the world. Unfortunately, all too often, academics feel overburdened, tired, isolated and excessively busy, struggling to meet escalating performance metrics and be good teachers, productive researchers and efficient administrators (Clark and Sousa, 2018). There is often very little time to discuss, share and reflect on the strategies, ideas and initiatives that might enhance work productivity, reduce the likelihood of burnout and foster work–life balance, with only the lucky few having wise mentors to advise them.

Our vision for this book is that it fills this void, playing the role of a generous mentor and lifting back the veil to show, in detail, the day-to-day work lives, choices, challenges and achievements of design academics. By openly sharing their unique career paths, both conventional and unconventional, each chapter author provides unprecedented insight into their personal and professional experiences in academia. To date, despite the growing number of books exploring different aspects of academia, from how to complete a PhD (Mewburn, 2013), be a productive and "happy academic" (Clark and Sousa, 2018), thrive as a female professor (Murray and Mifsud, 2019) or negotiate the transition to retirement (Burge, 2015), to our knowledge, there are no books targeted specifically at *design* academics.

Design academics have a unique experience of academia for several reasons. Firstly, unlike most other disciplines, until very recently, it was rare for a design academic to hold a doctorate. A vocational field where professional practice is highly valued, design had a "history of a limited research culture, professional practice rather than research and few staff who are qualified at the doctoral level" (Durling, 2002, p. 80). Even in our own School of Design, where research is highly prized, approximately a third do not currently hold PhDs (ten years ago, that figure would have been 50%), and are completing their PhD while working as an academic. This is a uniquely challenging experience that Williams and Satherley explore in Chapter 5.

Secondly, by nature, design is a very experiential and practical discipline. Seen by practitioners as a process and seen by society as the result of or creations from that process, the word "design" has many different meanings. Design, Redström (2017) stated, is first of all a process, with Don Norman defining it as the "deliberate shaping of the environment in ways that satisfy individual and societal needs" (2009, p. 171). Design embraces a broad range of disciplines, including, but not limited to: architecture, industrial design, interior design, fashion design, interaction design, design management, design education, digital and visual communication and landscape architecture. What unifies these diverse disciplines is how they think and their approach to problem-solving: regardless of their specialist disciplinary training, designers share an iterative design process that privileges a unique way of problem-setting and problem-solving, often through visualisation and realisation (that is, making). While many of these skills are transferable to academia, as Swann and Rezayan document in their chapter on the new academic's mindset, all too often, other skills (such as writing and conforming to traditional norms in academia) are harder for some to master.

And, like all academics, design academics must also learn how to navigate and negotiate the culture of academia – how to be great teachers and researchers, manage their own workloads, stay resilient, learn and evolve from early career academics into leadership roles. All universities have their own unique culture and ways of doing things, but typically, academics are expected to be independent and manage their own workload. Some design academics may be able to benefit from a supportive culture and colleagues and/or an engaged, hands-on line-manager, who provide advice, mentoring and clear direction. However, all too often, academia is a "sink-or-swim" culture, with very little tangible guidance for staff. This book is purposely written – in an accessible, conversational and informal style – to fill this gap and guide the reader through the challenging but highly rewarding career of being a design academic.

1.2 HOW TO READ THIS BOOK

The book is organised into two parts, with six specific sections. While you can read the book from start to end, you can also focus on a specific issue and read that chapter as a stand-alone narrative. Each chapter in this book covers an important issue, and the relevance will vary depending on your unique career trajectory. If you are seeking your first job or have just secured your first academic position, then *Part 1. How to Be a Design Academic: Starting Out* (Sections 1–3) will be especially pertinent, while mid-career and senior academics will be inspired by the chapters on leadership in *Part 2. How to Be a Design Academic: Becoming a Leader.* In the overview that follows, we describe the purpose of each section and briefly summarise the main contributions of each chapter.

1.2.1 Part 1. How to Be a Design Academic: Starting Out

Part 1 (containing Sections 1, 2 and 3) is most useful for new- and mid-career design academics, providing rare insight into their day-to-day experiences, challenges and learnings. Elsewhere, there is relatively little written about how to navigate the first few years and develop the teaching, research and administrative skills needed to survive and thrive in academia, especially as a design academic. Part 1 addresses all three of these, with a focus on the teaching and research skills most likely needed at this stage. Early and mid-career design academics share their unique career paths, and outline how they have navigated the challenges and opportunities of academia.

Part 1 explores a range of relevant issues covering professional identity, teaching and research, including: crafting your unique career path as a design academic and becoming a successful design researcher and educator.

Section 1: Crafting Your Unique Career Path

This section contains three chapters documenting how different design academics have approached their careers in very individualised ways, based on their diverse backgrounds.

We start with a thought-provoking chapter by Evans, Kelly and Kerr, who discuss the various identities that a design academic may assume, depending on their background and career priorities. This chapter proposes that the three constructs of design process, design practice and design thinking are useful for navigation within the profession, distinguishing between academics who have expertise in a design discipline or design cognition/science/method and those who do not (non-designers). This chapter is useful for understanding the priorities and points of view of the various academics who may teach within a design school, or teach some aspects of design from outside a design school. As well as helping readers to frame their own identity as academics and/or designers, it can also help them to understand and value colleagues who may have different backgrounds and priorities – as discussed in the next two chapters.

Lindgren explores the concept of the "pracademic" – an industry professional who has decided to enter academia, as a "practitioner-academic." Using examples of how he has embedded practice into his teaching, he exemplifies why universities often seek to employ those with industry experience, especially in teaching roles. A fashion pracademic, Lindgren outlines how his extensive industry experience with the design, making and marketing of fashion has informed and enhanced his teaching practice, as he has translated his creative practice into a teaching tool.

Changing tack, Miller outlines how she has found her place in a design school – where she has worked for more than a decade, even though she was not trained as a designer. Her chapter describes the pros and cons of this type of scenario, giving insight into how this approach can enrich the culture, research and teaching of the school. She gives advice on how an academic who finds themselves in a workplace that may not naturally seem to be their "home" can nevertheless carve out a niche for themselves and make a valuable contribution.

Section 2: Becoming a Successful Design Researcher

Section 2 focuses on design research, from the PhD experience to the early career researcher stage. Williams and Satherley used semi-structured interviews to explore the experiences of those who complete a PhD whilst working as academics. This is a scenario that is relatively common in design as many people entered teaching after a stint in industry, and, while they were often recruited based on their industry experience (Lindgren highlights why this is important, especially in design), they found themselves in need of a PhD in order to progress their academic careers. Their findings include issues with identity as a staff member who is also a student, problems with work–life balance, mental health and supervisor relationships. Some of these issues are common to all PhD cohorts, but are exacerbated by the experience of being a part-time student and working full-time in a demanding role. Williams and Satherley offer recommendations to help academics in this situation to navigate this period and have a more successful and rewarding PhD experience.

Swann and Rezayan explore the mindset of "always learning," which they have found beneficial as early career researchers in the post-PhD stage. This entails developing new skills as well as actively seeking to transfer existing skills into academic work contexts, and adopting an attitude conducive to identifying and making the most of opportunities that are presented. Using a combination of self-reflection and semi-structured interviews conducted with eight early career academics, they provide insights into the experiences of new design academics, and their perspectives on skill development and transfer, and offer skills maps to provide cues to spark potential opportunities.

Section 3: Becoming a Successful Design Educator

This section explores the common issues design academics might face in teaching design courses and designing a curriculum, and documents some innovative approaches to engaging students, both in person and online. At our institution, most students take two semesters per year, and a full-time load within each semester is four "units." A unit is a self-contained course of study (although it will often have prerequisite units required to have been previously completed depending on level, skill and knowledge required) – sometimes called a course, subject or a module at other institutions. Each standard unit is worth 12 credit points, with some capstone units (such as final year projects) being double or even triple credit points. Ninety-six credit points (equivalent to eight standard units) per year for three years is required for a Bachelor's degree. The academic staff member who creates, leads and teaches a unit is called the "unit coordinator" – sometimes called "course convener" at other institutions.

Scott and Thomson start this section with a detailed description of an innovative teaching practice: running design intensives "outside the studio." For almost four decades, first-year design students at our institution started their design education with an overnight camping excursion intentionally designed to foster team collaboration and the creation of peer support networks – many students form lasting friendships "on camp," which endure throughout their studies and beyond into professional practice, and foster students' enthusiasm and motivation for design. In this reflective chapter, Scott and Thomson outline the practical, logistical and cost considerations of camp, alongside the desired pedagogical processes and outcomes. In 2019, for example, students collaborated to prototype ideas for a hypothetical event, designing and building a series of "pop-up micro-communities." Despite being labour-intensive and administratively challenging for the teaching team, Scott and Thomson remind us that intensive, immersive project-based activities – such as off-site camps – provide memorable learning experiences for design students and a wonderful foundation for accelerating socialisation with the design process, peers and teachers.

Shifting gear, in the next chapter, Thomson and Scott describe how they complimented the same first-year introductory design unit by integrating interviews with design practitioners, using online video materials derived from a wide range of design professionals, creating a truly blended approach. Chapter 8 outlines their step-by-step process, documenting challenges and lessons learnt. They describe the pros and cons of this approach, including the many practical considerations in making videos in the field, and how it can enhance a large interdisciplinary unit.

Moving even further into the digital realm, Seevinck and her colleagues document how the staff members of two design disciplines (interaction design and visual communication) took all their units online at short notice during the COVID-19 global pandemic. As online learning becomes more mainstream, and we investigate a whole range of new ways to approach teaching post- COVID-19, these chapters provide a source of inspiration, practical guidance and possible strategies for making your face-to-face and online teaching more engaging and relevant.

1.2.2 Part 2. How to Be a Design Academic: Becoming a Leader

Part 2 of the book is ideal for those who are interested in leading in academia in the design fields, developing their own careers and those of others. It is intended to guide readers in growing as academics and in leading, mentoring and developing others. This is important because design has the potential to have a huge impact on the world, but designers are not traditionally seen as potential leaders in academia. Therefore, opportunities exist to fill this gap, and, to grasp them, design academics need to strategically focus on leading their own careers, and leading various teams of their colleagues, students, industry partners and academic peers.

Section 4 starts with an internal focus on academics leading themselves in order to make the most of their time, energy and talent. Section 5 builds to a more external focus on developing a reputation and profile by leading others, both close to home and around the world. Finally, in Section 6, the focus shifts to leading a whole group or school. Part 2 contains contributions from a diverse range of experienced design academics in various formal and informal leadership roles. Focused on providing practical, real-world advice, the case studies, examples and ideas included are intended to guide readers in developing a leadership portfolio and, if desired, undertaking a substantial role as an academic leader. This will enable readers to move from being a design academic to an academic leader in their field, whether they wish to be a global research leader, Head of a School or Faculty, or leader of a group of teachers or researchers, peers or students.

Section 4: Leading Yourself

This section highlights how, in contemporary academia, one of the most important skills any academic must learn is how to manage their own workload and prioritise as appropriate. Miller describes the importance of learning to be resilient, discussing the practical strategies she has engaged in to try and maintain work–life balance and the importance of embracing a growth mindset, as well as being intentional and strategic in academia. In their chapter, Chamorro-Koc and Caldwell use the analogy of "running the academic marathon" to delve deeply into the tactics and techniques mid-career and senior academics use to manage their workloads, documenting how work approaches and activities differ. Based on their findings from self-reflection and semi-structured interviews, they offer a set of strategies for running the academic marathon.

Section 5: Leading Others

Section 5 focusses on leading others, whether they be students, colleagues, peers or industry partners. Two chapters in this section are devoted to research leadership, outlining how and whom you might lead as a growing or senior design academic. Two are dedicated to leading in the teaching sphere – both students and other staff in developing mentoring schemes and curricula.

Blackler outlines how to turn your research passion into an international influence, as well as some practical strategies to lead within your school at a local level. First, she outlines how she has developed an international reputation and global network in her field of intuitive interaction, giving practical advice on the steps to take in order to create influence and impact at a global level. Next, she shifts to the local level, and leading within your own institution. As well as championing a weekly *SUAW* (shut up and write) group for colleagues, she is involved in a Women's Writing Accountability Club (co-founded by Miller and Blackler) and Promotion Club, for design academics intending to apply for promotion to Associate Professor or Professor.

The next chapter switches gear, looking at leading industry–research partnerships. Design academics usually have strong industry connections, and the chapter by Zelenko, Gomez and Kelly describes how they have leveraged these partnerships for impact. Drawing on three case studies – one each in industry, government and not-for-profit sectors – this chapter outlines the value of engaging in a co-design approach to deepen research engagement.

The last two chapters explore examples of teaching leadership. There is relatively little published about how design schools navigate the process of large-scale curriculum development and change. Meth, Brough and Thomson use thought-provoking vignettes to vividly share their experience negotiating the institutional approval processes and varying visions and degrees of buy-in and ownership from stakeholders. Curriculum development is a tough job, and this insightful chapter shares multiple practical points, thought questions and specialist reference works designed to assist and empower anyone undertaking such a curriculum journey. Finally, in the last chapter in this section, Beaver discusses the benefits of students building peer support systems. She documents her learnings from setting up a student peer mentoring group, how she addressed the various issues she encountered along the way and how such initiatives helped to positively transform the culture for both students and teaching staff.

Section 6: Leading a Group or School

Leadership requires a group of followers, a team and something that brings people together, and these chapters outline different approaches. We start with a chapter by Foth and Rittenbruch, which uses a far-ranging interview format to reflect on how to build and lead a research group: the Urban Informatics group, which started in 2006. As well as describing the group's origin and formation, Foth and Rittenbruch outline some of the opportunities and challenges that helped and hindered the group's progress early on, before explaining how they built group identity and branding to promote visibility and establish a sense of belonging amongst researchers and research students. From developing the brand, securing space and developing positive research culture through online communication channels, regular weekly meetings and an annual retreat, this chapter offers critical guidance and advice for other design academics working to show research leadership.

Finally, Scharoun and Muratovski compare their respective experiences of leading design schools in Australia and the US. They highlight the challenge of leading in a university where change is almost a constant, exacerbated by the COVID-19 pandemic which occurred during the writing of this book. They start off by describing their schools and then use an interview format to compare their leadership styles and philosophies, how they connect with their staff, foster diverse and inclusive environments, and communicate and facilitate change. The chapter finishes with advice on how to lead in a crisis, focussing on the challenges of COVID-19 in the tertiary sector.

1.3 CONCLUSION

There is a variety of contributions made by the various chapters in this book. As well as real-life reflections and case studies, with practical how-to advice on teaching innovations (see, for example, Scott and Thomson and Thomson and Scott, Beaver, Seevink et al., Lindgren, and Meth et al.), leading your own career (e.g. Miller) and navigating leading a design school or research group (Blackler, Foth and Rittenbruch, Scharoun and Muratovski), some chapters provide explicit recommendations and strategies on career management issues (e.g. Williams and Satherley, and Chamorro-Koc and Caldwell), and some also develop frameworks or models to help conceptualise the design academic experience (e.g. Evans et al., Zelenko et al., Swann and Rezayan).

In editing this book, our aim has been to bring together and foreground the diverse experience of design academics. We benefit from the experiences of our authors; in each chapter, they openly share their diverse personal and professional experiences and career paths, documenting highlights, challenges, learnings and approaches that worked for them. While almost all authors work together in one School of Design at an Australian university, what is quite striking is how different our day-to-day activities, experiences and priorities are. That is the joy of academia: we can follow our passions and craft our own unique career path, whether as a pracademic, design educator and researcher or by leading curriculum renewal, or leading a research group or school.

As you read this book, our hope is that you will be inspired by one of the ideas, processes, concepts and frameworks outlined, and integrate that approach into your work practices. Our work is emotionally and intellectually challenging and, as you will see, there is no one way to be a design academic. However, by recognising and following your ambition, your skills, your networks and your strengths, you will craft your own academic identity – and make a unique impact as a design academic. Wherever you are on your academic journey so far, this book will help you plan the next step – enjoy!

REFERENCES

Burge, G. (2015). *Mapping your academic career: Charting the course of a professor's life.* InterVarsity Press.

Clark, A. & Sousa, B. (2018). *How to be a happy academic: A guide to being effective in research, writing and teaching.* Sage.

Durling, D. (2002). Discourses on research and the PhD in design. *Quality Assurance in Education*, 10, 79–85.

Mewburn, I. (2013). *How to tame your PhD.* Lulu.

Murray, R. & Mifsud, D. (2019). *The positioning and making of female professors: Pushing career advancement open.* Palgrave Macmillan.

Norman, D. (2009). *The design of future things.* Basic Books.

Redström, J. (2017). *Making design theory.* MIT Press.

Part 1

Starting Out

Part 1 is mainly for new and mid-career design academics, providing rare insight into their day-to-day experiences, challenges and learnings. Despite a growing number of books on how to secure that elusive first academic job, there is very little written about how to navigate the first few years and develop the teaching, research and administrative skills needed to survive and thrive in academia, especially as a design academic.

How to Be a Design Academic: Part 1: Starting Out explores a range of relevant issues, including:

- Crafting your unique career path as a design academic, whether that is navigating the path from industry to academia, finding your discipline, or focussing on design process, design practice or design thinking approaches
- Becoming a successful design researcher, from getting that PhD to embracing the "always learning" mindset
- Becoming a successful design educator, from designing design intensives and using video and blended learning, to teaching in the time of COVID-19

Part 1 helps new and early career design academics think strategically about their own careers and the different ways in which they can take control of and navigate the turbulent waters of academia with confidence and impact.

Section 1

Crafting Your Unique Career Path

In this section, design academics reflect on how they understand design. This is an important issue as design is a term often overused and misapplied and, as design academics, we need to be clear about how we see ourselves. Also, there are multiple design and related disciplines with which design academics are affiliated, and it is useful for them to consider their alignment and contributions within the wider sphere of design.

Chapter 2 focuses on the relationship between design discipline knowledge, our understandings of design methods and design science, what it means to have non-designers teaching "design" and non-designers studying it, and vice versa. Chapter 3 explores the experience of "pracademics" or "practitioner-academics" (an industry professional who has decided to enter academia), while Chapter 4 documents how a non-designer has crafted a role for herself in design.

Section 1

Crafting Your Unique Career Path

2 Being a Design Academic
Design Process, Practice, Thinking

Richard Evans, Nick Kelly and Jeremy Kerr

CONTENTS

2.1 THE ROLE OF DESIGN ACADEMICS

The widespread adoption of *design thinking*, and its embrace within the public sphere, has served to create some confusion around the role of design academics in the contemporary university. In our own university, for example, there are design thinking subjects being taught within areas of information technology, business, law, education and, of course, design. This chapter aims to provide some clarity around the question of: what is the role of the design academic in the contemporary university in which non-designers teach "design" and non-designers study it? Even the terms used in constructing this question are contested territory, where the distinction between designers and non-designers is more of a continuum than a dichotomy. Further, the idea of *what it means to teach design* is itself a long-identified, but still-unresolved question: what is the relationship between design discipline knowledge and our understandings of design methods and design science?

Some of these distinctions have been identified by Cross (2001) in his discussion of the need for design as a discipline to be "studied on its own terms, and within its own rigorous culture" (p. 54). His position, while predominantly concerned with design research, is that there is a need for study of "the techniques of the artificial" (p. 54) as well as the "sciences of the artificial" (in reference to Simon, 1969/2019). Here we build upon the work of Cross to consider this same question from the perspective of the professional identity of the modern design academic: who do they teach, what do they teach, and how are they different to non-designers who are teaching and studying design?

Schools of Design (and here we are limiting the discussion to those schools within universities) are diverse in nature, and they typically offer a range of different disciplines such as architecture, urban planning, landscape architecture, fashion design, interaction design, graphic design, industrial design and interior design. Many Schools of Design have only been around since the 20th century, and some disciplines (e.g., interaction design) have only had a widespread presence since the 21st century. Each of these *design practices* has its own traditions, knowledge, skills and values.

Sitting alongside this discipline knowledge is the fruit of half a century of research into design science, design cognition and design methods – attempts to understand how designers think and what designers do, which we refer to collectively here as the *design process* – as typified by the work of (Cross, 2011; Dorst, 2015; Lawson, 2005). This knowledge has found its way into the pedagogy of design. For example, scientific knowledge about the differences between expert and novice designers has led to some changes in how and what we teach novice designers in the different disciplines (Crismond and Adams, 2012).

Finally, overlaid on top of this, is the recent "packaging up" of this knowledge about how designers operate as *design thinking*: as a series of processes, tools and mindsets that non-designers can use to be able to solve problems like a designer (Dorst, 2011; Kimbell, 2011).

In the contemporary university, the professional identity of a design academic has multiple facets. To what extent does an academic identify with their discipline knowledge and expertise? To what extent do they identify as an expert on design science/cognition/process? And to what extent do they identify as an expert in design thinking or in teaching/facilitating design knowhow to/with non-designers? The contribution of this chapter is to provide some further clarity around this ambiguous notion of what it means to be a design academic by focusing on the question: who and what design academics teach? We argue here that these three different domains – of design process, design practice and design thinking – serve as useful landmarks for navigating the confusing territory of being a design academic in the contemporary university.

2.2 DESIGN PROCESS, PRACTICE AND THINKING

2.2.1 Defining Design Process

The meta-level analysis of designers working, thinking and carrying out design activities has led to an understanding of what is referred to as *design process* (Bayazit, 2004, p. 16). The origins of this perspective on design – as a science, method and way of thinking – can be traced over the last 50 years, with significant debate centring around the tension between creative and scientific perspectives (Papanek and Fuller, 1972; Cross, 1999; Gero and Kannengiesser, 2004; Lawson, 2005; Lawson and Dorst, 2013).

There is no consensus around the series of activities that designers, across the range of different disciplines and practices, undertake during design activity – even the start and end points of the design process remain the subject of argument. Dorst and Dijkhuis (1995) suggest that there have been two fundamental ways of approaching the design process. The first is typified by the problem-solving theories introduced by Herbert Simon (1969) – similar to the "systematic methods for designers" proposed by Archer (1964) before him – who rationalised the design process as a discrete sequence of stages, with each informed by research and knowledge. The second is typified by the notion of design as reflection-in-action (Schön, 1983) and characterises the design process as a reflective conversation with the design situation, in which doing and thinking cannot be separated but rather are complementary and interdependent. From this perspective, designers are improvisers who are active in restructuring the problem, often evaluating their own actions as a means for solving the problem – a perspective which is not compatible with any universal sequence of stages that might be conceived (Dorst and Dijkhuis, 1995. p. 263).

These two paradigms, one positivist (design as problem solving) and one constructivist (design through reflection on action) are problematised in Cross's (2001) juxtaposition of "design science" and "design discipline" as different ways of understanding the design process. Design science is characterised as an "explicitly organised, rational, and wholly systematic approach to design; not just the utilisation of scientific knowledge of artefacts, but design in some sense as a scientific activity itself" (Cross, 2001, p. 53). In contrast, what Cross refers to as "design discipline" aims to develop "domain-independent approaches to theory and research" based upon the reflective practice of design. In recent decades, these two approaches have converged through scientific approaches that recognise the situated nature of cognition and the social nature of knowledge.

The term *design process* can be used to refer to this entire body of knowledge that seeks to understand what constitute "designerly ways of knowing, thinking, and acting" (Cross, 2001, p. 55). This description recognises that knowing, thinking and acting are three intertwined and inseparable aspects of the design process.

2.2.2 Defining Design Practice

The term "design," as an industry, has been applied to describe a range of disciplines or practices, each of which has specialist skills and represents an established career pathway. These professions have named roles, such as architect, fashion designer, interior designer, graphic designer and industrial designer, amongst others (in this book, for example, Lindgren reflects on his experience as a fashion designer who became an academic; something which, in his terms, is a "pracademic"). Accordingly, the traditional model for design education has been organised to align with these separate areas of specialisation, with a focus on students developing expertise in their unique discipline. Significantly, while design education entails teaching theory and application within each specialist field, the idea of the *design studio* operates an overarching educational approach that occurs across all design practices, a "signature pedagogy" of design education (Crowther, 2013).

Within Design Schools, the design studio has come to describe both the unique learning environment and the delivery model itself. The term references the professional place of employment and the process of educating designers attempts to replicate this context. In design studios, students respond to discipline-specific "design briefs," which are representative of industry projects, and centred on learning specialist design skillsets and integrating theory into practice. The learning is experiential and semi-structured, with the studios/classrooms serving as informal creative environments where students regularly develop and present work, receive critical feedback from both peers and teaching staff, and then iterate work further. Turnbull, Littlejohn and Allan (2010, p. 128) identify the need for authenticity in design briefs within this context as being integral for "learning through experience."

The design studio approach is both a form of project-based learning and "problem solving." Casakin and Goldschmidt (1999) observe that within this studio context, design education is about learning to apply practical methods, knowledge and natural ability to solve new problems. Evaluation of student learning occurs primarily through drawings, models and prototypes that reflect the design solutions that students have developed (Morton and Obrien, 2005). Crowther (2013, p. 19) states that as students produce these original designs, they "create their own learning experiences" and that this "flexible pedagogy is in response to the flexible nature of the design process itself."

Central to the development of expertise in a design practice within studio classes is the dialogue between the tutor and student. Both Schön (1983) and Biggs (1999) highlight the significance of this dialogue and of face-to-face learning across design education. This approach has its origins in the master and apprentice relationship that was central to older design disciplines when they first emerged (Schön, 1987). The skills and knowledge in a traditional design practice, such as architecture, were imparted by an "experienced practitioner" to a "protégé" as they worked on the practitioner's project (Littlejohn, 2017). In the 19th century, as part of a larger movement centred on professionalisation across industries (Klein, 1990), institutions began developing this early apprenticeship model into a formal educational pedagogy.

Subsequent developments in design education have occurred at iconic schools such as the Bauhaus. In the case of the Bauhaus, this led to the introduction of a series of fragmented, discrete subjects that concentrate on specific skills development and knowledge within schools (Littlejohn, 2017). For example, first-year students being educated on introductory principles about design and colour theory, while experimenting with a variety of materials and approaches. As design education has continued over the decades, academics have also contributed to an expanding body of knowledge for each design practice and integrated this into curriculum. This has entailed a process of "archiving, documenting, critically writing, and publishing history, as well as engaging with social, cultural, and political contexts" within each discipline (Triggs, 2010, p. 3).

Over the last decade, as rapid technological change has transformed many disciplines, there is now renewed focus on evolving disciplinary design education further. Kiernan and Ledwith (2011) note how technology and the changing nature of design have seen the borders between design disciplines blur and extend into one another. Kullmann (2016, p. 31) identifies that there exists a "inter-disciplinary anxiety" occurring across spatial design disciplines in particular, wherein landscape architecture, architecture, urban design and other design practices "contest similar disciplinary territory." Bremner and Rodgers (2013) have also outlined how the boundaries of disciplines such as product, graphic, textile and fashion design have increasingly dissolved. Kiernan and Ledwith (2011) observe that this has occurred within industrial design through a "crossing over" with interaction design as a result of the embedding of digital technology within products, creating product service systems. As fluid employment resultantly becomes commonplace between traditional design practices due to these changes (Rodgers and Bremner, 2017), educators are increasingly introducing trans-disciplinary design projects that allow students to develop skills across a wide range of design disciplines. In some cases, this has led to "hybrid courses," which consist of multiple disciplines. Kiernan and Ledwith (2011, p. 182) see a challenge in such moves away from established design practices, because while they broaden the skills of designers, they also run the risk of designers becoming "generalists without specialist skills." Yet for the time being it seems, anecdotally, that the majority of design academics clearly identify with a discipline and have teaching that is predominantly focused within that area.

2.2.3 Defining Design Thinking

Design thinking has its origins in design science, in our understanding of how designers think. One way of seeing design thinking is as applied design science, to areas outside design. As defined by Kimbell (2011), "[t]he main idea [of design thinking] is that the ways professional designers problem-solve is of value to firms trying to innovate and to societies trying to make change happen" (p. 286). The term *design thinking* comes from design scholars such as Cross (2011), Dorst (2011) and Lawson (2005) who sought to understand the cognitive processes that designers use. However, more recently, the term design thinking has become overloaded, with Kimbell (2011) and Johansson-Sköldberg, Woodilla and Çetinkaya (2013) reviewing the literature in the area and finding respectively three and five different definitions. There is thus some potential confusion here, due to the fact that the work of those scholars who coined the term design thinking are categorised under "design process" in our taxonomy (and design science or design cognition in others' taxonomies), while design thinking has become its own derivative subdomain.

Design thinking has been widely adopted (e.g., by different faculties within a university, by businesses, by governments and by schools) but has also been criticised on the basis that it oversimplifies design, separates out *thinking* from *doing* and removes the nuance of design as an holistic activity, a situated process that takes place in the physical context of a design practice (Kimbell, 2011). Despite these criticisms, the sheer popularity of design thinking has led to widespread acceptance of the term as a useful referent for the design mindsets, skills and tools that can be used independent of both design experience and design context. In this respect, design thinking is a way of referring to the translation of design know-how into non-design contexts; that is, the application of designerly ways of knowing, thinking and doing outside of the disciplines. Examples of this can be seen in the use of design thinking for teaching business executives how to innovate (e.g., Brown, 2008), or the use of a design-thinking framework in facilitating co-design workshops (e.g., Kelly, Wright, Dawes, Kerr and Robertson, 2019).

Design thinking, defined in this way, has been widely adopted within education and research – and sits both inside and outside of established Schools of Design. The naming and packaging up of ideas from design process research as "design thinking" has allowed ideas from design education to bubble over into the public sphere, in large part due to the work of organisations like d.school (at Stanford) and IDEO (a design consulting firm), which developed and promoted frameworks for design that are accessible and useful to laypeople (non-designers) and businesses alike. Brown's (2008) article in the *Harvard Business Review* had a significant impact in leading businesses and government organisations to realise that:

> … as economies in the developed world shift from industrial manufacturing to knowledge work and service delivery, innovation's terrain is expanding. Its objectives are no longer just physical products; they are new sorts of processes, services, IT-powered interactions, entertainments, and ways of communicating and collaborating—exactly the kinds of human-centered activities in which design thinking can make a decisive difference (p. 2).

The success of design thinking has thus been to demystify the design process and make it accessible; one no longer needs to have a design discipline to gain the benefits of *thinking* like a designer. As the IDEO website frames it, design thinking "allows people who aren't trained as designers to use creative tools to address a vast range of challenges" (IDEO, 2019). Design thinking does this by clarifying a *mindset* that involves certain dispositions towards experimentation, empathy, optimism and collaboration (e.g., Rauth, Köppen, Jobst and Meinel, 2010); with certain *steps* that get carried out when designing; and where a range of *tools* can be deployed to carry out these design steps (refer to Cuthbert, Vaugh and Chambers, 2018 for a comparison of steps and tools in different models). The result is a framework for approaching even complex problems in a designerly way; one that is grounded in design science, but – crucially – divorced from any specific discipline. This is the definition that we refer to when we speak of design thinking, as being something separate from the design process.

2.3 ORIENTATION AS A DESIGN ACADEMIC

The metaphor that we adopt in this chapter is one of design academics orientating themselves within the domain of design in relation to three kinds of design academic expertise: design process, design practice and design thinking. As a design academic in the contemporary university, it can be easy to feel drawn in different directions. For example, cross-disciplinary research projects, out-of-faculty teaching and industry partnerships are common occurrences that can require a design academic to develop new kinds of expertise. The mastery of each of these three areas of design process, practice and thinking (See Figure 2.1), can be considered as a "landmark" that can be used by an academic for orientation.

The first of these locations is *design process expertise*, typically developed through scholarship and a deep knowledge of the literature. Foundational or applied research into design science, design methods, or design cognition and design pedagogy relating to the design process can all be found here. The second of these is *design discipline expertise* in which the history, skills, techniques and industry knowledge of a specific design practice is understood and perhaps even mastered.

The third of these, building upon the way that design thinking is used outside of Schools of Design, can be considered as *design thinking expertise*, or perhaps more accurately *design translation expertise*. This is knowledge of how to use design with non-designers and might involve diverse skills such as teaching design thinking to non-designers, facilitating a co-design process or using design-led pedagogy to teach traditional subjects.

2.3.1 Design Academic Personas

Three design academic personas (where a persona is a fictional assemblage based upon an amalgam of anecdotal cases from informal research) can be used to provide a demonstration of the use of these landmarks in practice: (1) the design academic with a focus on practice; (2) the design academic with a focus on both practice and

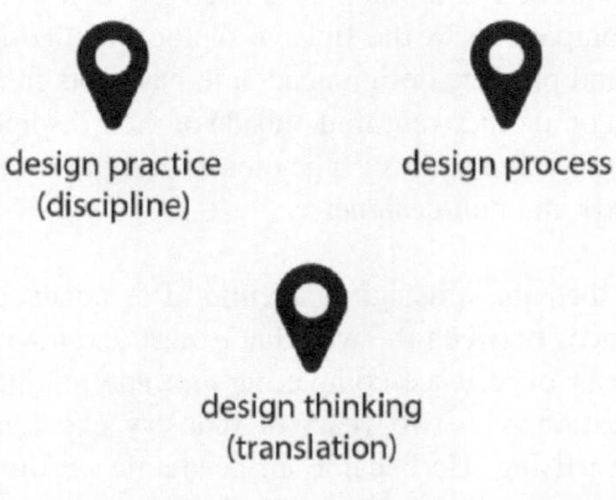

FIGURE 2.1 Design practice, design process and design thinking expertise represented as landmarks

process; and (3) the non-design academic (see, in this book, Miller's experience). The aim of articulating these personas is not to suggest that all design academics will necessarily relate to one of the three personas as there are certainly many other types; rather, the aim is to use these personas as archetypes to demonstrate the use of these landmarks for orientation purposes and perhaps show how they might be used as a tool for career navigation.

2.3.2 Persona 1: Design Academic (Practice-Focused)

The design academic (practice-focused) has been at the very heart of traditional design education for many years. They teach their craft to their students. That is, they are an expert in their discipline area and primarily teach students studying in that discipline area. An example of this is somebody who studied industrial design and became an academic in industrial design, and now teaches industrial design to industrial design students.

Persona 1 (P1): Sarah became a design academic after practising architecture for a decade with one of the larger firms in the city. She now teaches architecture students, running first- and third-year design studios. She is currently writing a book about one of the more significant architects from the 20th century.

This persona is an archetype of a design academic who is dedicated to spending their career located near the landmark of design practice (See Figure 2.2). Whole academic careers can be spent developing expertise in understanding and teaching discipline-specific knowledge and skills; there is no intrinsic need to venture into other spaces.

2.3.3 Persona 2: Design Academic (Process and Practice)

The design academic (process and practice) is relatively new in the design education teaching space, in comparison to the lineage of the practice-focused design academic. The process and practice design academic has roots in a certain discipline, but has for one reason or another ventured outside of their discipline, with a growing expertise in the area of design process. They teach design process and practice to a wide range of designers and non-designers.

Persona 2 (P2): Joe became a design academic after conducting a PhD focused on differences between the way that expert and novice designers sketch. Joe had prior experience completing an undergraduate degree in visual communication with two years of industry experience as a freelancer within advertising. He became an academic on the basis of his visual communication skills, responsible for teaching students in this area, but is increasingly gaining recognition for his foundational research on design cognition and is spending ever less time teaching students within

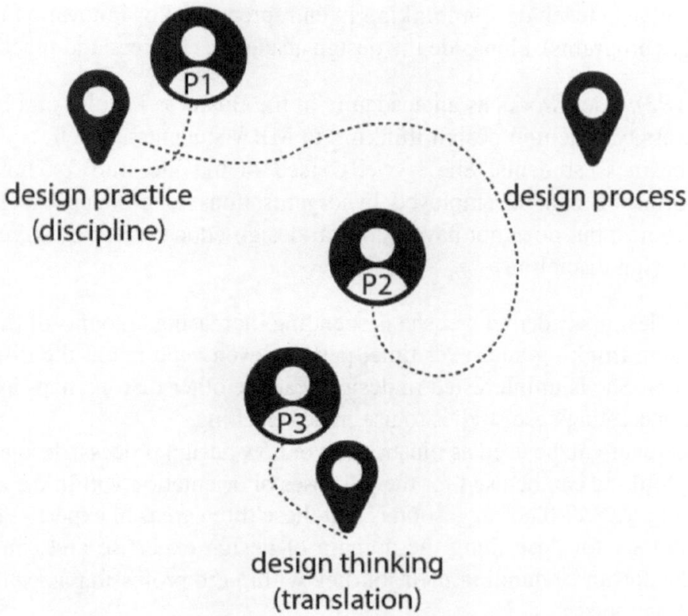

FIGURE 2.2 Academic journeys of personas represented in relation to three kinds of expertise

his school in favour of researching and teaching non-designers (often for consulting fees).

While Joe and Sarah might be very similar kinds of academics in their early career stages, they quickly diverge. Joe considers, perhaps, that he has "transcended" his discipline and has much broader interests than he used to have, spending his time reading journal papers about the latest design process research; while Sarah might consider that he has given up the possibility of retaining mastery of his discipline over time. In terms of territory, Joe has journeyed away from design practice towards design process, with perhaps some "tourism" in the direction of design translation.

2.3.4 Persona 3: Design Translation Academic

The non-design academic primarily teaches design thinking to non-design students. This academic does not have a design discipline and therefore would not be teaching either design practice or process. Their expertise is in teaching non-design students from areas such as information technology, business, law, education, etc., to think and act like a designer. This type of academic does have a discipline that lies outside of design; for without their own discipline they would not be so adept at contextualising design thinking into the discipline vocabulary to teach their non-design students. This also highlights that academics from outside design typically teach design thinking to their respective discipline students. Some do go beyond their non-design discipline and also teach design thinking in entrepreneurial or innovation incubator type settings (programs), alongside the design academic (process and practice).

Persona 3 (P3): Alex works as an academic in the Business Faculty of a large university, teaching design thinking to MBA students as well as third-year business students. She is well-versed in the literature on how design thinking can be employed by organisations to produce excellent outcomes, but does not have a formal design education or background in a design discipline.

Alex is not a design academic, yet she is spending increasing amounts of time in the space of design thinking and, over time, perhaps even ventures in the direction of design process. She is uninterested in design practice other than perhaps as a "tourist," to find interesting case studies to use in her teaching.

These personas can be used as illustrations of how design process, design practice and design thinking can be used for the purposes of orientation within the academic landscape (Figure 2.2). The suggestion is that these three areas of expertise compose a useful ontology for describing the domain of design expertise and can be used by academics for understanding their journey within the profession as well as their desired destination(s).

2.3.5 Design Academic Teaching Vignettes

This metaphor of orientation within a landscape can be used to address the question of who and what design academics are teaching, by referring to three vignettes of teaching from within our own university. These three vignettes show design academics teaching in contexts of three different units of study, where each has complexities in terms of who is being taught and the expertise of the teacher.

2.3.6 Emerging Design Technology (Taught within the Design School)

This unit (subject) was designed specifically to teach first-year undergraduate students about existing and emerging technology and how it applies to design. It is a common foundation unit taken predominately by fashion design, industrial design, interaction design and visual communication students, as well as by double degree students that might combine business and design (which is becoming increasingly popular). It is a core, multi-disciplinary unit for design degree students at our university in which they learn how to do the following:

- Apply appropriate technology to the design process
- Demonstrate knowledge of existing and emerging technology and how it constrains design and creates design opportunities; and
- Apply existing and/or emerging technology to design outputs

The main theoretical constructs used in the unit are drawn from design process literature and include technology adoption and futures thinking. Students use these constructs to help explain and predict how and why designers adopt existing and emerging technologies and their effects.

The kind of design academic that would be best suited to teaching this type of unit needs to be familiar with the territory of both design process and design practice (e.g., Persona 2, Joe). While the unit clearly has a design practice dimension, that practice is multi-disciplinary – in this example, it crosses four very different design disciplines and includes students who have no association with any design discipline who originate from faculty areas such science and engineering, business, law and education, etc.

Teaching this unit would be extremely challenging for any academic who is not familiar with the territory of design process. An academic such as Sarah who is focused on practice might not want to spend her time getting to grips with the theoretical depth (in the design process literature) or cross-disciplinary breadth (to speak to the wide range of students in the unit) required to teach this unit well. In contrast, an academic such as Joe is likely to get a lot of satisfaction from having to prepare and teach this unit.

2.3.7 Design Thinking (Taught within the Design School)

This undergraduate unit focuses on teaching design thinking to design and non-design students. It introduces them to the fundamentals of design thinking, methods and creative problem-solving techniques. It is aimed primarily at students who are passionate about design and want to become change agents, social entrepreneurs, strategists. and innovative thinkers. They learn how to do the following:

- Display a working knowledge of design-thinking methods and processes, and an understanding of the connection between design, creativity and innovation
- Demonstrate creative and analytic modes of thought within design processes, including an ability to think critically and reflect upon personal performance and design outcomes
- Apply design-thinking strategies to a real-world project and communicate outcomes via multiple media forms; and
- Demonstrate collaborative skills through participation in a multi-disciplinary team

While design thinking is just one part of what a designer does, it is a relatively new phenomenon, arising within the last decade or so, used to isolate this design thinking as a separate offering in Design Schools, specifically for design students.

The notion of non-design academics teaching design thinking to design students is a curiously provocative notion that is worth discussion. Given the popularity of design thinking outside of Design Schools for social entrepreneurs, strategists and innovative thinkers, etc., it would be remiss not to acknowledge that a non-design academic who has specialised in teaching design thinking to non-designers could, in principle, teach this type of unit. This does happen already outside of Design Schools.

Non-design academics (specialised in teaching design thinking, such as Persona 3, Alex) do teach design thinking within entrepreneurial or innovation incubator type settings (programs) – and design students occasionally enrol in these kinds of programs seeking complementary skills and expertise to become future entrepreneurs.

2.3.8 Problem Framing for Creative Action (Taught outside the Design School)

This unit is part of the Master of Business Administration (MBA) and exists to teach design thinking to non-designers. It forms part of a comprehensive suite of units for the MBA program in which students learn how to do the following:

- Analyse workplace problems using problem framing techniques to identify the real issues to be resolved
- Apply creative problem-solving techniques, design-thinking and analogical-thinking principles to generate solutions to organisational problems
- Evaluate creative problem-solving tools and techniques to determine the most appropriate tools and techniques to be used in a given situation
- Work collaboratively with others to achieve a specific outcome; and
- Recommend strategies for fostering creative and alternative thinking by individuals and teams to create and capture opportunities for organisational improvement

For an academic to teach this type of unit, our conceptual framework illustrates that it would be most suited to two of the personas defined: the design academic in practice and process (Persona 2, Joe), as well as the design translation academic (Persona 3, Alex). This vignette is important in illustrating one of the reasons why design thinking has had such popularity outside of the domain of design education. Design thinking taught by non-design academics has been a major factor in businesses, organisations and government departments adopting design thinking.

2.4 DISCUSSION

The role of the design academic is more challenging to identify than it has been previously, due to the reach that design now has across universities, businesses and organisations. The recognition that design has been receiving within society more broadly creates a range of different pulls upon the modern design academic.

Design methods are now widely adopted as a means to solve complex human problems through innovative solutions, and they are especially relevant in an era where competitive advantage is increasingly focused on the "experience" of the users or customers. The approaches of change by design and design-led innovation, for example – in which design leads the way for new products, services, experiences and environments – have been widely successful beyond the traditional domain of design. "Design" has become a language spoken by many (although perhaps with varying dialects), in what might be termed the "democratisation of design."

Design academics, and non-design academics with design-thinking expertise, are currently in high demand, with their skills and expertise widely sought, in part due to this democratisation. The rising trend of undergraduate students combining "design" as a double degree option with areas such as information technology or business (at least within our own university and, anecdotally, in others within Australia) is another indicator of how demand for design education is evolving. As these non-design students increasingly seek design expertise as part of their repertoire, if this is taught in Schools of Design, this will continue to impact design academics: on what they teach and who they teach it to.

While it is unclear what the future holds, it is important to recognise that the teaching of design practice is still as relevant today as it has been in the past and does form the mainstay of teaching for many design academics. We recognise that, as design practice evolves over time, so does the teaching of that practice. It is a relational continuum. We also acknowledge the pedagogical difference in how design and non-design academics establish their own unique professional identity and how they design the curriculum (Wrigley and Straker, 2017). We have intentionally separated "how they teach" from "who and what they teach" to focus on design expertise and not upon pedagogical expertise.

The demands and shifts we have outlined serve to illustrate that some design academics are moving beyond the traditions of their practices to become what we have termed "practice and process academics." With this comes new opportunities, such as "design thinker in residence" or "design expert" on large-scale projects or addressing societal problems (e.g., environmental challenges, linear to circular economy transition, etc.). However, these changes also present new challenges, and new question arise. For example, could some design academics evolve into purely design-process academics, teaching units similar to the vignettes described and therefore no longer be linked to a discipline other than "design" in general? And, if so, where would they belong within a School of Design, perhaps leading to a basis for design process becoming its own discipline?

2.5 CONCLUSION

In this chapter, we have proposed that design academics in contemporary universities can be viewed almost as victims of the success of design – they are pulled in many directions because, at least since the last decade, design speaks to so much of the world beyond the traditional disciplines. We discussed the important question of who and what they now teach, recognising the shift in which non-designers increasingly teach design translation (under the name design thinking) to non-design students. We have given a brief review of three key concepts that can be used by any design academic to orient themselves. Design process speaks to the broad scientific literature on what we know about what designers do, think and know in a way that speaks across the different design disciplines. Design practice speaks to the individual disciplines and their proud and continuing traditions, whereby design academics come to have mastery within a specific domain of knowledge. Design thinking refers to what we suggest might be more aptly called design translation – the application of what we have learned about design outside of the design disciplines.

This chapter has proposed that the three different concepts can be considered as landmarks that a design academic might use to orientate themselves within the profession. When considering where they have come from, where they are now and where they would like to go – in terms of teaching, service and research – these constructs might serve as useful landmarks that can perhaps be considered by each design academic in establishing their own unique professional identity.

REFERENCES

Archer, B. (1964). *Systematic method for designers*. London: Council of Industrial Design.

Bayazit, N. (2004). Investigating design: A review of forty years of design research. *Design Issues*, *20*(1), 16–29.

Biggs, J. (1999). *Teaching for quality learning at university: What the student does*. Buckingham, UK: Open University Press.

Bremner, C., & Rodgers, P. (2013). Design without discipline. *Design Issues*, *29*(3), 4–13.

Brown, T. (2008). Design thinking. *Harvard Business Review*, *86*(6), 84.

Casakin, H., & Goldschmidt, G. (1999). Expertise and the use of visual analogy: Implications for design education. *Design Studies*, *20*(2), 153–175.

Crismond, D.P., & Adams, R.S. (2012). The informed design teaching and learning matrix. *Journal of Engineering Education*, *101*(4), 738–797.

Cross, N. (1999). Design research: A disciplined conversation. *Design Issues*, *15*(2), 5–10.

Cross, N. (2001). Designerly ways of knowing: Design discipline versus design science. *Design Issues*, *17*(3), 49–55.

Cross, N. (2011). *Design thinking: Understanding how designers think and work*. Oxford: Berg.

Crowther, P. (2013). Understanding the signature pedagogy of the design studio and the opportunities for its technological enhancement. *Journal of Learning Design*, *6*(3), 18–28.

Cuthbert, B., Vaugh, T., & Chambers, F.C. (2018). What is design thinking? Learning to Mentor, in F. Chambers (Ed.), *Sports coaching: A design thinking approach*. New York: Routledge.

Dorst, K. (2011). The core of 'design thinking' and its application. *Design Studies*, *32*(6), 521–532.

Dorst, K. (2015). *Frame innovation*. Cambridge, MA: MIT Press.

Dorst, K., & Dijkhuis, J. (1995). Comparing paradigms for describing design activity. *Design Studies*, *16*(2), 261–274.

Gero, J.S., & Kannengiesser, U. (2004). The situated function–behaviour–structure framework. *Design Studies*, 25(4), 373–391.

IDEO. (2019). *Design thinking defined*. Retrieved from https://designthinking.ideo.com/.

Johansson-Sköldberg, U., Woodilla, J., & Çetinkaya, M. (2013). Design thinking: Past, present and possible futures. *Creativity and Innovation Management*, *22*(2), 121–146.

Kelly, N., Wright, N., Dawes, L., Kerr, J., & Robertson, A. (2019). Co-design for curriculum planning: A model for professional development for high school teachers. *Australian Journal of Teacher Education*, *44*(7), 84–107.

Kiernan, L. & Ledwith, A. (2011). The effect of the merging of design disciplines and its implication for product design education. *Design Principles and Practices: An International Journal*, *5*(4), 173–185.

Kimbell, L. (2011). Rethinking design thinking: Part I. *Design and Culture*, *3*(3), 285–306.

Klein, J. (1990). *Interdisciplinarity: History, theory, and practice*. Detroit, MI: Wayne State University Press.

Kullmann, K. (2016). Disciplinary convergence: Landscape architecture and the spatial design disciplines. *Journal of Landscape Architecture*, *11*(1), 30–41.

Lawson, B. (2005). *How designers think: The design process demystified*, 4th ed. Oxford, UK: Architectural Press.

Lawson, B., & Dorst, K. (2013). *Design expertise*. Oxford, UK: Architectural Press.

Littlejohn, D. (2017). Disciplining the graphic design discipline: The role of external engagement, mediating meaning, and transparency as catalysts for change. *Art, Design & Communication in Higher Education*, *16*(1).

Miller, E. (2021). A "Non-Designer" in a Design School, in Blackler, T & Miller, E. (Eds), *How to be a design academic*. Taylor & Francis, 45–55.

Morton, J. & O'Brien, D. (2005). Selling your design: Oral communication pedagogy. design education. *Communication Education*, *54*(1), 6–19.

Papanek, V., & Fuller, R.B. (1972). *Design for the real world*. London: Thames and Hudson.

Rauth, I., Köppen, E., Jobst, B., & Meinel, C. (2010). Design thinking: An educational model towards creative confidence. Paper presented at the *DS 66-2: Proceedings of the 1st international conference on design creativity (ICDC 2010)*.

Rodgers, P.A., & Bremner, C. (2017). The concept of the design discipline. *Dialectic 1*(1), 19–38.

Schön, D. (1987). *Educating the reflective practitioner*. San Francisco, CA: Jossey-Bass.

Schön, D.A. (1983). *The reflective practitioner: How professionals think in action*. New York: Basic Books.

Simon, H.A. (1969/1996). *The sciences of the A (3rd, rev. ed. 1996; Orig. ed. 1969)*. Cambridge, MA: MIT Press.

Triggs, T. (2010). Graphic design history: Past, present, and future. *Design Issues*, *27*(1), 33–51.

Turnbull, M., Littlejohn, A., & Allan, M. (2010). Creativity and collaborative learning and teaching strategies in the design disciplines. *Industry & Higher Education*, *24*(2), 127–133.

Wrigley, C. & Straker, K. (2017). Design thinking pedagogy: The educational design ladder. *Innovations in Education and Teaching International*, *54*(4), 374–385.

3 Industry to Academia
The Enhanced Academic

Tim Lindgren

CONTENTS

3.1 BEING A PRACTITIONER-ACADEMIC

An industry professional who has decided to enter academia is in many ways continuing their practice through a new lens of inquiry. "Pracademics," or "practitioner-academics" are best thought of as retaining the ability to move between industry and academic boundaries in either direction (Posner, 2009). Their ability to take the knowledge earned on the "frontline" of industry, and, through a process of knowledge conversion and reflective practice, create authentic teaching experiences in the classroom is the characteristic that makes the pracademic valuable in the classroom.

Consequently, pracademics are a relatively rare kind of academic. In this chapter, the ways that a pracademic is able to shape curriculum are examined from the perspective of an academic who has been a professional fashion designer and business owner, and who has moved into the creative industries at University. In the process, dynamic links that are cohesive between industry and academia are formed. Importantly, the pathway from industry to academia does not mean dissolving the ties between one's academic career and one's previous work in industry but leveraging them for greater graduate outcomes. Consequently, the ideas in this chapter are universally applicable across a wide range of disciplines and simply amount to a methodology of "teaching by doing."

By examining specific examples of fashion practice from industry and their translation into studio activities, new ways of identifying key graduate needs can be explored. Furthermore, implementing innovative methods of teaching can be made more precise, which provides an authentic, real-world perspective as a key theme in the classroom or studio. As Barbara Larrivee (1999) argues, when a teacher becomes a reflective practitioner, "they move beyond a knowledgebase of discrete skills to a stage where they integrate and modify skills to fit specific contexts, and eventually, to a point where the skills are internalized enabling them to invent new strategies." Encouraging the expression of creativity is important; however, embedding some of the principles of capitalism is equally important in the making of fashion because students must be able to utilise their creative practice to earn a living in the real world.

As the fashion pracademic is usually a person who is familiar with the design, making and selling of fashion, in this chapter, the examination of this practice is conducted by considering a variety of moments of reflection from my own career in industry, which have allowed me to develop a suite of teaching methods that draw the world outside the University campus and academia closer to each other and provide experiences for students that are interesting and engaging.

3.2 REVISITING THE FASHION ARCHIVE

One method I have used effectively in the process of translating one's creative practice into a teaching tool is an exercise where a garment from the pracademic's fashion archive is bought into the fashion studio. In this reflective context, this garment can be analysed by students according to the design aesthetic, as well as the commercial and cultural parameters that have informed the construction of the garment. It is also a tactile activity as the garment can be handled and turned inside-out if needed to see the often-hidden construction methods. If the course curriculum includes theoretical content, this also can be examined in the context of the garment at hand.

In some cases, archival garments may be many years or decades old and so can be reimagined by students in the context of new technologies, new textiles and new construction techniques. The distinguishing feature in this case is the alignment of the pracademic's authentic narrative with the activity that is made more compelling by the use of theoretical concepts to support the passage of the creative process. The activity is not unlike a museum visit, however there is a distinct advantage by including the personal, storytelling element. The ability to introduce real world outputs from industry practice into the theoretical space for deconstruction and for reimagining based upon new technologies and techniques is a unique way to conceptualise a design practice.

In my case, a specific example of this kind of activity was the introduction of a dress embroidered with paillettes (flat oval or circular pieces of shiny plastic) made in my workrooms during the late 1990s into a University studio concerned with examining the future of fashion. Students were asked to deconstruct the garment using a process of reverse engineering (which is done by taking numerous measurements and rebuilding the patterns used to construct the garment), and to reimagine the garment in alignment with contemporary cultural and academic themes. There were some notable outcomes.

One group of students decided to use laser cutting to recreate the paillettes in a variety of shapes, sizes and materials that were not commercially available at the time of construction. A key theme in the fashion curriculum was sustainability, and, in one team, paillettes were cut from repurposed wood veneer. Another group used 3D printing to form words in the size of paillettes that formed a short narrative that ran around the waist of the garment. Yet another very ambitious group decided to ask for grant funding from the University to redesign the paillettes as small linked solar collectors that could be linked to a supercapacitor for energy storage.

The diverse outcomes from this single exercise meant that students were stimulated to think for themselves using an interdisciplinary approach to problem solving that was directly applicable to the real world, which is necessary for "graduates who must be able to proactively navigate the world of work and self-manage the career building process" (Bridgstock, 2009). In reality, we discovered that, generally, the sun is not inclined to shine in the evening, and the idea was transferred to a garment that might be worn during the day (Figure 3.1).

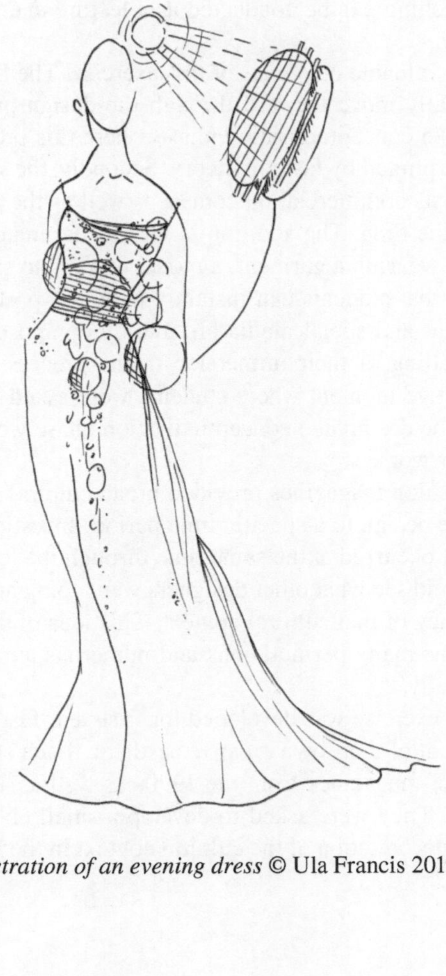

FIGURE 3.1 *An illustration of an evening dress* © Ula Francis 2019

3.3 A SOLAR COLLECTING EVENING DRESS?

Another exercise that has proven popular in the studio is to examine the contextual material around the creation of a new idea or design. Examining the original resources that were used as source material can provide clues to the development of the particular design; however, this process can also provide alternate pathways to consider how there might be different outcomes from the early stages of the design process. A particular example of this exercise was an activity in which I brought to the fashion studio a number of fashion magazines from the 1980s and a finished archival garment. The magazines were the original source for the garment and contained detailed notes I had made in the margins at the time of the garment's construction, as well as the small fabric swatches and trims that were under consideration at the same time.

My original hand illustration was bought into the classroom, with the patternmaker's notebook to provide the technical information. Students were able to gain a glimpse of the entirety of the creative process, from concept to final design and relate this to the development of the designer's aesthetic vision, and to see the tangible manifestation of this process in the realised garment in one studio setting. Once again, this kind of teaching can be conducted as a lecture and a classroom activity that follows the theory.

There were several valuable outcomes of this exercise. The first was that the students could intellectually move forward through the design process as well as in reverse, returning to the conceptualisation phase where this process could take the form of a lecture underpinned by fashion theory. Secondly, the success of the design could be measured by its commercial outcomes, as well as the attention it garnered in the print media at the time. The addition of rich visual material, such as a photograph of a customer wearing a garment, a model in a photo shoot or a newspaper article are all aspects that students can instantly relate to, particularly when ideas in contemporary digital and social media are mainly composed of visual images. The designer's storytelling of their immersion in the process provided authenticity, as well as a reflective moment where students were asked to consider whether varying the inputs to the design and conceptualisation phase would have altered the outcomes in a better or worse way.

Importantly, the fashion magazines provided broad cultural context and isolated the development of the design to a specific time period. Investigating other cultural or social activities that occurred at the same time through the lens of the magazines meant that students could see what other designers were doing and how their designs related to the immediacy of that cultural context. This idea of surveying the field is a valuable skill that has many permutations and numerous applications during the on-going process of study.

An iteration of this exercise was developed for first-year fashion students as part of the early development of their own creative aesthetic. Each student was provided with a small number of magazines from the 1990s as well as a fabric swatch from the designer's archive. They were asked to develop a small collection of garments that drew upon their interpretation of the cultural context in the magazines, that also incorporated the fabric swatch in some shape of form. The reason for providing magazines from this time period was that the early 1990s were largely pre-digital. This highlighted the importance of print material as a resource for designers and provided students with an understanding of their use of digital media as a design resource.

An interesting discussion among students that emerged as a side-line from this activity concerned changes to the relationships between celebrity culture and fashion. Many of the celebrities wearing or promoting fashion in the magazines were much younger at the time, however they were still well known to the students. One student commented specifically on the quality of the graphic design in the layout, which serves to highlight the close relationship between fashion and the entertainment industries as well as associated design industries. In terms of course curriculum, I was able to link this phenomenon to later courses in the degree program that examined the relationship between digital media, fashion and the rise of the social media influencer who has usurped the seats of the traditional fashion editor in the front row of the fashion runway.

3.4 PAST MISTAKES MAKE GOOD LESSONS

Teaching from experience by deconstructing mistakes or miscalculations that have been made in professional practice is also an important, real-world approach in the classroom. Using this process, a poor decision can be isolated from the pracademic's experience and reconsidered with the luxury of time in the teaching studio. Here, the decision to follow a particular path or choice, as well as the ramifications of the outcome of the decision, can be explored in rich detail.

Often there has been significant cultural change that sheds light on the decision, and, with the benefit of hindsight, alterations to the decision-making process can be considered in terms of potential outcomes. In industry, the fashion designer must often interact with disciplines that are outside the scope of their understanding. This is a normal part of a career in the creative industries, according to Ruth Bridgstock (2011), who has investigated the preparedness of graduates in managing their careers.

For instance, the process of designing might be considered as a way of taking an intangible concept and making it tangible. The generation of ideas in this case means that the designer must think about protecting their new intellectual property, which is likely unfamiliar territory for an early stage fashion student. Adopting a distinctly commercial approach to design is also often outside the realm of the traditional fashion education, yet this attribute is vital for the viability of the new brand a graduate may start, which means that some knowledge of the need to make a profit from your work in necessary. At the intersection of these different disciplines lies the distinct possibility for a costly mistake, however a pracademic may be able to avert this in their classroom.

3.5 THE STOLEN DENIM HALTER NECK

The story of the stolen denim top is a good example of this process. At some point in my early career, when I was developing a range of denim, my workroom produced a small run of denim halter-neck tops (Figure 3.2). Unfortunately, the patternmaker had cut the pattern for the tie at the rear of the neck about two inches too long, and so there was a problem where the tie began on the neckline seam when the garments were constructed. My solution was to simply form a knot in the tie to shorten it and turn it into an impromptu design feature. We did this before the halter tops were sent to be garment washed, which meant that it appeared to be part of the design and blended in well. We only made about 10 or 15 pieces and they sold very quickly in one of my shops. This kind of problem solving corresponds with Donald Schon's viewpoint, which is that professionals engage in a reflective conversation with their problems by reconsidering them in a variety of ways (Schon, 1983).

A month or so later, a staff member was travelling through the wholesale fashion district in my city and noticed a copy of my denim top in a wholesaler's window, complete with the faulty knot/design feature. When the owner of the shop was asked about it, they denied knowing anything about the origin of the design and said they sourced all their products from Guangzhou in China.

As a relatively young and naïve fashion designer, I was a little confused about how this could have happened. However, a member of my retail staff was compiling

FIGURE 3.2 *Illustration of a dark-haired person wearing a halter top* © Ula Francis 2019

a customer data base so we could refine our product ranges. In the process, she had captured the details of a transaction for a denim top with the customer details that led directly back to the wholesaler. As it turned out, the wholesale company was simply buying a size eight garment at retail from the current ranges of various designers in the city, sending them to China and reverse-engineering them with a very quick turnaround.

The clever part was that the wholesaler was using established designers who knew the city's clientele well and had carved out their own niches, to build a business using the intellectual property generated in those designer's workrooms without any cost to themselves, knowing well that the designers had little capacity to protect their creative works.

3.6 FINAL ASSESSMENT – HANGER APPEAL AND THE RETAIL WINDOW

In the University studio, this example from the frontline of industry is rich with potential for robust discussion about the ethical, moral and commercial implications of the wholesaler that inevitably leads to a conversation about how a young fashion designer can protect their creative work. The creative industries are fundamentally concerned with the generation of various forms of capital through the production of ideas (Hartley et al., 2013), and here the process of design, ownership and the provisions of the Designs Act and the Copyright Act might form an entirely new unit as the pracademic builds curricula. Teaching students about the generation of intellectual property in fashion is generally understated; however, it is a relatively straightforward field that, if addressed early in the curriculum, will provide a strong foundation upon which students can base their work (Figure 3.3).

Furthermore, there has been much discussion about the effects of stolen designs. In particular, Appel et al. have developed a framework that includes "acceleration," where the presence of the pirated design increases awareness of the design;

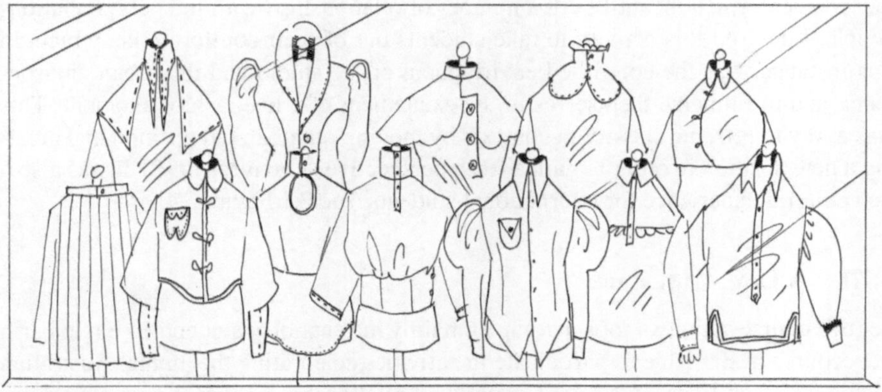

FIGURE 3.3 *Illustration of a rectangular window with a display of shirts* © Ula Francis 2019

"substitution" where the original brand owner loses sales because customers purchase the fake product; and loss because of overexposure of the design in the market (Appel, Libai and Muller, 2017). I consider it very important that a graduate designer is made aware of these types of effects when designing an original work because there are other viewpoints, notably that of Raustiala and Sprigman (2009), who argue that fashion piracy is paradoxically beneficial to fashion organisations as it induces rapid turnover while fostering innovation. Importantly, the designer can make material choices that are difficult to replicate early in the design phase that make it difficult for someone to replicate the garment as advocated by Janssens and Lavanga (2018) in their investigation into the rights of fashion designers.

This train of thought can be taken further. Fashion is not only about the design and making of a product; it also includes the presentation and retailing of the product. The opportunity to strengthen the course content by asking students to design and make their garment in the studio, to protect it with design registration and to put it in the shop window is a complete process, which begins with asking the students to critique their design by imagining they have just walked past it in a shop window, or viewed it online.

> I often point out to my students that, in this context, a good measure of success is whether they would buy their garment.

Here, it is important that students become impartial about their creative work if it is to be a commercial product offering so as to become more objective about the capacity for a return on the capital that has been invested into their career.

3.7 OUTSIDE THE COMFORT ZONE FOR ASSESSMENT

In my years in industry I was often involved with seasonal fashion events at horse-racing carnivals, mainly judging fashion-on-the-field contests, but also being involved in photo shoots, or supplying garments for models for other goings-on at racecourses, which meant I built a number of relationships with the public relations people. I devised this activity to take students out of their comfort zone, which, in this instance, was the controlled environment of the studio and the urban environment, and to immerse themselves in the excitement of a live fashion moment. This was a way to reframe classroom contexts by moving out of an old position and creating a new position in order to gain a fresh perspective (Larrivee, 1996). It was also a rich cultural experience for international students (see 2.1 Figure 3.4).

3.7.1 A Day at the Races

For this course activity, I took a group of mainly international students to Eagle Farm Racecourse in Brisbane to participate in activities celebrating the annual Australian Melbourne Cup Race. In a joint promotional relationship with racecourse management, we set up an illustration tent near the fashion-on-the-field contestant stage. As contestants waited in line before walking down the runway, one of my students

FIGURE 3.4 *An illustration of a horse racing track with people* © Ula Francis 2019

picked out a single contestant and sketched their outfit, which was presented to the contestant on behalf of the racecourse and my University. There were numerous outcomes that were all extremely positive. For instance, each student had to dress appropriately in race wear and many students had never been to the races so this was tremendously exciting for them.

Furthermore, this activity formed a strong bond among the group who took great delight in exploring the boundaries of the racecourse tents full of celebrities, as well as coming to grips with the Australian affinity for having a bet. An important benefit for my international students was their introduction to an event that had strong cultural significance for some Australian students. Many of the international students had never seen a racehorse before, and to have the horses thundering past our illustration tent was a great and colourful spectacle that created great opportunities for photographs. Over the course of the day my students were mentioned in various blogs, which provided other opportunities for social media exposure for all parties involved.

The assessment I set for this activity was a reflective multimedia assignment. Students were to document their day at the races from a fashion perspective and provide updates to their illustration journal that included a small piece of reflective writing. There was rich material to draw upon, and because this was an assignment based upon an authentic cultural fashion experience, the student response was enthusiastic.

3.7.2 Shopping

Another slightly more commercial activity I created was to take students on an excursion to a mall. While this activity might seem a little close to an everyday experience, it is the change of perspective of participant observation that is important (Garner, Wagner and Kawulich, 2009). Once there, they were broken into teams of two or three and asked to compare the activity underway in some of the retail stores. Among fixed costs such as rent, electricity and the minimum wage, they measured the number of garments on sale racks, staff numbers, the size of the shop and how many shopping bags were seen leaving each store over an hour.

The assessment task was to determine whether each store was viable, and to propose solutions to remedy underperforming stores, which might include changing the store layout, the kind of fashion stock on display and perhaps even removing the designer from their job.

The inspiration for this activity came from a period of time in the 1990s when I wished to have a retail store in London. I had no connections, so each time I visited, I would adopt a routine that was exactly as I asked my students to do. I would simply sit outside shops that I considered would be competitors in the market and observe them at regular times throughout the day. Over the course of a number of visits to London, I gathered rich data that showed me that having a shop in London was a bad idea, and I turned my focus to China. This kind of activity can take the "fashion" out of the aspiration and provide a real, measured and genuine appraisal of the situation without the qualitative aspects that an emotional decision might entail.

3.8 REFLECTIONS OF MANAGING MY TEAM

For the pracademic, teaching from the experience gained by making a mistake in industry has another benefit. The academic environment provides time for contemplative reflection of your activities with a view towards improving your teaching and creative practice. In my case, much reflection has been undertaken as I stand in front of my class giving a lecture about what appears to the perfect textbook scenario, wishing I could impart as strongly as possible that the perfect scenario does not exist. However, in industry, and specifically in small businesses, practitioners usually do not have the time for reflection.

Here the decision-making process is quick because of the commercial nature of industry, and there is often very little peer review, particularly if you are a manager. If you are the owner of the business, there is likely no checking of the decision-making process or the validity of the strategy you are about to enact. Pirtea, Nicolescu and Botoc have expressed that "what a strategic plan can do is shed light on an organization's unique strengths and relevant weaknesses, enabling it to pinpoint new opportunities, or the causes of current or projected problems." Yet, the impact of a poor decision reverberates very quickly and is sometimes difficult to correct. Here, important insights into industry can be discussed with students in terms of why a decision was made and what changes could be made to improve the outcome, should the situation arise again.

This is very important when considering how team-based exercises or assessments are placed in a curriculum, because it highlights the reliance of team members upon one another while drawing attention to the team mission; or, in industry, the common goal of the business. A key point relating to this activity is that University graduates in creative practices often enter industry by setting up their own businesses or begin working for a small business where they must multitask. Here, the employability of the graduate is enhanced as students are a little more prepared for the reality of taking responsibility for their decisions in the workplace.

To further this point, I often recall some of the moments in my very early fashion career that related to how I treated my staff when I was a young designer and business owner. The nature of my business meant that I was constantly under immense pressure to keep the cash flow growing. My business at the time included several retail stores with the associated retail staff and a workroom that housed about five or six staff members. I realise now that there are many moments where I made decisions without any strategic imperative that had a poor impact on the business and affected staff morale. At times, staff simply left my employ.

3.8.1 See You Later!

I particularly recall speaking in confidence with a senior staff member, whom I thought was sympathetic to the volatile nature of the industry, about a poor sales period we had encountered at retail and how we would have to reduce costs, perhaps by reconfiguring machinist hours in the workroom. The next day I received a resignation letter. The staff member had misread our discussion and had decided to jump ship, fearing the business was going under. Unfortunately, the person was a senior patternmaker and I was left without the ability to create new garments until I could find a new patternmaker, a process that took several weeks and cost the company a lot of money (see Figure 3.5).

As I stand in front of my students in the University studio, I have reflected upon this moment often, particularly of the naïveté of my decision to share my problem and how this might have been counteracted by having more experience. It is clear now that I was working in an organisation (my own small business) where there were certain values, behaviours and political meanings to the structure I had created; however, it is also clear when I consider the very clear structure that Iuscu et al. (2012) propose: that my grasp of organisational leadership was limited.

Barbara Larrivee (2000) describes how "the process of becoming a reflective practitioner cannot be prescribed. It is a personal awareness discovery process." In hindsight, I was really asking for some feedback without being very clear about it. However, I had received no formal training of fashion design or people management at the time, and knew only to keep pushing forward. In this context, what I think about most is how to construct teaching methods and other tasks that allow my students to not make this kind of mistake when they set up their own business or work for a larger company.

It is difficult to build a workplace culture of transparency by teaching alone; however, one of the ways that I have utilised this experience is to build assessment tasks that are conducted in teams where each member is reliant on the team effort to move towards the best outcome. The task also teaches students that no man is an island, and that a single person cannot do everything required to build a viable business.

FIGURE 3.5 *Two people and a resignation letter* © Ula Francis 2019

The point of this task is to build resilience through adversity (Deveson, 2003), and the task can put this in place quite effectively. There are two pressure points to consider. As the team becomes familiar with one another and the activity is underway, the lecturer can remove a portion of time from the activity, thereby speeding up the deadline for delivery, or the lecturer can remove a team member without notice and substitute someone else. I often think that this is the way a fashion collection is made, by a constant editing and re-editing of ideas and the people involved until the deadline of time means the ideation is halted and the production of the garments must begin. The difficult part to teach is that people improve their capacity for this task over time, which is often in short supply in the classroom. A way to frame this is to ensure that you have imparted the importance of understanding the process more than the particular practical outcome required by the assessment.

This activity works well in teams of four, where three people can reorganise themselves to pick up the slack and to integrate an unfamiliar person into the activity. It simulates the reality of the workroom quite well; students acquire skill sets that complement their own by having to react quickly, whilst also learning to manage their professional presence. An additional layer of pressure is to constrict the parameters of the assessment task by some sort of mock commercial outcome. For instance, this could be coming in on-budget, or achieving the most sales in a pop-up shop where students can sell their creative work. Linking the activity to the process of making money is a key learning outcome.

3.9 FASHION, CAPITALISM AND DESIGN DNA IN THE STUDIO

It may seem strange to converge several seemingly diverse topics in the studio; however, the production of a creative work, especially a product such as fashion, relies on fundamentally imparting an understanding of the role of financial capital and the creation of value derived from human capital to the student cohort. In many ways this is a cross-discipline approach, which demonstrates the importance of interaction among diverse disciplines. At the outset of determining the key themes of the curriculum, I have tried to instil exercises that allow a student to begin to understand the commercial value of their creative activities, which provide a way of measuring one's individual worth.

Most students will be aware of the minimum wage, so one way to do this is to allow each student to determine their hourly pay rate and to use this as a measuring stick throughout the course delivery. However, the primary outcome of this activity is to centre the student in their creative practice and to equip them with a series of tools that allow them to place a value on their work, which includes a profit from their work that, in turn, means they can build the skills to determine how to invest in themselves as they move through their career.

Once again, this teaching method is drawn from my own industry experience when I reflect upon how I determined the price of something I had created. It is difficult to place a value on one's creative work. In the fashion industry, the use of a costing sheet is standard practice, which is a method of aggregating all material, labour and shipping costs so that a final manufactured cost price can be determined. If software such as Adobe Illustrator is used for the creation of tech packs, the work is date- and time-stamped as well. Yet, often, the costing sheet is a generic document that fails to capture the value attributed to the aesthetic value of a design, especially when a young designer is on the path to establishing their own creative DNA which they will refine over the course of their career.

In my opinion, this is the most important value that must be identified and accounted for because it comprises the very essence of a brand and possesses a tangible value. Each graduate fashion designer has the potential to develop a design signature that will form the basis of their design career. If they are passionate about developing this aesthetic, the process will sustain them when they question their progress. However, in the early stages of a design career this is difficult to identify, and is often only made manifest when a designer begins to understand who they are in the context of their life experience and cultural immersion.

A simple method for identifying this emerging design DNA is to build a component into the assessment tasks that requires the student to identify "who they are, where they have been and where they are going" at regular times throughout the semester, and to have a one-on-one discussion with each student about how this is reflected in their journals. Generally, their journal activity will show a process of refinement as they move through their first year at University. It is also important to ensure that each member of the teaching team who is responsible for connected courses across the three years of the fashion degree is cognisant of each student's positional DNA so that this can be supported at various stages. Here, the important point is to teach to the strength of each student.

Leading on from the concept of team-based activities is the ability of the pracademic to expose students to a professional network. In the fashion industry students who are exposed to as many aspects of the supply chain as soon as possible are more prepared for the experiences they will encounter when they enter the workforce.

3.10 GLOBAL FASHION – CONTEMPORARY CONTEXTS

Fashion is a global industry, with design and supply chains often geographically separated. The classroom or studio is also an international space, with students from many countries, often separated from their families. In Australia, many fashion students are from Asia or India, and English is a second language. When I was in industry, my workrooms were in Shanghai, China, and I was the outsider in an unfamiliar culture. I spent many years trying to understand how the fashion industry in China worked and how young Chinese fashion graduates would build their businesses.

I am often reminded of this when I stand in front of my class at the beginning of a new semester with fresh faces in front of me. I have learnt that, while there are sometimes vast geographic, political and cultural divides, the process of making fashion remains the same. However, there is also a great tension arising from the environmental impact of fashion and the widening chasm between fast fashion and the movement towards slow fashion (Binotto and Payne, 2016). This is often discovered early by a student in their design education and often differs depending on where the student has come from; however, this is also an important aspect of design.

Learning how to adapt quickly to changing conditions, to become resilient and to move forward with confidence are important elements in one's design practice that are difficult to teach until they are encountered in reality. Consequently, I try to ensure that there is equality in the content of my course, particularly in the case studies we unpack and much of the visual imagery. I do this by trying not to assume that ideas about identity, language and meanings are shared by all, and by exploring new and reflexive perspectives in creating content (Bouten-Pinto, 2016). This can be done by including as many Asian designers as European designers in case studies, and by providing the opportunity for international students to study design practices from their own countries. Generally, the feedback has been positive. For instance, students are given permission to explore their own backyard in the context of the theoretical underpinnings delivered in the course: "This new insight that I got from today's lecture was extremely educational and informative" (John, 2017).

Another activity I have encouraged, and often built into assessment tasks, is to ask students to be aware of fashion events that are occurring around the world. Dominant European fashion capitals now have competitors, particularly in Asia where a rising middle class is learning quickly about conspicuous consumption. In the past, this kind of contemporary research was conducted by reading fashion magazines; however, with the decline of traditional media, this activity is increasingly found in the digital realm. Here there is a wealth of information generated each day that adds to the shifting form of the global fashion system, which means rethinking fashion according to its everyday practice (Thornquist, 2018).

When developing course content, I ensure that a requirement to read prominent online newspapers is included. Each newspaper has a fashion section that is updated daily, and this provides a simple yet effective way to impart an understanding of the connections between actors in the fashion system in a way that a layman might understand. Most importantly, this is a way to have a dialogue or discussion with the student cohort in which the pracademic can gain real-time feedback about the direction of the course content. Lawrence Susskind (2013), who possesses a legal background, speaks specifically of this aspect; of finding problems in the field in order to work collaboratively with stakeholders to generate solutions. This kind of levelling mechanism enables genuine discussion and fosters a sense of inclusion in the process of learning that really should be a back-and-forth exchange of new information with an accompanying feedback loop.

In this chapter, I have tried to use this principle to illustrate my point literally. The illustrations that form part of the storytelling aspect that is a unique attribute of a pracademic are mostly drawn from photographs of each moment or incident. I asked a fashion student whose work I noticed in her reflective fashion journal to use a little bit of creative licence with the images and, in this way, I have included her perspective and gained a little feedback in the process. We have used the principles of teamwork to build something a little better than we could as individuals and, because the student was paid for her work, she has learnt a little about how to value her creative output. I am reminded by Larrivee (2000) that the "route to critical reflection cannot be pre-planned – it must be lived."

3.11 CONCLUSION

In conclusion, this chapter is a personal reflection upon some of the unique attributes that an industry professional might bring to academia, which in turn may create opportunities to construct teaching and learning outcomes that are conducted with the real world in mind. While the anecdotes in this chapter are drawn from the fashion industry, there are many other applications of these ideas that may be suitable in diverse, cross-discipline settings from any industry. Key aspects of the teaching methodologies that have been discussed include the importance of teaching "process." It is clear that there is not enough time in the University curriculum to seek perfection in the studio or the classroom, and fashion design is a constant lifelong refinement of one's personal aesthetic; therefore, learning the concept of a process is vital for a career in the creative industries. Finally, design is concerned with the creation of value as well as the protection of that value in its form as intellectual property.

REFERENCES

Appel, G., Libai, B., & Muller, E. (2017). On the monetary impact of fashion design piracy. *International Journal of Research in Marketing*, *35*, 591–610.

Binotto, C., & Payne, A. (2016). The poetics of waste: Contemporary fashion practice in the context of wastefulness. *Fashion Practice*, *9*(1).6–7 doi: 10.1080/17569370.2016.1226604

Bouten-Pinto, C. (2016). Reflexivity in managing diversity: A pracademic perspective. *Equality, Diversity and Inclusion, 35*(2), 136–153.

Bridgstock, R. (2009). The graduate attributes we've overlooked: Enhancing graduate employability through career management skills. *Higher Education Research and Development, 28*(1), 31–44.

Bridgstock, R. (2011). Skills for creative industries graduate success. *Education and Training, 53*(1), 9–26.

Deveson, A. (2003). *Resilience.* Sydney: Allen & Unwin.

Garner, M., Wagner, C., & Kawulich, B. (2009). *Teaching research methods in the social sciences.* Farnham: Ashgate Publishing.

Hartley, J., Potts, J., Cunningham, S., Flew, T., Keane, M., & Banks, J. (Eds.). (2013). *Key concepts in creative industries.* Los Angeles, CA: Sage.

Iuscu, S., Catalin, N., & Zguba, F. (2012). Organizational behaviour and leadership. *Metatlurgia International, 17*(11), 207–210.

Janssens, A., & Lavanga, M. (2018). An expensive, confusing, and ineffective suit of armor: Investigating risks of design piracy and perceptions of the design rights available to emerging fashion designers in the digital age. *Fashion Theory. 24*(2) doi: 10.1080/1362704X.2018.1515159

John, Z. (2017). Tim Lindgren: Contemporary Chinese fashion design and the European fashion system. Retrieved from https://zoyajohn.wordpress.com/2017/09/10/tim-lindgren/

Larrivee, B. (1996). *Moving into balance.* Santa Monica, CA: Shoreline.

Larrivee, B. (1999). *Authentic classroom management: Creating a community of learners.* Boston, MA: Allyn and Bacon.

Larrivee, B. (2000). Transforming teaching practice: Becoming the critically reflective teacher. *Reflective Practice, 1–3*, 293–307. doi: 10.1080/713693162

Posner, P. (2009). The pracademic: An agenda for re-engaging practitioners and academics. *Public Budgeting and Finance, 29*(1), 12–26.

Raustiala, K., & Sprigman, C. (2009). The piracy paradox revisited. *Stanford Law Review, 61*(5), 1201–1226.

Schon, D. (1983). *The reflective practitioner – How professionals think in action.* New York: Routledge.

Susskind, L. (2013). Confessions of a pracademic: Searching for a virtuous cycle of theory building, teaching, and action research. *Negotiation Journal, 29*(2), 225–237.

Thornquist, C. (2018). The fashion condition: Rethinking fashion from its everyday practices. *Fashion Practice 3.* 291–292 doi: 10.1080/17569370.2018.1507147

4 A "Non-Designer" in a Design School

Evonne Miller

CONTENTS

4.1 BEING A NON-DESIGNER IN A DESIGN SCHOOL

In 2007, I joined QUT's School of Design as a lecturer/research fellow. What was quite unusual about this appointment, however, is that I do not have any formal qualifications in design. My educational background is in psychology. I hold a BA (Hons) and a PhD in Psychology, as well as a BCom in Marketing Management (for the record, I rarely disclose this qualification to colleagues, fearing that this might make me the key person for any marketing, branding and communication work), all from the University of Otago in New Zealand. In this chapter, I reflect on my experience being a non-designer in a design school and how I have a crafted a unique role for myself, culminating with promotion to Professor of Design Psychology in 2019 (I have essentially created this title to reflect my expertise, inspired by others across the globe). Before I reflect on the origins, advantages and disadvantages of this transdisciplinary appointment, and my learnings and misadventures in crossing disciplinary tribal boundaries, I will start with a discussion about the changing role of disciplines and disciplinary knowledge, and why transdisciplinary appointments such as mine may become more common.

4.2 THE CONTEXT – UNDERSTANDING AND CROSSING TRIBAL BOUNDARIES

Becher's (1989) book was the first to deeply explore discipline-based differences in values, norms, behaviours, ideologies and operating paradigms. Becher introduced the tribes and territories metaphor to explain how knowledge structures of disciplines (academic territories) strongly influence the values and culture of academics, who share a framework for understanding the field's subject matter and live in disciplinary tribes. As Becher and Trowler (2001) explain, "being a member of a disciplinary community involves a sense of identity and personal commitment, a way of being in the world, a matter of taking on a culture frame" (p. 47).

In expanding on what is and is not considered a discipline, Trowler (2012) recently argued that, while he viewed nursing studies as a discipline, he felt design was too broad a term given the differences in the field of knowledge and disciplinary backgrounds. As a disciplinary field, design is diverse – ranging from fashion and product design to multimedia graphic design, to architecture, landscape and interior design. My own School of Design, for example, consists of seven design disciplines grouped under two broad clusters: spatial design (architecture, landscape architecture and interior architecture) and experiential design (interaction design, fashion design, visual communication and industrial design). Despite sharing a broad commitment to creating and making, each of these unique design disciplines is underpinned by differing intellectual frames, assumptions, processes and methodological approaches.

Rodgers and Bremner (2016) have even suggested that design disciplines have dissolved, arguing that "design today is characterized by fluid, evolving patterns of practice that regularly traverse, transcend and transfigure disciplinary and conceptual boundaries" (p. 22). As many of the complex challenges of the 21st century (climate change, sustainability, health, ageing populations) warrant fresh interdisciplinary and multidisciplinary approaches (Land, 2012), situations such as mine – being located outside your academic tribe – may become more mainstream. Thus, this chapter critically reflects on my learnings from this unusual experiment in transdisciplinary research, teaching and practice.

At the outset, I must acknowledge my privilege; while my path has been windy and unexpected, I am now a full Professor. The most recent statistics in Australia suggest less than a third of PhD graduates will secure an ongoing academic job, and the competition for these roles is intense. For example, in an auto-ethnography of failure during the academic job search entitled *"I know I'm unlovable,"* Herrmann (2012) outlines, in detail, how he lost his confidence after multiple rejections. While Herrmann ultimately landed an ongoing academic position, the most recent Australian statistics suggest approximately half of employed doctorate holders are working in tertiary education and research, with the remainder in business (~23%) and the public sector (~24%; ABS, 2016). Of course, while not all PhD graduates want an academic role, such stories and statistics remind us that we must have open conversations with the higher degree research students we supervise about their expectations, different career paths and strategies that will help them achieve their dreams – and to make them aware that securing an academic job is no easy undertaking.

4.3 HOW DID A NON-DESIGNER END UP IN A DESIGN SCHOOL?

You may be wondering how exactly a non-designer ended up in a design school. The story, like many in academia, is one of accident, happenchance and serendipity – that is, university restructuring! I originally joined QUT in mid-2004, as a fixed-term Postdoctoral Research Fellow in the Centre for Social Change Research. In 2008, the Centre and associated faculty, the Faculty of Humanities and Human Services, was restructured and closed, with some staff dispersed throughout QUT. With several colleagues, I was moved into the School of Design in the Faculty of Built Environment and Engineering, in part because we had been collaborating with them on teaching and research into sustainability (a core research interest of mine).

As a discipline, design straddles engineering, built environment, and creativity. And so, another restructure in 2012 saw the School of Design move into the Creative Industries Faculty, where we remain today - co-located with the School of Communication and School of Creative Practice. The fact that Design has been based in both engineering and creative industries faculties highlights a critical and unique distinguishing feature of design: it crosses multiple disciplinary bases. And, in fact, the move away from our built environment and engineering colleagues has not been easy for some, illustrating the many disciplinary differences that exist with schools, departments, faculties, and disciplines.

4.4 LEARNING FROM EARLY MISSTEPS IN TRIBAL TERRITORIES

In writing this chapter, and reflecting on my experiences, I am reminded of how I have learnt from the mistakes I made as a young undergraduate psychology student in the late 1990s. I studied at the University of Otago, a prestigious "sandstone" university in New Zealand. At the time I trained, the Department of Psychology had a very strong quantitative, behavioural, positivist research orientation. The quantitative experimental approach to research was unquestionably presented as superior to qualitative research – and I drank this quantitative cool-aid! As part of my education, I dissected rats' brains, designed and ran experiments, and crunched numbers in SPSS. I was, as I have described elsewhere, very strongly "indoctrinated with the experimental method" (p. 385, Miller, 2018).

What changed my view on the value of qualitative research was my study outside the Department of Psychology. While joint and double degrees are now commonplace, at the time, it was relatively rare to compete a double degree. I was the only person in my cohort who was endeavouring to do this, and I vividly remember very complicated conversations at the beginning of each year when our course plans were discussed and approved. This was before online course plans, and so required a scheduled 10–15 minute face-to-face conversation about your course plan with the subject area coordinator, and getting formal approval each year to complete a Bachelor of Commerce alongside a Bachelor of Arts (Hons) in Psychology was always difficult. While these battles taught me resilience, determination and grit (personality features that have stood me in good stead as an academic, as I discuss in Miller (2021), in this book, my introduction to the value of qualitative research in my Commerce degree was a game-changer.

I vividly remember, as a third-year student, confidently telling my consumer behaviour lecturer (and the class) that "qualitative research was a waste of time, and of much lesser value than quantitative." To her credit, she heard me out, replying: "let's see what you think after your assignment" – which was to design, conduct, analyse and report on a small qualitative research project. Drawing on Russell Belk's concept of possessions and the extended self, I used an ethnographic approach to explore the meaning people attach to possessions – and I was hooked forever on qualitative research, and design and arts-based research has become my passion (see Miller, 2018). While my doctorate used experimental approaches to explore women's body image and the media, over the last decade, my research focus has moved from consumer culture to the design of aged care and how we might redesign our lives, homes and cities to be more resilient, connected and sustainable.

Other academics have described a similar journey in their research careers, as they have moved away from their training in positivist science towards experimentation with qualitative and arts-based research methodologies. Hodgins and Boydell (2014), for example, recall being scientists whose "roots are (were) firmly planted in a positivist tradition—training in a faculty of medicine and particularly in community health and epidemiology," and then "branching out" into the interesting yet murky waters of arts-based research.

While I am not aware of any academic publications reflecting on a similar situation to mine (having a tenured role in a department of Design, very different from disciplinary training in Psychology and Marketing Management), I know of several high-profile design academics with diverse non-design disciplinary origins; for example, Don Norman, with a background in engineering and psychology, was Director of The Design Lab at University of California, San Diego (and renowned for his best-selling book, The Design of Everyday Things). At Monash University in Australia, Sarah Pink, with a background in anthropology and visual culture, is Professor of Design and Emerging Technologies, and Director of the Emerging Technologies Research Lab, exploring the design, use and futures of new and emerging technologies. Yet, while there is a growing number of publications on "learning the ropes" in academia, from the job-hunt to reaching the professoriate and becoming Dean (see Kelsky, 2015; Justice, 2019), to date, publications on how to approach working in a different discipline from your original training are scarce.

There is one notable exception. Kalekin-Fishman's (2019) recent auto-ethnography outlines how she "didn't make it" at the university, reflecting on how "a series of inept choices" led to her not fitting in with "desirable discipline" (p. 185). Kalekin-Fishman's doctorate explored music in early childhood education, yet she found "the terminology, the methods, and the questions that were important to her as a sociologist were almost like a foreign language" (p. 186) to her colleagues in the School of Education, who had been trained in psychology. She also describes ignoring organisational politics to her peril, rejecting an assigned mentor (due to workload demands) and becoming an outsider who missed out on critical tacit knowledge and organisational learning. Kalekin-Fishman reflects on how she missed out on an appointment to professor, because, she "repeatedly mis-read structures" and located herself outside "ongoing conversations" in her department (p. 183).

There are several critical learnings and lessons in Kalekin-Fishman's honest account. First, having personally worked closely with a sociologist on research projects, there is no doubt that, as disciplines, sociology and psychology are very, very different. But what stands out to me most is that Kalekin-Fishman actively disengaged from organisational politics, and simply did not seem to realise the vulnerability of her position. She acknowledges making limited efforts to learn and fit into the organisational culture and philosophy; the reality is that, whether we like it or not, learning to navigate workplace politics and personalities is a critical part of our career success or failure.

While I'm definitely *not* the most measured or politically astute person, on reflection, I believe I brought a very different mindset and approach to navigating my academic role than Kalekin-Fishman. By nature, I am quite pragmatic and grateful to have an academic role in this competitive marketplace. Additionally, the mission, culture and purpose of each institution differs, valuing different competencies, approaches and achievements; my institution has been an early champion of inter- and trans-disciplinary perspectives and partnerships.

4.5 MY TRANSITION – FROM EXPERIMENTAL SOCIAL PSYCHOLOGY TO DESIGN PSYCHOLOGY

In reflecting on how and why my transition differed from Kalekin-Fishman's experience, I am drawn to transition theory. William Bridges (2004) describes change as situational events (e.g., moving to a new city, having a child, starting a new job), but transition as a slower, more complex psychological adjustment that involves a reorientation of our identity – an "inner reorientation and self-redefinition that you have to go through in order to incorporate any of those changes into your life … Unless transition happens, the change won't work, because it doesn't 'take'" (p. xii).

Bridges identifies three stages: (1) an ending, (2) a neutral zone and (3) a new beginning. The first stage, ending, means "letting go" of the familiar, and starting to see exciting new opportunities for growth. This was perhaps easier for me, as I was forced to embrace new directions in design teaching and research, which involved a "deep dive" into a new world of design theory and practice. In explaining what this means, Bridges draws on folk-wisdom to explain how "you can't steal second base with your foot on first." The "neutral zone" involves disengaging from a past identity to adopt a new identity, which leads to "the new beginning" – behaving in a new way, with a new sense of self and greater clarity about what is possible. While I would never call myself a designer, I am now comfortable labelling myself a Professor of Design Psychology – an acknowledgment of both my disciplinary origins and the last decade I have spent in a design school.

4.6 ACKNOWLEDGING AND OVERCOMING "IMPOSTER SYNDROME" – AND THE IMPACT OF GENDER

For me, one of the biggest challenges has been learning a new disciplinary language; a topic discussed in detail by Evans et al. (2021) in this book. That has involved frequent bouts of imposter syndrome, which Clance and Imcs (1978) succinctly define as the internal experience of being an "intellectual imposter" in acaemdia. Given that most honestly self-reflective academics report experiencing "imposter syndrome" when they have trained in the discipline, imagine, for a moment, my experience.

Luckily, there is much conceptual overlap between psychology, marketing (consumer behaviour) and design, and my knowledge in environmental psychology (which focuses on the interplay between people and their surroundings) has created a convergent space for collaboration. Indeed, one of my proudest career moments was in 2020, when I co-authored a book squarely on design issues with a Landscape Design colleague entitled *Creating Great Places: Evidence-based urban design for health and wellbeing* (Cushing and Miller, 2020). Given the title of my 2003 PhD in Psychology was "*Media and body image: Exploring the discrepancy between body perception and satisfaction*," there is no doubt that the last 17 years have been a wonderfully diverse time of learning and exploring new territories.

This transition has not been always easy, but I know that I am not alone in struggling to navigate the complicated terrain of academia. As well as differences in disciplinary background and knowledge, a large body of literature has documented gender differences in confidence. Over 40 years ago, Clance and Imcs (1978) wrote how high-performing female professors "fear their stupidity will soon be discovered." One department chair noted how "*obviously, I'm in this position because my abilities have been overestimated*," while another woman professor shared she felt she was "*not good enough to be on the faculty here. Some mistake was made in the selection process*" (p. 1, 1978).

What this body of research on imposter syndrome reminds me of is that everybody feels insecure, and we must all learn to overcome imposter syndrome (or at least silence that self-critical voice). It is critically important that we talk about such beliefs in academia because, as Clance and O'Toole (1987) remind us, the experience of anxiety, doubt and fear that accompanies imposter syndrome means many women (1) do not achieve all they are capable of (potentially turning down or not applying for opportunities, settling instead for what seems certain rather than reaching for their dreams); and (2) do not enjoy their successes, always fearing being revealed as a fraud and losing "the sense of reward and joy that ordinarily accompanies success" (Clance and O'Toole, 1987, p. 53).

Even today, many academics report experiencing imposter syndrome. Knights and Clarke's 2014 study of 52 business academics in the United Kingdom (ranging in level from Lecturer to full Professor) found high levels of guilt and anxiety, with respondents struggling to live up to their ideals about what it is to be a "proper" academic – someone who demonstrates "incredible intellect and eloquence." Being an inspirational, charismatic teacher; a successful, grant-winning researcher; and a competent administrator were described as impossible to do well, with one professor explaining feeling constant guilt because "the job is never done; it's never done properly and it's never done well enough. You're always feeling terribly guilty (p. 342). Participants explained "you need to be fucking amazing" because "a good, successful academic … requires a day to have 48 hours not 24," with one querying whether one person could actually do all the required tasks:

> Can you do all of these things in one professional label? We're asked to teach students, engage with students, have assessment strategies, feedback strategies, supervise MScs, PhDs, mentor people, mentor other members of staff, research, write research bids, write research papers, present at conferences, publish in high quality journals, administration and all aspects of pastoral care. I mean it's just never-ending but is it realistic? (Knights and Clarke, 2014, p. 342).

A debate about workload and performance expectations is beyond the scope of this chapter, although Chamorroc-Koc and Caldwell (2021) discuss some of the strategies design academics have implemented to manage workload. What is important to remember is that fretting about our own insecurities and engaging in imposter syndrome is time wasted and often self-sabotaging. As Facebook's chief operating officer Sheryl Sandberg explained in her best-selling 2013 book *Lean-in: Women, Work and the Will to Lead*, many women hold back when what they need to do is "*lean in*." In the book's first chapter, *The Leadership Ambition Gap: What would you do if you weren't afraid?* Sandberg reflects on how, in comparison to men, women have internalised negative stereotypes that restrict them in the workplace. Sandberg tells how Virginia Rometty (the first female chief executive officer of IBM) was offered a "big job" early in her career but worried she did not have *all* the required experience – until her husband asked her: "*Do you think a man would ever have answered that question that way?*" Sandberg (2013) herself describes experiencing imposter syndrome, and how critical it is to move past fear and doubt:

> I still face situations that I fear are beyond my capabilities. I still have days when I feel like a fraud. And I still sometimes find myself spoken over and discounted while men sitting next to me are not. But now I know how to take a deep breath and keep my hand up. I have learned to sit at the table.

I have slowly learnt to overcome my feelings of imposter syndrome in a design school. I have, however, engaged in a few conscious strategies to earn my seat at the table. I have actively engaged in teaching activities, ranging from postgraduate research method units to socio-cultural studies and a large first unit introducing sustainability to 1,000 students across the faculty. I have reached out to my design colleagues, collaborating on publications, workshops and research grants. I have deeply engaged with the design literature, reading widely across all seven disciples that are represented in our school. While I can now speak with confidence about being inspired by the beauty of Pallasma's work (influential architectural theorist), Richard Haag's contribution to urban design (landscape architect who redefined the meaning of industrial landscapes), and Linda Rosenberg's advocacy for thoughtful interior design for disability, as well as the power of co-design methods and the potential of design thinking for disruptive innovation, developing my confidence with the methods, theorists and mindset of design has taken time – and commitment.

If, like me, you also experience imposter syndrome, Clance and Imcs (1978) recommend a homework assignment of keeping a record of positive feedback (my colleague Professor Jillian Hamilton recommends academics set up a "brilliance folder" on their computer, and save all positive feedback there). The advice in a humorously titled blog post in the Times Higher Education Supplement "*Think like an impostor, and you'll go far in academia*" also resonates with McMillan (2016), reminding us that self-delusion is a common feature of academia, and that even if you feel "completely and hopelessly inadequate," you are in the best position to excel – "surrounded by experts, equipment, and relevant literature." I agree. We must all make the most of being an academic, challenge ourselves to grow by continually engaging in lifelong learning, and extend our disciplinary expertise by reading books, listening to podcasts and attending workshops and lectures outside our unique field of expertise. That is the essence of lifelong learning, an attribute we work hard to foster in our students and one we should all embrace – as Swann and Rezayan (2021) do.

4.6.1 Finding My Disciplinary Home

While other chapters in this book (for example, Evans et al., 2021) have discussed design thinking and design education, I want to reflect briefly on the disciplinary origins of design – and how and why I have been able to adjust so well here. Firstly, I think it helps that psychology and marketing (and my specific focus of consumer behaviour) is the study of people, as is design. I found very early on that, while I was less aware of specific design theorists and approaches, my design colleagues and I shared some common frames of references – the literature in environment psychology, behaviour change and consumer behaviour focus on people–product interactions and provided a strong platform for shared communication and collaborations. For instance, just last week, I had a conversation with an Industrial Design colleague about product attachment and design for durability; we discussed consumer behaviour theorist Russell Belk's concept of possessions and the extended self – seminal work I had read and referenced in my undergraduate Marketing degree. Similarly, in the context of healthcare design, I share with my Architecture and Interior Architecture colleagues an appreciation of Rodger Ulrich's classic study of how patients whose hospital beds had a view of nature healed quicker, and complained less, than those viewing a brick wall.

In many ways, my research interests have centred around issues of social justice, health and wellbeing, and these are shared by my design colleagues. And, my unique transdisciplinary perspective is valued and sought – for example, last year, I chaired the most higher-degree research panels in my faculty (12), with the next highest academic chairing 7. I have worked to be a productive, engaged and supportive colleague in design, always aware of how fortunate I am.

It also helps that design is a young and diverse discipline, which has a strong tradition of respecting, engaging with and building on the contributions of other disciplines, including engineering, geography, anthropology, sociology and, of course, my disciplinary backgrounds of psychology and marketing. I often ponder how a psychology department might have reacted in the same situation: would they have embraced a young, somewhat naive designer the way my colleagues have supported me? I am not so sure. What I do know is that I have *never* been made to feel out of place or disrespected by my design colleagues, who have instead displayed the upmost openness, collegiality and support for the female psychology scholar in their ranks.

In response to their generosity, as described earlier, I have endeavoured to learn the language and norms of being a design academic – accepting and engaging in a purposeful process of listening, learning and respecting different domains of knowledge. In many ways, Swann and Rezayan's chapter in this book, *Always Learning*, resonates, as that is how I feel – but luckily, I enjoy learning! Gary Burge, in his book *Mapping Your Academic Career*, makes a similar argument about the value of continual growth throughout our career. Burge (2015) outlines how each of us can fail, at any stage, if we are not self-aware – we must, he argues, make the conscious choice to continue to grow and not "fossilize" in the form we had when we "were thirty-five" (p. 67).

I also have endeavoured to be a "good academic citizen," engaging in what has been labelled "institutional service" as a local health and safety representative, on the ethics and space planning committees, and serving as higher degree research director for the faculty. While many academics do not want or seek out these roles, believing that the only route to promotion is through research and that service is a distraction, I believe there are many benefits to strategic service. As Pfeifer (2016) notes, these roles enable people to learn more about institutional processes, thus broadening perspectives, and they are also a critical way to meet and connect with different colleagues. Networking, and developing strong interpersonal relationships, is critically important to succeeding in academia, and being willing to actively and happily engage in "service roles" has helped build my reputation as a collegial colleague.

In a field where, as Hearn (2008) notes, "competitiveness, intellectualism, achievement-orientation, hierarchy, and evaluativeness [may give rise to] all manner of high emotions, anxieties, defences, denials, deceptions, and self-deceptions, rivalries, insecurities, threats, vulnerabilities, [and] intimacies" (p. 190), I endeavour to be genuinely authentic and kind. After all, as the popular saying goes, everybody in academia is smart – so distinguish yourself by being kind. Additionally, in reflecting on why I have felt so welcomed in a design school, I need to acknowledge my strengths: I am an engaging teacher, with a very strong research track record in the topical areas of design for health, population ageing and climate change. Someone who was failing on the key metrics of teaching, research, leadership and service may well have had a less positive experience.

4.7 FROM IMPOSTER TO PROFESSOR

In closing, I am reminded of how Land (2012) pondered if and how the pace of change this century would accelerate "interdisciplinary and even transdisciplinary ways of thinking and practicing among academic tribes" (p. 185), or if academic incentives would mean we remain separated within our disciplinary territories. Even disciplines which share conceptual and theoretical origins (a focus on people and their behaviours) inhabit vastly different territories. However, I believe that my experience – as a non-designer who has crafted a successful career in a design school and reached the rank of full professor – is evidence that transdisciplinary appointments can work.

However, it requires openness and willingness to learn, on both sides. Personally, I am grateful for the university restructuring that altered the path of my career, moving my research from body image and the media into issues of design. I feel I have found my home, and my tribe. But, as Land reminds us, interdisciplinarity remains rare and difficult, as "disciplinary specialists tend to find it difficult to leave the comforting Eden of their discipline, to cross the boundary into interdisciplinarity" (p. 183). Working outside your disciplinary training requires a willingness to tolerate uncertainty, to be comfortable with different perspectives, and be open to constant learning. While interdisciplinary appointments such as mine remain rare in academia, I hope my account illustrates their value – and might help encourage other design departments to take a chance on employing a non-designer.

REFERENCES

ABS. (2016). *Microdata: Education and work, May 2016.* Cat. 6227.0.30.001. Australian Bureau of Statistics.

Becher, T. (1989). *Academic tribes and territories: Intellectual enquiry and the culture of disciplines.* Open University Press.

Becher, T., & Trowler, P. (2001). *Academic tribes and territories.* The Society for Research into Higher Education and Open University Press.

Bridges, W. (2004). *Transitions: Making sense of life's changes.* 2nd ed. Da Capo Lifelong Books.

Burge, G. (2015). *Mapping your academic career: Charting the course of a professor's life.* Westmont: InterVaristy Press.

Chamorro-Koc, M., & Caldwell, G. (2021). Running the academic marathon: Planning and executing as planned. In Blackler, A., & Miller, E. (Eds), *How to be a design academic.* Taylor & Francis, 189–207.

Clance, P., & O'Toole, M. (1987). The imposter phenomenon. *Women & Therapy, 6*(3), 51–64.

Clance, P.R., & Imcs, S.A. (1978). The impostor phenomenon in high achieving women: Dynamics and therapeutic intervention. *Psychotherapy: Theory, Research and Practice, 15(3)*, 241–247.

Cushing, D., & Miller, E. (2020). *Creating great places: Evidence-based urban design for health and wellbeing.* New York: Routledge.

Evans, R., Kelly, N., & Kerr, J. (2021). Being a Design Academic: Design Process, Practice, Thinking. In Blackler, A & Miller, E. (Eds), *How to be a design academic.* Taylor & Francis, 13–28.

Hearn, J. (2008). Feeling out of place? Towards the transnationalizations of emotions. In Fineman, S. (Ed.), *The emotional organization: Passions and power* (pp. 184–201). London, UK: Blackwell.

Herrmann, A.F. (2012). "I know I'm unlovable": Desperation, dislocation, despair, and discourse on the academic job hunt. *Qualitative Inquiry, 18*, 247–255.

Hodgins, M., & Boydell, K. (2014). Interrogating ourselves: Reflections on arts-based health research. *Forum Qualitative Sozialforchung/Forum: Qualitative Social Research, 15*(1), online.
Justice, G. (2019). *How to be a dean*. Johns Hopkins University Press.
Kalekin-Fishman, D. (2019). Mis-making an academic career: Power, disciplines, structures, and practices. In R. Murray & D. Mifsud (Eds), *The positioning and making of female professors: Pushing career advancement open* (pp. 177–200). Cham, Switzerland: Palgrave Macmillan.
Kelsky, K. (2015). *The professor is in: The essential guide to turning your PhD into a job.* New York: Three River Press.
Knights, D., & Clarke, C. (2014). It's a bittersweet symphony, this life: Fragile academic selves and insecure identities at work. *Organization Studies, 35*(3), 335–357.
Land, R. (2012). Crossing tribal boundaries: Interdisciplinarity as a threshold concept. In P. Trowler, M. Saunders and V. Bamber (Eds), *Tribes and territories in the 21st century: Rethinking the significance of disciplines in higher education* (pp. 175–185). London: Routledge.
McMillan, B. (2016). Think like an impostor, and you'll go far in academia. Retrieved 26/11/2019 from: https://www.timeshighereducation.com/blog/think-impostor-and-youll-go-far-academia.
Miller, E. (2018). Breaking research boundaries: A poetic representation of life in an aged care facility. *Qualitative Research in Psychology, 15*(2/3), 381–394.
Miller, E. (2021). Developing the resilience to 'dare greatly' in academia. In Blackler, A., & Miller, E. (Eds), *How to be a design academic*. Taylor & Francis, 177–188.
Pfeifer, H. (2016). How to be a good academic citizen: The role and importance of service in Academia. *Journal of Criminal Justice Education, 27*(2), 238–254.
Rodgers, P.A., & Bremner, C. (2016). The concept of the design discipline. *Dialectic, 1*(1), 19–38.
Sandberg, S. (2013). *Lean in: Women, work, and the will to lead*. New York: Random House.
Swann, L., & Rezayan, L. (2021). Exploring the mindset of a new academic. In Blackler, A., & Miller, E. (Eds), *How to be a design academic*. Taylor & Francis, 83–111.
Trowler, P. (2012). Disciplines and interdisciplinarity: Conceptual groundwork. In P. Trowler, M. Saunders, & V. Bamber (Eds), *Tribes and Territories in the 21st Century: rethinking the significance of disciplines in higher education* (pp. 5–29). London: Routledge.

Section 2

Being an Effective Design Researcher

Section 2 documents the experience of becoming a successful design researcher. In design, unlike most other disciplines, until recently, it was rare for a design academic to hold a doctorate or have much experience doing research. Chapter 5, therefore, explores what it is like to complete a PhD while also working in academia, a uniquely challenging and often stressful experience. Chapter 6 focusses specifically on the experience and mindset of a new academic, documenting the value of being open to continually learning. Both chapters highlight how design researchers grow in confidence as they learn the language of research, and outline the critical importance of developing peer support networks and mentors who support them on this journey.

5 Getting a PhD – "How Hard Can It Be?"

Tim Willams and Shannon Satherley

CONTENTS

5.1 GETTING THAT PHD WHILE WORKING FULL TIME IN ACADEMIA

This chapter discusses the challenging situation of gaining a PhD whilst working in a full-time Design academic position within a university. Today, this is a relatively common situation in the Design fields in which the MFA or equivalent was until recently the terminal degree. The days when a Designer without a PhD could start an academic career based on professional practice experience alone now seem long gone, and in recent years many Design academics have suddenly found their employers requiring them to achieve the Doctoral qualification. This chapter draws on a combination of the authors' personal experiences and those of seven other Design academics completing, or having recently completed their PhDs while working full time. It unpacks their experiences and identifies the challenges and benefits, offering suggestions for survival, and even for having a rewarding experience whilst completing this milestone.

Our study is based on semi-structured interviews with nine Design academics from one Australian university, with reference to the scant literature on this topic, and on that relating to the challenges of PhD study in general. The authors make no claim that the experiences of staff within a single institution represent those of all Design academics. However, there are many similarities and differences identified with the challenges faced by all PhD students, and we consider this chapter a valuable historical snapshot of a group of individuals caught in the transition between the practice-based traditions of Design teaching and the research-based traditions of academia.

What emerges from our thematic analysis of the interviews is a clear picture of this PhD experience as an extremely challenging juggling act, but one from which there are benefits. Thematic challenges unpacked in this chapter include the type of thesis to pursue; the pressure to publish during candidature; the usefulness of the PhD process as a method for learning; the relationship with professional design practice experience; the varying forms – and lack – of support during and after candidature; relationships with supervisors; the challenges of time pressures, both professional and private; identity issues; and, ultimately, the emotional experiences and costs of the process of PhD completion. Synthesising our participant responses, we offer some suggestions to help those now undertaking this journey to navigate these challenges. For brevity, we refer to full-time Design academic PhD students as "staff PhDs," as distinct from "regular PhDs."

5.2 BACKGROUND

Higher education in Australia and the United Kingdom underwent significant restructure in the late 1980s and early 1990s. Institutional mergers transformed a range of tertiary educational institutions, including institutes of technology and colleges of advanced education, into "new" universities. Hitherto "applied" disciplines such as Design became part of the university sector, ushering in a new requirement for teaching staff to undertake research (Brew, 2001; Alison Lee and Boud, 2003; Melles, 2011).

In the Australian context, Lee and Boud (2003, p. 189) identify the 1999 Federal Government White Paper on research and research training (Kemp, 1999) as the first of a series of government initiatives directing university research priorities. It recommended universities report on many aspects of their research performance, including "research active members of staff and their research outputs and achievements. The last point will be important in ensuring institutions' research claims have a strong grounding in the activities and performance of their staff" (Kemp, 1999).

This shift in emphasis prompted universities to adopt strategies to develop their research, including the development of the creative, applied and Design research methodologies widely accepted today, the rise of the professional doctorate, and the engagement of academics with research experience from other fields (Melles, 2011). The focus of this chapter is on the strategy Melles refers to as the "doctoring" of staff (2011, p. 451). While the 1999 White Paper did not explicitly instruct that all university academic staff must gain PhDs, the response of universities to require this is unsurprising.

Several authors discuss the emotional dimension of academic "identity" and the challenges individuals can face when significant identity change is required (Alison Lee and Boud, 2003, p. 130). Henkel and Vabø describe disciplines as "cultures, centred round different knowledge traditions, which shape how individuals define themselves" (2006, p. 130). Swann and Rezayan explore the mindset of early career Design academics (2021). Rosch and Reich found that new academic recruits in the mid-1990s tended to have preconceived ideas of their future academic role "complementing past experiences and individual values … not necessarily consistent with the actual performance expectations" (1996, p. 125). This experience was shared by many of our study participants, as were the findings of a case study of staff at a "new" UK university, initially employed to teach, but who found themselves required to become researchers. Sikes identifies that such radically changing work requirements

> impact upon one's workload, working conditions, job content, sense of self, and personal and professional identities. How people feel about their work has implications for how they approach and go about doing it, which has consequences for both the individual and the institution (2006).

The shift towards research and new requirements for academics to gain PhDs was challenging for those in applied fields such as Design, which, as Davis describes, has "a short history of research and a long tradition of training in know-how, in the craft of solving problems with the information immediately at hand" (2008, p. 71). Melles identifies a tension arising between the more "creative" interests of academics in Art and Design, and the "academic traditions of research" predominating in universities at the time (2011, p. 453). Rowland describes this as a "fault line" between teaching and research, and that academics needed to "re-integrate" in the face of their changing circumstances (2002, p. 53). Lee and Boud describe formerly teaching-only staff as having "struggled to redefine the nature of their work and to adopt new practices for which they [had] not been systematically prepared" (2003, p. 189). Even into the late 2000s, many Design academics had commenced their careers in industry after completing diploma and undergraduate degrees, and later moving into the university sector, attracted to the idea of teaching. A relatively recent survey of Design educators and practitioners found limited understanding of the difference between "academic and vocational research and practice-oriented teaching" (Melles, 2011, p. 457). As a result, they brought with them varying "degrees of understanding and antagonism towards the institutional demand to be academically research-active" (Melles, 2011, p. 457).

Lee and Boud describe the new demands teaching-only academics faced in the late 1990s and early 2000s as creating "fear and anxiety, as well as offering the possibility of new forms of satisfaction and pleasure in a richer and more complex environment" (2003, p. 189). The PhD is the key that should open the door to this environment. Yet Lee and Boud argue that, within the Australian context, while many staff from non-traditional backgrounds (such as Design) have completed PhDs, their experiences in doing so, usually part-time and with competing personal and full-time work-related demands, have not been sufficient to make them into "competent and confident individual researchers" (2003, p. 188). As will be discussed, all these experiences and concerns were shared by our study participants.

5.3 METHOD

We invited 15 participants from 3 universities to be interviewed for this study, each undertaking or having completed a PhD while working as a full-time Design academic. A total of 9 invitees from 1 university agreed to participate (including the 2 authors), and whilst this is a small sample, the total population of Design academics currently completing a part-time PhD in Australia is estimated to be between 100–200 and so our sample represents 4.5%–9%. There are 20 accredited tertiary design "schools" in Australia offering Bachelor level degrees. Our school has 15 staff currently completing PhDs. We have a relatively big school so we estimate it would be about 10 in a smaller school. So, assuming most schools in the country have the same ratio of PhD completing staff as us, that gives us 150, plus or minus 30% gives us 100–200. As the criteria for selection are well defined, a minimum of 6 respondents would be sufficient (Guest et al., 2006; Mason, 2010).

During a series of individual interviews, we asked participants semi-structured questions about their experience of gaining a PhD while working full time as a Design academic. The questions were organised into two groups. The first aimed at getting participants talking by starting with direct, easily answered questions about the practicalities of their PhD. For example, "Was/is the thesis by monograph or publication?" "What was the length of candidature?" Often, these questions were followed with open-ended ones such as "why was that?" which added context to responses. The second group of questions were open-ended and aimed to explore emotional responses to the experience of their PhD journey. For example, "How did/do your colleagues regard your PhD student status?" and "What do you think were/are the biggest challenges of managing working full time as an academic, doing a PhD part-time, and maintaining the rest of your life? What were/are the greatest benefits?"

Some who declined to be interviewed were concerned their comments would reflect badly on their supervisors and possibly their university. Therefore, we acknowledge that some strongly held views may not appear in this study. To protect the confidentiality of participants, all names have been removed, and some quotes have minor modifications to remove identifying elements, without changing intent. The promise of confidentiality encouraged the remaining participants to express their views openly.

Leech (2002, p. 665) states that an unstructured interview should be more like a conversation than an interview, and that the topic may even change as this conversation progresses. Accordingly, in addition to our set questions, further ad hoc questioning encouraged conversation, resulting in what were more "collegial chats" than "interrogations." We chose this approach as we were not testing a hypothesis but aiming to gain insight into participants' experiences. The interviews were recorded then transcribed using Otter.ai and analysed using Atlas.ti for thematic coding. The emergent themes form the sub-sections of our findings. In the spirt of this conversational approach, we use quotes from the interviews to primarily present the narrative in the words of the participants.

5.4 FINDINGS

5.4.1 Employer Expectations

While most of our participants felt there was a strong expectation from the university employer to complete a PhD, this was largely implied, not stated. Only one stated that a higher degree was part of their employment agreement, but for the rest there was a perception that if they wanted to retain their job, and certainly if they wanted to progress, a PhD was essential.

> … quite a few times, people said things like, 'the universities are changing. In the past, you could get away with not having a PhD. But that's no longer the case'… So I did feel that if I didn't complete my PhD, that my longevity as an academic … was at risk.
>
> … the position was that if you wanted to progress in your role that you need a PhD.
>
> I realised to have an academic career I needed a PhD.
>
> … the PhD was just something I needed to do, because I was on borrowed time.
>
> Oh, these kinds of grants were not meant for people without a PhD.
>
> I think the most negative thing about the PhD is that if you don't have the PhD, you don't get into research centres or anything like that.
>
> I never set out to do a PhD, it was kind of something I had to do for my job. I didn't ever want to do a PhD necessarily. And even now that I's finished, I don't really care that I've got a doctorate, it was never something I thought 'I'd love to be doctor, I'd love to have a floppy hat.'

However, not all participants felt pressure to complete a PhD. This seemed to depend on their length of employment at the university. Those who joined the university within the last 12 years felt pressured, and those who had been there longer did not. Although it is hard to draw a firm conclusion from such a small sample, this suggests a change in emphasis.

5.4.2 Type of Thesis

In Australia, there are three formats for a PhD thesis. The first is the traditional thesis by monograph: a single document of up to 100,000 words, generally following a well-defined structure. The second is the thesis by publication: a more flexible format incorporating a series of published and submitted papers with linking text. The third is by practice-based creative works with supporting explanatory text.

None of our participants had or were completing a thesis by practice-based creative works. Of those interviewed, two participants began their PhDs by publication and stated they still preferred that option; the rest started their PhD by monograph. Of those who began by monograph, two changed to publication part-way through, and two others were continuing by monograph but stated a preference for thesis by publication. The remaining three did not express a preference.

> I would have liked to have done it by publication because I realise now [that] it's not the PhD so much, it's the publications, and I don't have any publications, I haven't got to that.
>
> I wanted to do it by publication. But because my topic was so vague and I did grounded theory, so even more vague. It wasn't anything that you could write into papers on the way. You really didn't know anything until the end.
>
> And, you know, doing it by publication definitely helped because each publication had deadlines. Without that, I just don't think I would have made any progress whatsoever.

As an academic, there is an expectation to publish research. Therefore, to focus on the PhD thesis for several years at the expense of publishing is a significant problem, one increasingly shared by regular PhDs who wish to become full-time academics (Atwell and Buchan, 2017). The old adage "publish or perish" is still very much alive.

> I think it's difficult when you know that you have to produce outcomes for your academic position. To do that, whilst you're doing a PhD, although I think I've found a way, but I think also that you'll often hear people say 'just finish your PhD.' And the reality is, is that sometimes when you just focus on PhD you can't do outputs, necessarily. You're in a catch 22 if you don't do outputs, and you can't progress. But if you're told to just focus on your PhD …
>
> And so it was four years without any publication … And then how's it going to be the next year? I won't have anything either, neither the following year. So I'll have a couple more years of nothing … So that's when I … changed it to by publication.
>
> I did it by publication. And the reason being is that I was publishing anyway. As an academic, there was an expectation that I was publishing.

According to most participants, publishing papers during the PhD journey was important, but they did not all feel this to be the best or easiest approach.

> I found it very difficult to string the papers together to form a cohesive whole. Had I had a clear direction at the start, I think that would have been much easier.

> Because I did mine by publication there were quite disparate, and sometimes repetitive, publications that had to be wound into some sort of coherent narrative. And that's really hard … And of course you don't want to do it by monograph because you're doing nothing for all the time you're doing it. And you need to be publishing.
>
> I think writing a traditional dissertation would be a whole lot easier. And I think the reason for that is the individual publications were targeted to specific book chapters, or journals or conferences, where they're very specific themes, and was quite difficult to weave all these quite disparate concepts together into a final publication. And that was quite a difficult task to do. And indeed, was way harder than writing any of the papers.

Only one participant was happy to have chosen the monograph, and published each section of it as a series of papers. For example, at the Confirmation stage of the PhD process (a milestone which takes place at the point equivalent to 1 year in full-time time, normally expected at the two year mark for part-time students), students have normally completed a significant review of relevant literature, which can be adapted for publication.

> I think it is a good way to do it [publishing papers from a monograph] because a lot of what you've put into your confirmation is literature grounding. And that's helpful for the research realm, I guess. And to put a position out there. That means that you probably are starting to publish a lot earlier on. And that's helpful for a lot of things. It's helpful for your position as a full-time academic. And just in terms of getting your ideas out there and getting them acknowledged externally. And I think it's helpful psychologically as well.

The importance of publishing along the way is clear, but not only because of employer expectations. It also helps set benchmarks, an experience shared by regular PhDs (Atwell and Buchan, 2017).

> I remember really early on really struggling with knowing whether I was performing at the right level … I really struggled with that because I didn't know whether what I was producing [was] what I should be producing, I had no way to compare it with anything. And I wasn't getting that feedback from the supervisors and I didn't ask them for [it] either. I probably should have. So it wasn't till later on that I … had drafts for them to read that I … started to get that feedback.
>
> Yeah, I mean there's a whole imposter syndrome thing that I think everyone seems to suffer from at one point or another. And by publication at least you can go 'well, you know what? That was peer reviewed. So at least a couple of people thought it was OK.'

5.4.3 Supervision

It is perhaps unsurprising that the topic of supervision revealed some of the most strongly held views. The importance of an effective supervisor–student relationship for regular PhDs has been well scrutinised (Bui, 2014; Gill and Burnard, 2008; Jones, 2013; Mainhard et al., 2009). Typically, this is structured as a Master/Apprentice relationship, which some go so far as to describe as Master/Slave (Grant, 2005). However it is perceived, there is inevitably a power imbalance in which the supervisor holds the knowledge and power and the student is expected to be obedient (Lee, 2008, pp. 271, 272).

These power dynamics are an important part of normal student–supervisor relationships, as supervisors should feel comfortable guiding and pressuring students when required, and students should expect this. However, when students and supervisors are peers, colleagues and even friends, this "power balance" can disappear or become awkward or dysfunctional, depriving the student of the necessary guidance and pressure.

> So some of my supervisors were colleagues and others were both colleagues and friends. That had challenges in itself … it's that changing hats where you walk into a space and you're sitting down adjacent to someone that you consider a friend who is now your supervisor, and then you walk out of the meeting and have a debrief in the office as a friend …
>
> I think part of the issue with my supervisor was that I think we felt like peers, and which is not a bad thing. But I think he felt he didn't demand things of me. You know, he didn't treat me as someone who needed the guidance or the structure, perhaps thinking, you know, that perhaps he [the student] is big enough and ugly enough to do it himself?
>
> I possibly went into it with more confidence than I should have had. It certainly got kicked out of me in the end of it, but I went in going, 'How hard can it be?' And my supervisor went, 'Oh good, that means I won't have to help you much.' I think we probably learned over the course of it that that was complete rubbish and I knew nothing, and I needed a lot of help.
>
> Make sure you have a compatible working relationship with your principal supervisor, and know thyself … be upfront about what you need. Don't be shy about talking about the meta process. Not just the actual topic.

5.4.4 Learning Outcomes

All our participants felt they learnt something through doing a PhD, but some questioned the efficiency and benefits of the process. In two prominent handbooks for regular Australian PhD students, the question of whether students will learn from the process does not arise, and the onus of efficiency is largely on them (Denholm et al., 2012; McMaster et al., 2017).

> I don't feel like I have learnt a huge amount doing the PhD, it was a thing I had to do, and I did it. It was like a project to me. So I learned different skills along the way but, you know, at the end of the day, I don't feel like my life changed.
>
> The PhD did give me a degree of personal intellectual satisfaction, but not as much as I had hoped.
>
> I really didn't feel that it was an efficient way of learning what I needed to learn, many people say that a PhD is just a research apprenticeship … But before I started the PhD, I was already publishing, I was doing research. Plus I did research as a practitioner.
>
> Though I certainly learned how to do research much better through doing the PhD, I just think I could have learned that a lot quicker in other ways … I think the other thing that a PhD demonstrates is the ability to handle large projects over an extended

period of time, and I've been doing that all my professional life, running large projects and managing them. So, I didn't need to do a PhD to prove that.

I just think it was a lot of effort, a lot of stress, for not much gain.

Since I have got [the PhD], I don't think it's been a benefit.

I'm not sure whether the PhD really has added anything to my ability to research.

But, in terms of my ability to research, I'm not sure it added an enormous amount that I wouldn't have gotten any way, just through my publishing.

… now I feel like, why did I do it?

I have experienced managing, you know, multi-million dollar projects. So it's not as if I need experience running projects … Yeah. What I did need experience on is, I guess, how to research. But I still had to write papers, so I had to learn that on the fly anyway.

Whilst most of our participants found the PhD process inefficient, some nonetheless felt they gained a lot from it. However, the main benefit was not in research but in teaching and supervising post-graduate students, as we see in the last few quotes below. Many participants had already published and undertaken other research activities before starting the PhD, and queried whether completing the PhD had actually "added anything to my ability to research."

Before I started, I didn't know much at all about research. I didn't even know the difference between quantitative and qualitative research. I didn't even know what a PhD was, really. So, I have learnt a lot.

The PhD has changed me, changed the way that I think and the way I view the world. Hopefully for the better.

But, I think … what it has added is my ability to supervise PhD students, because I think I understand the process now, having gone through it.

I haven't seen one benefit. Except that I have a PhD student – he's really interested in my topic and I feel like I can work with him in the future.

I'm now supervising PhD students. So, my PhD experience obviously helps that.

And, I guess, those feelings of inadequacy and self-doubt … it's a great insight to undergraduate students, you know, the ones that don't come to class. Yeah, the ones that turn up after five weeks not having anything, I've got a lot of insight into how they feel. Digging yourself into that hole.

I'm better equipped to supervise students, because I've been there, and I can talk from experience, and obviously I've learned skills that are of value to a PhD student. So certainly I wouldn't have had that without the PhD.

5.4.5 Professional Practice Experience

Most of our participants had significant professional practice experience prior to taking on an academic role; something we did not find discussed in the reviewed handbooks for regular Australian PhDs. Some participants felt a practice-based PhD would have been a better option than either of the text-focussed formats, with others suggesting that their professional design work had been equivalent to a PhD.

> I ended up doing a thesis about the design of [topic removed]. I think it would [have been] better for me to design something and write about that … To actually propose some models. I think that probably would have been a better contribution.
>
> I would push them to start a new pathway for PhDs and do it as a practice-based PhD.
>
> You know, I would say … that I have effectively done a PhD in practice on a project that needed considerable research that went for three years. There's an equivalence there, I think.
>
> So in practice, you've got a potential project, you put in a proposal, you articulate all these things that you need to do to deliver it. You have a program, you have to develop a position, then you go and do the work. And being in practice, you actually do a lot of research in order to achieve whatever you need to do. It's just that it's not articulated as research.

Whilst their professional practice experience helped secure their academic jobs, most participants felt this experience was then undervalued compared to having a PhD. Though the PhD might be essential for career progression, participants felt it was their professional practice experience which informed their teaching much more than their research could.

> From a teaching perspective, I think my professional practice experience far outweighs my PhD research.
>
> I had quite a significant practice career before moving to academia, and I have kind of kept one foot in that world. So I don't feel like I've left that world behind. So for me, I think it'll be a natural place for me to kind of sit somewhere in between.
>
> [Name redacted] is a university that wants students to be workplace ready. But they don't seem to acknowledge the importance of practitioners within the academic sphere, unless they're sessional tutors. And that's a problem.
>
> And so I teach from my practice experience.

Despite this perception of professional practice experience as undervalued by the university, participants agreed that it was the combination of research qualifications and practice experience that was recognised as a strength.

> Because now, the university sector and the research sector is looking at partnerships with industry. They want that applied knowledge and the ability to apply their research within industry. So I think that there's a mutual benefit for industry teaching into university, and university enabling industry through research-led practice. And I think that's the way that's it's going to go.
>
> They [an external partner] were very interested when they heard that I … have had quite significant practice experience … before moving into academia … And I think that the whole package is what has made me attractive to them to collaborate with. So

> I need to probably remind myself that it is a bit of an edge that I have over other colleagues … I have to remind myself as I don't feel that's really respected much.

> And I really do believe that my professional practice experience has contributed to my ability to be a good academic just as much as my ability to research. And so now at least I have the benefit of having both. Whereas my colleagues will never have that.

> Professional Practice? … I don't think it's valued whatsoever. I feel like when I started at … I had to go 10 years back in my career to actually start being [an] academic.

5.4.6 Institutional and Peer Support

Almost all of our participants spoke favourably of the level of institutional support they received, including reductions in teaching load, access to a significant amount of leave to complete the final stages of their PhD, as well as some financial support. Some, however, felt there were inconsistencies in the application of that support.

> I mean, there was a certain amount of funding that was given to me when I enrolled in my PhD … which was wonderful … I was provided with time away to go on writing retreats, I did access professional development leave. So I had one semester of leave to try to complete [and] tie up the final thoughts of my PhD and get it in. So I think all of those sorts of things were very supportive. So … as an organisation, yeah, I would say. Absolutely [supportive].

> There is no sort of direct, 'if you're going to get a PhD, this is the direct line for you to do it. And you know, we've cleared all the path for you,' you negotiate your way to it …. they were more supportive of it earlier on, and then there was a period when it just dropped off.

While support during the PhD was generally seen as very good, that support seemed to dry up once the PhD was complete:

> I was dismayed to discover that completion of a nearly 100,000-word dissertation would not be counted as a 'research output,' and that I would therefore have to keep conducting research purely in my 'own time' in order to be granted time in my job to undertake research.

Support from other colleagues varied from positive to ambivalent.

> So, incredible support from colleagues who are friends. Other colleagues, no, Probably just little interest and, and not much support.

> Yeah, there's some colleagues that will ask about it. And then the ones that are actually in it or have just recently completed, and they gently ask. It's this kind of unspoken thing that you don't talk about it too much. You know, because for some reason, it's a touchy subject.

> I don't think I was ever in a situation where I was supported by staff colleagues who are doing PhDs and that's because we're all so busy.

Support from non-staff, regular PhDs varied depending on the level of contact. Most participants had little to no contact with regular PhDs, whereas some experienced a lot. All, however, saw the benefits of being part of a PhD cohort – which they often missed out on.

> I didn't feel like part of a group, peer group or anything. I've heard people talking about the peer support group for PhD students, talking to each other, egging each other on and those dynamics as being important, and I've always … wistfully observed those. Yeah, it's pretty isolating.

> So there were lots of PhD students in that group that were not my colleagues … it was incredibly helpful. Yeah, good, really, really good because I could talk to them as a PhD student and they saw me as a PhD student. They didn't see me as a lecturer or any of those other things. So … that group has been very good at supporting and nurturing the HDR students. So, yeah, it was an excellent experience for me.

5.4.7 Being a Student

The situation of being a student again at the same time as being a full-time academic is quite unusual. As already touched on, it can lead to tensions and confusion in relationships with colleagues, supervisors and regular PhD students. This occurs in particular when staff PhDs have already been teaching and even researching for 10, 15 or even 20 years before undertaking their PhD. Neither "hat" seemed to quite fit them properly.

> I never felt like I was allowed to be a PhD student because I was an academic … So I never felt a connection with PhD students, other than my colleagues who were also PhD students.

> I didn't feel like I was a real student, you know, never really part of the student cohort … But then I never felt like a fully fledged academic either because I didn't have a PhD, couldn't apply for grants, promotion and so on. So, it was a state of limbo for 8 years really. No wonder I struggled with the imposter syndrome!

> … part of my frustration was that I was never really able to be a student when I was doing my PhD. And I didn't have that indulgence … and the PhD was just something I needed to do, because I was on borrowed time … So I didn't have the kind of joy of just being able to immerse [myself] in the literature and, you know, bounce around and enjoy learning about research, I … kind of always felt like I was on borrowed time, or [like it was] something I just had to do.

> I guess it's different from a traditional research or PhD student's path where they complete the PhD and that's what equips them to go and apply for jobs. I was already in the job. So I don't want to sound ungrateful, but I think [it] was just a different journey when you're already actually in the job. And it's a matter of actually just doing what you probably should have done before you had the job anyway.

5.4.8 Work–Life Challenges

We cannot tell from this study whether the process of undertaking a PhD while working full time as a Design academic is any more or less challenging than for regular PhDs. By definition, the Design academic already has the job, so there is not the pressure of what happens afterwards. There is also support from the university to which regular PhDs may not have access. However, the Design academic does have to manage their PhD around a demanding full-time job and is often at an age when it is common to have many responsibilities apart from work and study. Overall, though, the work–life challenges which arise are not unique to Design academics, as is clear in Australian PhD handbooks with chapter titles such as "Who am I? Surviving the battle of the roles" (Stone, 2017), "Parenting through a PhD" (Eades, 2017) and "Managing part-time candidature with work and family commitments" (Evans, 2012).

> I had quite significant home pressures as well … and that was quite challenging … on top of trying to complete a PhD … So there's this constant [of] being pulled in so many different directions at once.
>
> Why did that have to be so hard?
>
> And it [is] just difficult to juggle it all. Yeah, you know, in retrospect, would it be easier to do this [at] the age of 22? Yeah, probably, if I'm still living with my parents and having laundry and meals made for me, but would I have bought the same maturity to my study? No, because I've got life experiences.

A recurring theme among our participants was the challenge of finding time to devote to the PhD amidst the specific demands of an academic job.

> The biggest challenge is just time. There are so many things that are more urgent than the PhD. The PhD always gets put at the bottom of the priority list. Because there are other demands, things that need to be done that have deadlines. The PhD just didn't have deadlines. It was just something you had to keep working and working on.
>
> The demands of a heavy teaching load are always more immediate.
>
> Because everything else was urgent. There was always, you know, you had to get a lecture ready for next week … you had to go to this meeting, you had these things [that were] always more urgent, perhaps not more important, but certainly more urgent, and therefore they were done at the expense of the PhD.

Finding suitably long, uninterrupted blocks of time to work on the PhD was identified as the hardest aspect of this challenge.

> The absolute minimum, [to] do any useful work was half a day, and ideally a whole day. So you need to just start in the morning, and just take an hour or two to actually re-engage with the content, get into the flow, and [then you can] be productive for hours on end.
>
> And I'd have bursts of productivity and then take a break and just not get back to it you know, let us find other things. You know, all the procrastination things, incidental

> cleaning, you know, all the stuff that you know, email. Email was the single best procrastination thing.
>
> It's not something I felt that I could quickly just get in the zone and write for an hour and progress, something I needed at least a day, and ideally, two or three days, because sometimes it took a day just to get into the headspace of writing again. So I think having writing retreats, whether they were away from home or at home, they were incredibly useful to me. So you know, sometimes I'll just block out two or three days in my calendar, and I'll just say I'm on a writing retreat, and I'll just be home. And I'll just write, and I'll try to try very hard to turn off the email communication, and it normally takes about half a day to get into it.

This preference for writing in large blocks was stated by all but one participant, and is at odds with most advice to PhDs to develop the habit of writing in small increments of an hour or two every day (Golash-Boza, 2012; Grandy, 2016; Harkonen, 2010) rather than "binge-writing." The participant who trained herself to write in small daily increments preferred this "routine writing" as it kept the research advancing and "live" in her mind all the time. We can only speculate on the reasons for the majority preference which was so at odds with the advice to work in this "routine" way. It could be discipline-specific: large blocks of time may suit the topics typically written about by Design academics, or perhaps the way most Designers think about and tackle projects suits deep immersion better. Or, it could be an aberration due to our small sample size or simply a useful procrastination excuse as half or full days are rarely available.

5.4.9 Emotions

Completing a PhD can be a stressful experience provoking strong emotions. Over 50% of PhD students experience psychological distress, and a third are at risk of common psychiatric disorders such as depression (Levecque et al., 2017). These are challenges encountered by both staff and regular PhDs, as advisory handbook chapter titles such as "Countering isolation: joining the research community" (Conrad 2012) and "Peer support groups" (Fisher, 2012) attest.

Mental health is a sensitive issue, and one which arose several times during our interviews. Due to this sensitivity, details are not included here other than to say that our analysis aligned with the statistics. This issue is much discussed in the literature, as summed up by Stubb et al. (2011, p. 34): the burnout commonly experienced by PhD students is the product of emotional exhaustion, cynicism and a reduced sense of efficiency. These are caused by factors such as heavy workloads, social conflict, lack of resources and organisational injustices. The stress, anxiety and lack of affect characteristic of burnout can lead to withdrawal from both work and study.

From our small participant sample, we cannot accurately say whether the incidence or severity of psychological stress is greater or lesser among Design academics compared with regular PhDs. However, we note that regular PhDs often have the additional stressor of trying to find work post-PhD, but may be less likely to suffer the effects of ambiguous status during the PhD. Our participants work as full-time academics and have mostly come from professional practice backgrounds, leading to some unique challenges including the change in status from "successful professional" to "research apprentice." Watts (2009, p. 688) warns PhD supervisors against underestimating the significant psychological challenges faced by those from professional backgrounds "becoming a student again," especially if the student perceives their knowledge to be equal or superior to that of their supervisors. Resultant feelings of isolation or inferiority were common with our interviewees.

> [I] probably never felt that lonely in my whole life.
>
> I felt like an imposter, like I shouldn't be there, kind of worthless until I got that PhD.

Although not all our participants had yet completed their PhDs, those who had reported surprisingly similar emotional responses to actually finishing.

> 'I've got my PhD.' And he was like, 'why are you crying' … the emotion I felt was just enormous. I don't know, sorrow. It was bizarre. It was really bizarre.
>
> And that's kind of how it felt for me. It was sorrow. It was almost like grieving.
>
> It was almost like 'so what.' But when I did get it, yeah, I cried too … I just felt this was just relief. It was kind of just this horrible experience in my life that's finally ended.
>
> I thought I'd be jumping for joy. But no, not at all, it was just like, Okay, done, tick, you know, move on … So many people have said to me. 'I bet you're so happy it's finished.' And I said 'Yeah, yeah.' Because you tell them what they expect to hear. But in reality, actually, no, no joy.
>
> I think [I was] so overwhelmed with emotion, at the fact that I'm finally finished that I just couldn't even interact with anyone, so I just didn't really tell anyone. I didn't post on Facebook, when I graduated, I just I just kept it very quiet. And maybe a lot of it was just shame that I hadn't got it before. I don't know.

When looking back on the process of doing their PhD, many participants reported overwhelmingly negative feelings such as sadness, anger and disappointment.

> I can distinctly remember that feeling of waking up in the morning and thinking, you know, I've got a research day. And Oh f**k, you know, and yeah, instead of the delight of having that time to concentrate on it, It was just this foreboding.
>
> And you were on borrowed time. That was the kind of sense that I got.
>
> I kept grinding away with an increased feeling of futility, I guess. And you get to the point where you don't really make progress, you just kind of spin your wheels.
>
> So what did it give me? I don't know. I guess frustration, self-doubt.
>
> I do have some regrets. Because it was a long time of feeling guilty. It was a long time to be cut off from friends and family and to lose, you know, personal time. Yeah. And I feel like it was a waste of time.

These comments might suggest our participants did not enjoy any aspect of their PhD journey. However, when asked about their overall experience, responses were not so bleak. Participants were asked to rate their overall PhD experience on a spectrum from zero – totally negative – to ten – totally positive. All participants struggled to select a single number and felt the need to contextualize their response. Most felt the experience comprised very bad and very good experiences and so settled on a five. One participant gave a ten, and two participants gave a two.

> If you'd asked me at the two year mark, I would have said one or two. At the halfway mark, probably five or six. My experience got better as I went through, and I think that that grew with my confidence because I just didn't know what the hell I was doing early on.

> Yeah. Well, I'll probably have to say five, because it was some things that were eights and nines. And there were definitely some ones and twos.

> It's not that it was a negative experience, it was intense … I enjoyed being a student, I enjoyed learning from others, including my supervisors, who I am very fond of and still work closely with them, and am very proud of the outcome. It did set me up quite well for more exciting projects to eventuate and I think an interesting career as an academic. So, for me, I would have to rate it as a totally positive experience, a ten.

All participants had strong feelings about the PhD journey, both negative and positive. We sensed they wished to share their experiences in order to contribute to improvements in the process for others. Some participants found their interviews cathartic, making comments afterwards along the lines of "well, that was a good therapy session." This reflects the sentiment that undertaking a PhD can be a lonely journey not understood by non-academic friends and family.

5.5 DISCUSSION

In this section, we distil the main points from our findings and offer some recommendations for surviving and even thriving while doing a PhD as a full-time Design academic.

5.5.1 Employer Expectations

30 years ago, a PhD in Design was rare, but has now become essential for a Design academic. Most of our study participants became full-time academics 10–12 years ago, at a time when having a PhD was not required to get the job, but a plan to get a PhD was expected. Most felt this as an implicit, rather than explicit, pressure to get their PhD, with reasons ranging from fear of losing their job to acknowledging the need to have a PhD to further their career.

5.5.2 Type of Thesis

There was a clear majority preference for doing a text-based PhD thesis by publication rather than monograph, as it provides quality benchmarks and feedback through external peer review and publication, satisfies the requirement to publish regularly, and provides a series of deadlines to progress the work.

The challenges of a thesis by publication are that "self-contained" papers written for different audiences and publications often contain repetitive content and explore tangential themes that may be difficult to weave into a coherent single document. For example, you may need to insert six papers with six different themes, literature reviews, discussions and conclusions into a thesis with its own literature review, discussion and conclusion. Careful planning at the early stages can avoid papers being too diverse, overlapping and repetitious. This is more easily said than done, as Design PhD topics are often fluid, changing emphasis over time.

A thesis by monograph does not suffer from these challenges, and it is possible to create publications from monograph chapters. This can require considerable rewriting but does prevent "publication drought" as well as maintaining the integrity and homogeneity of the monograph. Both options benefit from careful planning and both have strengths and weaknesses. However, a thesis by publication is probably the better career option for a full-time Design academic today. The option of a practice-based PhD will be discussed later.

5.5.3 Supervision

One very distinct challenge that staff PhDs have and regular PhDs do not is the nature of their relationship with their supervisor. The traditional "Master–Apprentice" model can become problematic when the "Master" is a peer, colleague or even friend of the "Apprentice." It can be challenging for staff to swap from those roles to those of supervisor and student. A danger reported by some of our participants is that the supervisor treats the student as a peer at all times and does not give enough guidance, or else assumes that, as the student is already an academic and may have published, they know more about the PhD research process and conventions than they actually do.

To avoid this, we recommend being clear from the outset regarding what the expectations on both sides of the student–supervisor relationship will be, and to discuss the structure and process of the PhD to establish a shared understanding. As is often the case with students; "you don't know what you don't know until you know it," making it difficult to know what questions to ask – but perhaps this state of unknowing is precisely what staff PhDs need to communicate to their supervisors. Asked whether they thought it better to have a supervisor who was a subject matter expert or a PhD process expert, our participants said both were important but, if they had to choose, it would be the process expert.

5.5.4 Learning Outcomes

Most of our participants felt they had not learned as much from the PhD process as they would have liked. Most thought the PhD an inefficient way to learn about research, and as they had often published academic papers before starting their PhD, felt there was not much the process contributed. If anything, it distracted many from writing papers, and reduced their university-measured research outputs. Most had significant professional practice experience in running more complex projects than the PhD "project" and so did not feel the need to demonstrate their ability to run a project through the PhD.

Most participants thought the main learning outcome was not about research per se, but about the formalities of the PhD process necessary to be able to supervise others. Again, most felt there would be quicker and more efficient ways to gain this knowledge than spending years doing a PhD. We recommend consideration of these issues and discussion with potential supervisors before submitting a PhD application to find the best type of thesis and research process to suit the desired outcomes.

5.5.5 Professional Practice Experience

Many of our participants were well-established design professionals before joining academia, often holding senior positions. A commonly cited reason for becoming an academic was having risen as far as possible in design practice and looking for a new challenge. Nearly all participants stated they felt the university undervalued their professional practice experience; that while a PhD was essential for career progression, their practice experience was deemed irrelevant. Yet this practice experience informed their teaching far more than their PhD research. Some participants did feel that having a PhD as well as practice experience placed them in a good position to leverage commercial research opportunities, giving them an edge over colleagues without this experience.

An interesting point is that some participants felt much of the design work they had done professionally involved research at an equivalent level to that of a PhD, though differently articulated. Some suggested a PhD by practice-based creative works might be more effective than a monograph or set of publications to leverage their existing skills, producing outputs more efficiently. While we recommend this option be considered by Design academics coming from substantial professional practice backgrounds, the academic requirement to continuously publish could make it a significant work–life challenge.

5.5.6 Institutional and Peer Support

The majority of participants felt well supported by the institution, including some having reduced teaching loads (although some stated they had a higher teaching load because they were not deemed "research active" while doing their PhD), extended leave to complete the PhD, occasional writing retreats, seminars, conferences and other organised activities to support their PhD journey. Some mentioned how privileged they felt to be supported in this way by their employer, but noted the level of research support dropped off once their PhD was complete. There were mixed levels of support from colleagues: while there was occasional verbal encouragement, it was more common to be reminded of all the things that were not possible until the PhD was complete, such as promotion and participating in grant applications. Overall, the PhD was mostly a topic of conversation not mentioned outside supervisory meetings, making for a lonely experience.

Only one of our participants actively engaged with a regular PhD peer support group, reporting it as an extremely valuable source of support. This participant had the highest level of overall satisfaction, as will be discussed in the next section. In summary, there was plenty of support during the PhD at the university level, but room to improve at a personal, collegial level and immediately post-PhD. We recommend the setting up of specific peer support groups amongst Design staff PhDs, to focus on research progress and related work–life challenges.

5.5.7 Being a Student

Two participants were enrolled externally (at other universities), and the remainder were enrolled at the university where they worked and had supervisors in the same faculty, mostly in the same school. It is challenging to be both a student and a staff member within the same institution, and particularly within a single Design school. It can lead to tensions and confusion in relationships with colleagues, supervisors and regular PhDs. Our recommendation is to wear two distinct hats. When wearing the student hat, be solely a student. When wearing the staff hat, be solely a staff member. It is important to have a clear conversation about this potential issue with supervisors early in the relationship.

Our second recommendation is, again, to join a PhD peer support group. However, only one of our participants enjoyed the experience of being part of a "regular PhD" group, while the rest thought they would be uncomfortable as they did not feel like true students, and would have little in common with the others, being at a different life and career stages. In addition, workload time pressures made attending group events difficult. Therefore, we recommend forming specific Design staff PhD peer support groups if they do not already exist. Since writing this chapter, we have done this at our institution as it has highlighted this need.

5.5.8 Work–Life Challenges

Completing a PhD is certainly a challenge to maintaining a good work–life balance, including in the domestic sphere. "If it were easy everyone would have one" is a commonly heard phrase, reflecting the notion that, amongst other things, a PhD demonstrates the ability to persevere with a difficult task over an extended period. None of our participants suggested completing a PhD should be easy, though some asked "why did it have to be so hard?" referring not to the research aspect of the PhD but to the administrative and logistical aspects. They felt the challenges should come from the process of research and new knowledge creation, but instead came from administrative tasks such as ethics approvals, and issues with supervision and time management.

The biggest challenge cited by our participants was making time to actually do the PhD work. While this is not unique to Design academics, completing a PhD while working in a demanding job without a regular workflow or working hours is difficult. The academic has recurring, urgent, non-negotiable teaching and other deadlines. It is not acceptable, for example, to stand in front of a full lecture theatre and say, "I haven't prepared a lecture for you today because I was up all night working on my PhD." Besides teaching, academics have other work-related obligations:

some of our participants held leadership positions while doing their PhD, and most were involved in committees or other service roles, all of which had to take priority over the PhD. The so-called "Eisenhower Principle" (Covey, 1989) states "there are two kinds of problems: the urgent and the important. The urgent are not important, and the important are never urgent." The principle is that you should focus on the important, not the urgent. The PhD is important but usually not urgent. Other aspects of an academic job are usually more urgent, and not necessarily less important. Regardless, the PhD always seemed to end up at the bottom of our participants' priority lists.

All but one participant felt they needed a block of time for PhD research to make meaningful progress: a minimum of half a day per week. "Binge writing" sessions over several days were favoured by the majority. This is in direct contrast to much "PhD advice" literature suggesting an hour or two a day of writing. However, most participants were adamant on this point, so it deserves further exploration. Our recommendation is to negotiate and set aside regular blocks of "protected time" such as writing "retreats" free from interruptions. While it is ideal for these to be held in a non-work environment, it could still be beneficial to book a room on campus, lock the door and write for half a day, or even one, two or three days. We recommend Design academics use their peer support group to establish a culture in which individual and group writing retreats are a regular, accepted practice whereby staff can go "off-grid" for half a day to a few days, rather than being "constantly connected" to their jobs.

5.5.9 Emotions

Mental (and physical) health problems occur at a significantly higher rate in PhD students than in the general population (Oakley, 2017; Storrie et al., 2010). All of our participants spoke about issues with emotional and mental health, some in depth. Some felt less worthy for not yet having a PhD; others felt resentment at a perceived undervaluing of their professional practice experience; many felt lonely. There was little joy in finally completing the PhD, mostly sorrow, guilt, anger and relief. One participant actually regretted completing the PhD, and others felt it was a waste of time. Some spoke of burnout and attributed physical and mental health problems to the PhD process. Only one participant was entirely happy with all aspects of the PhD experience.

As authors, we would prefer to paint a more positive a picture than this, but as researchers we convey our findings unvarnished. Clearly there is a problem, and the first step to solving a problem is to acknowledge it. Training for both supervisors and PhD students in recognising the causes and symptoms of burnout could help, along with jointly identifying and managing stressors and heavy workloads before they become problematic. Early in the PhD process, set in train practices to help retain a focus on the positive benefits of doing a PhD. Choosing a non-traditional mode of PhD such as a practice-based model may help some Design staff PhDs leverage and apply their professional practice experience, improving their sense of this being valued. These things may help in the short term, but we consider that in the longer term more research is required into the emotional effects of the PhD experience and the development of better institutional mitigation strategies and tactics.

5.6 CONCLUSION

In the 1990s, the "new" universities typically had Design staff who came from professional practice, with very few, if any, PhDs. There has since been a steady process of "doctoring" these staff (Melles 2011, p. 451). This chapter has explored the effects of this process on nine full-time Design academics from one Australian university, unpacking the challenges and benefits and elucidating the similarities and differences compared with "regular" PhD students. The themes emerging from our nine participant interviews ranged from concerns with the practicalities of balancing thesis type with the academic requirement to publish, management of professional relationships, to the emotional costs of personal and professional identity challenges, stress and burnout.

We are in no doubt that most Design academics love their jobs. The act of teaching, of helping people gain the knowledge and skills to start their careers, is very rewarding. Research, the finding of something interesting and exploring it to the point of expertise, is a privilege. Moreover, we are paid to do these things! Completing a PhD while working as a full-time Design academic can also be rewarding but is undeniably challenging. Resulting high levels of stress can lead to physical and mental health problems, and institutions and individuals being aware of these potential problems is essential to developing mitigation strategies and tactics.

As a result of this study, we have made a series of recommendations to our Design staff PhD colleagues, here recapped in brief:

- Do a thesis by publication if publishing is a requirement of employment
- A thesis by practice-based creative works may be a good option for academics joining the university after a substantial professional practice career
- Ideally, choose a principal supervisor with expertise in the PhD process and conventions over one without, and associate supervisors who are content experts
- Be clear with yourself and your supervisor about your different roles when you have your "student hat" on, and when you have your "staff hat" on
- Discuss your existing research and project management experience (or lack thereof) with your supervisor, and develop clear, joint expectations of the best PhD process to maximise your learning
- Be clear with your PhD and employment supervisors early on regarding the support you need to undertake the PhD part time whilst working full time
- Join or set up a peer support group with other Design staff PhDs, with a dual focus on research progress and work–life challenges
- Set up research/writing "retreats" individually and with your PhD peer group, working to normalise this as a regular, accepted practice in the workplace
- Identify emotional and workload stressors with the potential to cause burnout, and discuss mitigation strategies with your PhD and employment supervisors

- From the beginning of the PhD process, set in train personal practices to assist you to retain your focus on the positive benefits of doing a PhD

For most university Design schools in Australia, the academic staff "doctoring" process seems to be nearing completion at the same time as they are producing large numbers of young PhD graduates eager to gain academic positions. This makes it less likely that Design schools will continue to recruit practitioners without PhDs, and then wait for them to be "doctored." In this case, Design schools could become more research-focussed, moving away from applied practice, and this chapter merely an interesting historical snapshot. If, however, Design schools recognise the value of staff with professional practice experience and continue to recruit and "doctor" them, then this chapter may help inform the development of improved PhD programs. We certainly hope it helps our colleagues who are undertaking PhDs while working full time as Design academics to survive, and even have a rewarding PhD experience.

REFERENCES

Atwell, C., & Buchan, J. (2017). An academic apprenticeship: Publishing during doctoral candidature. In McMaster, C., Murphy, C., Whitburn, B., & Mewburn, I. (Eds), *Postgraduate study in Australia: Surviving and succeeding* (pp. 13–21). New York: Peter Lang.

Brew, A. (2001). Conceptions of research: A phenomenographic study. *Studies in Higher Education*, *3*(26), 271–285.

Bui, H.T.M. (2014). Student–supervisor expectations in the doctoral supervision process for business and management students. *Business and Management Education in HE.*

Conrad, L. (2012). Countering Isolation: Joining the research community. In Denholm, C., & Evans, T. (Ed). *Doctorates Downunder: Keys to successful doctoral study in Australia and Aotearoa New Zealand.* (pp. 38–44). Camberwell: ACER Press.

Covey, S.R. (1989). *The seven habits of highly effective people.* New York.

Denholm, C., & Evans, T. (Ed). (2012). Doctorates Downunder: Keys to successful doctoral study in Australia and Aotearoa New Zealand. (pp. 179–185). Camberwell: ACER Press.

Eades, Q. (2017). Parenting Through a PhD. In McMaster, C., Murphy, C., Whitburn, B., & Mewburn, I. (Eds), *Postgraduate study in Australia: Surviving and succeeding* (pp. 83–90). New York: Peter Lang.

Evans, T. (2012). Managing Part-time Candidature with Work and Family Commitments. In Denholm, C., & Evans, T. (Eds), *Doctorates Downunder: Keys to successful doctoral study in Australia and Aotearoa New Zealand.* (pp. 45–53). Camberwell: ACER Press.

Fisher, K. (2012). Peer Support Groups. In Denholm, C., & Evans, T. (Eds), *Doctorates Downunder: Keys to successful doctoral study in Australia and Aotearoa New Zealand* (pp. 45–53). Camberwell: ACER.

Davis, M. (2008). Why do we need doctoral study in design? *International Journal of Design*, *2*(3), 71–79.

Gill, P., & Burnard, P. (2008). The student-supervisor relationship in the phD/Doctoral process. *British Journal of Nursing (Mark Allen Publishing).*

Golash-Boza, T. (2012). Get a life, PhD: Ten ways to write every day. Retrieved October 26, 2019, from http://getalifephd.blogspot.com/2012/01/ten-ways-to-write-every-day.html

Grandy, T. (2016). Support your writing productivity and leave binge-writing behind | *GradHacker.* Retrieved October 26, 2019, from www.insidehighered.com/blogs/gradhacker/support-your-writing-productivity-and-leave-binge-writing-behind

Grant, B.M.B.M. (2005). The pedagogy of graduate supervision: Figuring the relations between supervisor and student. *Philosophy*.

Guest, G., Bunce, A., & Johnson, L. (2006). How many interviews are enough?: An experiment with data saturation and variability. *Field Methods*, *18*(1), 59–82.

Harkonen, J. (2010). Practical tips for doctoral students. Retrieved from www.researchgate.net/publication/274287777

Henkel, M., & Vabø, A. (2006). Academic identities. In B.I. Kogan M., Henkel M., Bauer M. (Eds), *Transforming higher education* (pp. 127–159). Dordrecht: Springer.

Jones, M. (2013). Issues in doctoral studies -forty years of journal discussion: Where have we been and where are we going? *International Journal of Doctoral Studies*.

Kemp, D.A. (1999). *Knowledge and innovation : A policy statement on research and policy statement*.

Lee, A. (2008). How are doctoral students supervised? Concepts of doctoral research supervision. *Studies in Higher Education*.

Lee, A., & Boud, D. (2003). Writing groups, change and academic identity: Research development as local practice. *Studies in Higher Education*, *2*(28), 187–200.

Leech, B. L. (2002). Asking questions: Techniques for semistructured interviews. *PS - Political Science and Politics*.

Levecque, K., Anseel, F., De Beuckelaer, A., Van der Heyden, J., & Gisle, L. (2017). Work organization and mental health problems in PhD students. *Research Policy*.

Mainhard, T., van der Rijst, R., van Tartwijk, J., & Wubbels, T. (2009). A model for the supervisor-doctoral student relationship. *Higher Education*.

Mason, M. (2010). Sample size and saturation in PhD studies using qualitative interviews. *Forum Qualitative Sozialforschung*.

Melles, G. (2011). Art, media and design research and practice: Views of educators in a "new" Australian university. *Research in Post-Compulsory Education*, *4*(16), 451–463.

Oakley, C. (2017). There is a large-scale crisis of mental ill-health at doctoral and postdoctoral levels that demands wider attention and action. Retrieved October 30, 2019, from https://medium.com/@catherine.oakley/there-is-a-large-scale-crisis-of-mental-ill-health-at-doctoral-and-postdoctoral-levels-that-demands-42fcad6b1525

Rosch, T. A., & Reich, J. N. (1996). The enculturation of new faculty in higher education: A comparative investigation of three academic departments. *Research in Higher Education*, *1*(37), 115–131.

Rowland, S. (2002). Overcoming fragmentation in professional life: The challenge for academic development. *Higher Education Quarterly*, *1*(56), 52–64.

Sikes, P. (2006). Working in a "new" university: In the shadow of the research assessment exercise? *Studies in Higher Education*, *31*(5), 555–568.

Stone, B. (2017). Who Am I? Surviving the battle of the roles. In McMaster, C., Murphy, C., Whitburn, B., & Mewburn, I. (Eds), *Postgraduate Study in Australia: Surviving and succeeding* (pp. 77–81). New York: Peter Lang.

Storrie, K., Ahern, K., & Tuckett, A. (2010). A systematic review: Students with mental health problems-A growing problem. *International Journal of Nursing Practice*, *16*(1), 1–6.

Stubb, J., Pyhältö, K., & Lonka, K. (2011). Balancing between inspiration and exhaustion: PhD students' experienced socio-psychological well-being. *Studies in Continuing Education*.

Swann, L., & Rezayan, L. (2021). Exploring the Mindset of a New Academic. In A. Blackler & E. Miller (Eds.), *How to Be a Design Academic:from Learning to Leading*. Boca Raton, FL: CRC Press, 83–111.

Watts, J. H. (2009). From professional to PhD student: Challenges of status transition. *Teaching in Higher Education*.

6 The Mindset of a New Design Academic

Levi Swann and Leo Rezayan

CONTENTS

6.1 INTRODUCTION

There is an adage that states "jack of all trades is a master of none, but oftentimes better than a master of one." It suggests that, although having a broad and shallow skillset has limitations, it might be preferable to being a specialist in only one area (Morris and Morris, 1988). In our experience in the context of design, this saying is often used in short form – "jack of all trades is a master of none" – to express the undesirability of having only moderate capabilities in a range of areas, without being particularly skilful in any one area. Having both trained and worked as designers, and now design academics, the dichotomy between skill generalisation and specialisation, as conveyed by this saying, has been of great interest.

As designers and early career academics, we have both made homes in diverse areas because of a willingness to engage in situations outside our discipline scopes. Through a mindset focussed on learning and development, we have been able to draw upon a varied skillset to adapt to different projects, teams and their needs. The importance of this type of mindset is also echoed by Miller (2021b, 2021a) in her chapters in this book, demonstrating its power not only for early career academics, but also in later stages to develop resilience and create spaces for opportunities and collaborations to flourish.

In this chapter, we explore the mindset of new design academics who, like us, have been required to find their place in diverse and changing work environments. To begin, Swann provides a brief overview of his experience in academia to provide context to the motivation for writing this chapter. We present a case for developing a generalist skillset and then discuss adaptability and its related factors. Then, the experiences of eight early career academics, gained from semi-structured interviews, are presented. We do not suggest the experiences of these participants generalise to all early career academics. Rather, they are personal accounts that serve to explore different perspectives and experiences of people with design backgrounds entering academia. They provide examples of how personal characteristics and skills have been developed and transferred to assist in adapting to academic tasks. The outcomes of the research are visualised in a Skills Map, which highlights common themes and informs discussion on important skills for new design academics to develop.

6.2 "I CAN PROBABLY DO THAT" – SWANN'S CAREER SO FAR

Against common professional advice you might read in career blogs, I tend to say yes to things. But it is not because I have difficulty saying no … It is because curiosity has repeatedly been emphasised in my education and professional experience. I completed a Bachelor of Industrial Design (Honours) in 2010 and, throughout this education, exploration of diverse problems and opportunities was met with enthusiastic encouragement. The synergy of research and design activity provided creative licence and the space to experiment with novel ideas. After graduating, I took opportunities as a freelance and contract designer, and developed skills in several areas: Industrial Design, Visual Communication, and Illustration predominantly.

Entry into academia had been on my mind since my Honours year and persisted throughout my time working as a designer. This eventually led me to start a PhD in design in the study areas of human factors and expertise. My academic career, from the start of my PhD up until the time of writing this chapter, is shown in Figure 6.1. I was fortunate enough that my PhD (1) led to a relatively stable Research Associate position (7), which has arguably been the closest thing to specialisation in my career so far; I applied and progressed very specific knowledge of human factors and expertise in airport security screening, which was initially developed during my PhD.

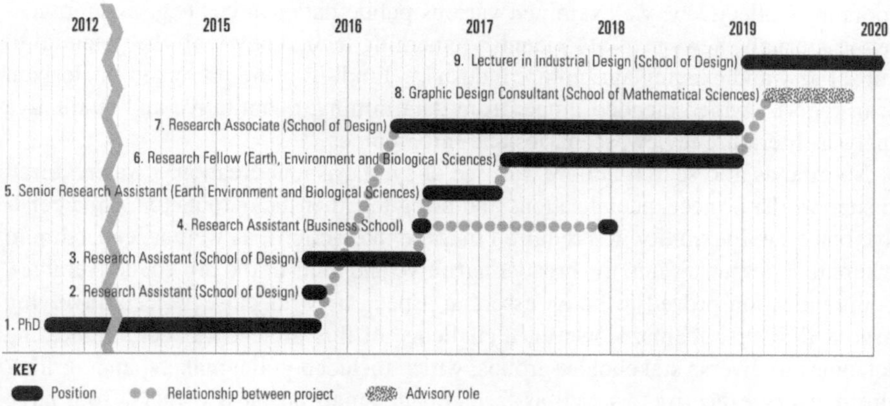

FIGURE 6.1 Summary of academic work experience

Although I eventually secured an ongoing academic position (9), the period between finishing my PhD in late 2015 and the start of 2017 closely resembled my time as a designer. I took opportunities that were presented and adapted to the needs of the projects I was working on. To varying degrees, each project required me to draw on the hard and soft skills I had developed during both my professional design and academic research experience. As a Research Assistant (3), I applied design knowledge of 3D printing, scenario development and concept design with literature review and trend analysis in a feasibility study investigating cotton as a feedstock for 3D printing and rapid customisation applications (Donovan et al., 2016). As a Senior Research Assistant (5), I drew on technical design skills of graphical data visualisation. This was blended with data collection and analysis skills to generate evidence-based scenarios for various Australian rural industry stakeholders (Hamilton et al., 2018). Short Research Assistant contracts (4) required illustration skills for the development of visual stimuli to assist with research activity. Each of these research projects led to successive opportunities, due largely to personal referrals and my ability to integrate design and research skills to generate diverse project outcomes.

Eventually, the Senior Research Assistant (5) role I was performing as a data visualisation specialist evolved into a Research Fellow (6) position on the same project. This provided a context for me to apply the varied skillset I had been developing over ten years of design and academic experience. The project, Detecting Opportunities and Challenges for the Australian Rural Industries, ran from 2017 to 2018 (Hamilton et al., 2018) and led to a successive project, Horizon Scanning: Opportunities for New Technologies and Industries, from 2018 to 2019 (Hamilton et al., 2019). Both projects involved a series of Horizon Scans performed at three-month intervals to identify opportunities and challenges facing Australia's rural industries. Horizon Scanning is a research technique used to systematically explore emerging issues. This typically includes the identification and examination of trends, drivers of change, emerging developments, as well as novel issues or wild cards (Miles et al., 2016). In our case, this required a multidisciplinary project team comprising 14 researchers across 9 domains. Collectively, we examined various public data sources (e.g. Twitter, academic journals, news media) to identify emerging technology with the potential to impact new and existing agriculture industries. Findings were presented in Horizon Scan reports, which blended inputs from data mining, quantitative and qualitative analysis, literature review, scenarios and infographics.

My role as a Research Fellow was one of synthesis and creation. I was required to support the project specialisations and bring together the outputs to form a cohesive outcome. Flexibility was required in these interactions, as well as leadership to ensure that outputs fit into the broader narrative and aims of the project deliverables. Communication proved to be an essential aspect, as with most projects involving large and diverse teams. However, a challenge of this project was communicating outcomes to diverse stakeholder groups, which included policymakers, and agriculture industry executives as well as farmers and small business owners. To a large extent, I drew on my design experience to develop and deploy diverse textual and graphic communication modalities to ensure appropriate and engaging outcomes for each group.

Having since moved into a role as Lecturer in Industrial Design, it has been instructive to reflect on these past experiences. The opportunities for collaboration and skill development that have come with each experience have shaped my diverse skillset and varied research interests. I believe they have also fostered a mindset of adaptability, which is core to my identity as a design academic. My experience serves as the motivation to write this chapter – a curiosity to explore the unique experiences and learnings that have shaped the path of other design academics in the early stages of their career.

6.3 A CASE FOR THE GENERALIST

As outlined in the introduction of this chapter, the dichotomy between a generalist and specialist skillset is a relevant discussion in the context of design. Designers are expected to apply broad skills and knowledge to various problems, while also demonstrating skills deep enough to be effective in a specific context (Kolko, 2011). This dichotomy is similarly relevant to the design academic. The standard of academic training, the PhD, suggests the requirement of a specialist skillset to produce a deep and significant body of work in a specific subject area. However, there is a tendency for design researchers to seek a creative contribution to broader social, cultural, political and environmental problems (Ehn and Ullmark, 2017). These require various threads of knowledge to be drawn together and understood to form a cohesive whole (Buchanan, 1992; Ehn and Ullmark, 2017). Designers must determine the subject matter and then draw it from the relevant domains. The integration of theory and practice that occurs during this process allows new knowledge to be generated for new productive purposes and to address various problems (Buchanan, 1992).

Underlying the integration, generation and use of knowledge described by Buchanan (1992) are two essential aspects of the design process: problem synthesis and problem solving (Cross, 1982; Gulari and Fairburn, 2013). Problem solving is at the core of design. After all, designers are solution-focussed and engage in incremental efforts to explore problems and create innovative solutions (Cross, 1982). Problem solving requires continuous solution finding and evaluation (Kolko, 2010b), but it also entails dynamic relationships with defining and revising problems (Cross, 1982; Kolko, 2010b; Smithers, 2002). This is because many design activities do not deal with a singular problem or problems with well-defined requirements (Buchanan, 1992; Cross, 1982). Therefore, before acceptable solutions can be developed, designers are required to define problems and set requirements. Here, problem synthesis is used to actively construct problems in order to explore what is and what should be possible (Cross, 1982; Kolko, 2010b; Smithers, 2002).

The process of synthesis is considered to be one of understanding and connecting various, and sometimes not obviously related, issues. It is a multi-activity process of both using and constructing knowledge (Smithers, 1998, 2002). Designers apply general context knowledge, which embeds knowledge of the domain, customers and the purpose of designing. Exploration knowledge for forming requirements and problems, defining problems and then finding and evaluating solutions is also applied. This use of knowledge enables problem definitions to be modified and revised as required (Smithers, 2002). Designers also construct different types of knowledge through the design process. Knowledge construction occurs through exploring problems and finding solutions, as well as through the learning that occurs during documenting, recording and presenting designs for evaluation (Smithers, 1998, 2002). Combining this constructed knowledge with existing knowledge serves to both make sense of problems and to create new information, which facilitates the generation of compelling and innovative solutions (Kolko, 2010b).

The ability to apply existing knowledge as well as construct other types of knowledge through process enables designers to adapt to new problem areas during their design activity. Given this, the transfer of this skill to an academic context might also present potential advantages for the design academic. Specialism is a pillar in academia as it fosters depth of knowledge and an authoritative voice. However, the value of interdisciplinarity is also recognised due to its potential for research innovation. Interdisciplinarity is considered to foster rich outputs and equip academics with experience across multiple disciplinary methodologies and research languages (Dalton et al., 2017). There is an organisational value inherent in interdisciplinarity as research and industry collaborations provide an opportunity to break new ground at the intersection of disciplinary expertise (Szostak et al., 2016).

Rather than argue one against the other, perhaps it is more constructive to adopt an approach that facilitates specialism to maintain the deep understanding of a research landscape and instil trust in collaborators, but also foster an interdisciplinary mindset to develop collaborative institutional structures (Dalton et al., 2017). The richness of such a mindset is explained by Miller (2021b), albeit in the case of her coming to design with a psychology and marketing background. Her ability to thrive in such a situation was in part attributed to the willingness of both sides to collaborate and share ideas.

In these situations of interdisciplinary collaborations, possessing the skill of synthesis can likely be of value; to draw specialist knowledge from domain experts to support one's own existing knowledge and assist the exchange of this knowledge across domain boundaries (Gulari and Fairburn, 2013). It might be particularly important for early career academics so that they can take advantage of diverse opportunities as they arise, while also maintaining a research direction and academic voice.

6.4 ADAPTABILITY

Several factors contribute to how adaptable a person is. Pulakos, Arad, Donovan and Plamondon (2000) used eight factors to describe individual adaptability in general contexts. These range from the ability to handle emergency situations, dealing with uncertainty and learning new tasks and interpersonal skills. Another scale specific to vocational contexts, the Career Adapt-Abilities Scale (CAAS), identifies four factors: Concern, Control, Curiosity and Confidence (Savickas and Porfeli, 2012). In an analysis of both scales, Hamtiaux, Houssemand and Vrignaud (2013) found there to be relationships between factors in each separate scale, as well as among factors in the same scale. Such overlapping relationships indicate that adaptability is not based on a set of simple and discrete constructs but involves the interplay of many factors (Hamtiaux et al., 2013). Due to this complexity, this paper will use a higher-level discussion of adaptability based on three broad concepts used to discuss how a person adapts (Savickas and Porfeli, 2012).

6.4.1 Capacity to Adapt

The first concept, a person's ability or capacity to adapt, is determined by the resources or capabilities available to them (Savickas and Porfeli, 2012). This concept is also referred to as versatility and includes traits of resilience, vision, attentiveness, competence and self-regulation (Alessandra and O'Connor, 1996), among other capabilities that enable problem solving in unfamiliar, complex or transitional contexts (Savickas and Porfeli, 2012). These traits are predominantly knowledge- and skill-based (Alessandra and O'Connor, 1996), and therefore develop over time with experience. As experience and skill are gained, a person's aptitude to adapt increases, thus improving their versatility (Alessandra and O'Connor, 1996).

6.4.2 Willingness to Adapt

The second concept is a person's willingness or readiness to adapt and affect change (Alessandra and O'Connor, 1996; Savickas and Porfeli, 2012). Unlike a person's ability to adapt, this has little to do with skill and more to do with a person's attitude, how flexible and how open they are to adjust their behaviour to new ways of working. This includes a willingness to engage with unfamiliar situations, as well as a willingness to work with others in diverse teams (Alessandra and O'Connor, 1996). Traits positively associated with willingness to adapt include tolerance, empathy, respect, confidence and a generally positive attitude towards people and situations (Alessandra and O'Connor, 1996).

6.4.3 Adaptation Environment

The third concept is the context or environment in which both ability and willingness to adapt are situated. Ability to adapt, including specific resource development, is tied to the environments in which experience is gained. Willingness to adapt, comparatively, is considered as a set of traits that reside in the individual rather than being influenced by the environment (Savickas and Porfeli, 2012). In either case, however, whether a person adapts to a situation or context is conceptualised in relation to how the environment is perceived and then reacted to (Hamtiaux et al., 2013). Therefore, a person's individual adaptability influences how they respond to an environment, whether it is changing or not. A highly adaptable person is likely to be more proactive towards change and more likely to anticipate and affect a change before it is required. They are also likely to be better able to frame the situation, select appropriate strategies and regulate behaviour to manage the process of adapting (Ployhart and Bliese, 2006).

6.4.4 Mindset

The concepts of a person's capacity, willingness and ability to affect their environment, as well as their contribution to adaptability relate to "mindset." Mindset encompasses these concepts as it describes the overall view a person adopts of themselves in terms of personal characteristics and capabilities relative to different applications (Dweck, 2006). Dweck (2006, p. 6) found that mindset profoundly affects how people lead their lives, including professional accomplishments, relationships and personally held aspirations. A distinction of two mindsets is made by Dweck: (1) a *fixed mindset* is the belief that one's qualities are set, and there is little or no room for growth, including a certain amount of intelligence, personality and moral character; and (2) a *growth mindset* is the belief that qualities can be developed and improved relative to effort and experience.

With a growth mindset, the belief that new skills can be developed facilitates a willingness to stretch oneself beyond what is already known and fosters a passion for learning. Based on this description, people with a growth mindset fall into the category of being proactive towards change, as outlined by Ployhart and Bliese (2006). They might perceive a change in the environment and consider it as a challenge and opportunity to grow. On the other hand, they might enact change for other reasons, such as for personal fulfilment or aspirations. This belief in change and desire for development recognises the value of being challenged and putting in the effort to achieve a desired outcome (Dweck, 2006).

6.5 EXPLORING THE EXPERIENCES OF NEW DESIGN ACADEMICS

The previous sections have discussed skills, characteristics and mindsets that inform a generalist skillset and enable adaptability in individuals. We argue that these benefit people entering academia and that individuals with design backgrounds are advantaged. Designers are often required to deal with ill-defined problems or problems without singular solutions (Cross, 1982). The capacity to adapt to these problems is facilitated by synthesis, during which information is actively constructed and problems are shaped and defined to facilitate solutions (Cross, 1982; Kolko, 2010b; Smithers, 2002). In professional practice, this has been referred to as problem setting and it is attributed to the effective adaptation of knowledge to deal with new problems and situations (Schön, 1983). Further to capacity to adapt, Savickas and Porfeli (2012) and Alessandra and O'Connor (1996) outline traits of willingness to adapt to include qualities such as empathy, tolerance and optimism, among others. As these traits are emphasised in the education and practice of design, it is likely that they also transfer to other professional activities, including academia.

To inform this argument, this research explores the experiences of new design academics relating to the development and transferability of skills and characteristics to assist in adapting to academic tasks. It investigates the research question, *What skills and characteristics enable new academics with backgrounds in design to effectively adapt to academic tasks?*

6.5.1 Method

To investigate the research question, perspectives from early career academics in design disciplines were gained through a series of semi-structured interviews. Semi-structured interviews were used as they enable a consistent set of questions across participants but a different set of prompts and varied sequencing provide flexibility and the opportunity for new information to arise (Donley and Grauerholz, 2012; Wilkinson and Birmingham, 2003). In this way, interviews can act more as conversations (Donley and Grauerholz, 2012). As the research is focussed on exploring the experiences and perspectives of different individuals, such qualities were considered important in the research methods.

Findings were used to develop a Skills Map, visualising skills and characteristics identified to be important to new design academics. The Skills Map is not intended to represent the general experience of all early career design academics. It represents the shared experiences of the eight academics interviewed in this research and therefore provides some potential cues for developing the skills and attitude to facilitate adaptability to academic tasks. The following sub-sections outline the research participants and procedure of semi-structured interviews, including the approach used to analyse the data. This research complies with the Australian National Statement on Ethical Conduct in Human Research and was approved by the Human Research Ethics Committee at Queensland University of Technology (Ethics approval number 1900000896). Informed consent was obtained from each participant.

6.5.2 Participants

Eight early career academics participated in semi-structured interviews. Each participant had completed their PhD within the previous one to five years. Participants' design disciplines and countries are outlined in Table 6.1.

TABLE 6.1
Participant Details

Participant Number	Design Discipline	Country
P1	Interaction Design	Australia
P2	Industrial Design	Australia
P3	Landscape Architecture	Australia
P4	Landscape Architecture	Australia
P5	Interaction Design	Canada
P6	Architecture	Australia
P7	Interior Architecture	Australia
P8	Landscape Architecture	Australia

Each of the participants had either been previously employed or were currently employed within an Australian university. One was currently working in an overseas university. They were recruited from the authors' professional networks. Although each participant had shared some experience of employment in Australia, they collectively represent diverse cultural, educational, and professional backgrounds. Participants 3, 4, 6, 7 and 8 each received their formal design training and professional design experience overseas, including Asia, Europe and Oceania. In the cases of Participants 3 and 6, their work and study spanned over multiple countries in Western Asia and Europe, and multiple languages. Participants 3, 4, 5 and 8 each had extensive professional design experience prior to entering academia. This ranged from mid-level to executive-level positions in organisations of varying sizes. In contrast, the design experience of Participants 1 and 2 was predominantly gained through their undergraduate and postgraduate studies. Participant 2 mentioned having "a few other lives before coming to design" referring to a career in another domain prior to beginning a Bachelor of Design (Honours) as a mature age student.

6.5.3 Procedure

Eight participants were recruited via email and nominated to take part in the semi-structured interviews. Each participant was sent a Consent Form to review, sign and return prior to conducting the interview. Once the Consent Form had been completed, a time for the interview was arranged via email. Participants who were employed at an Australian university participated in either face-to-face interviews at a private location on campus or interviews over Zoom video conferencing software. The interview with the participant who was employed at an overseas university was conducted over Zoom at a time convenient to the participant. All interviews were audio recorded to facilitate analysis at a later time.

The same set of questions was asked in each interview. Prompts were used to encourage participants to elaborate on their responses and explore experiences unique to each participant. The interview questions were designed to gain insight into participants' experiences and perspectives regarding their:

1. Professional design work experience
2. Academic work experience
3. Transition from professional design to academia (if applicable)

The focus on these three areas was to develop an understanding of each participants' design experience and to elicit the skills and characteristics seen to be important for transitioning to academia and performing academic tasks. Asking broadly about experiences and perspectives allowed traits and characteristics to emerge inductively from the data. These would form the basis of a Skills Map outlining important skills for new design academics, including potential avenues and opportunities for skill development and transferability to aid transitioning to academia.

6.5.4 Analysis

Analysis of interview data focussed on identifying themes describing skills and characteristics considered important for design work, academic work and the transition from design to academia. This was achieved using Affinity Diagramming, a tool used to break down data into small chunks, and then organise and create linkages between chunks across a dataset (Kolko, 2010a). Typically, this involves organising data (e.g. statements, keywords, etc.) recorded on Post-It notes into themes on a large wall. The resulting organised collection of data, the Affinity Diagram, is built from the bottom up. Labels are applied to the data based on its content rather than using predefined categories. This allows themes to emerge inductively based on the perspectives provided by the research participants (Holtzblatt et al., 2004).

To facilitate the Affinity Diagramming process in this research, statements made by interviewees relating to skills and experience were transcribed on digital Post-It notes using Miro visual collaboration software. Timestamps and participant numbers were recorded for each statement so that the concept and associated statement could be relocated in the data. Once all relevant statements had been recorded for all interviews on digital Post-Its, tags were applied to specify key concepts discussed in each transcribed statement. The identification of tags followed a process of open coding in which they emerged inductively based on the content of the interviews (Benaquisto, 2008). As tags were identified for discrete concepts, they were combined where possible to establish the most important concepts and streamline the presentation of results. The purpose of using an open coding approach instead of predefined codes was to represent the participants' experiences and perspectives when identifying important skills and characteristics. Figure 6.2 shows an example of transcribed statements on digital Post-It notes with tags applied.

Post-It notes were thematically arranged based on the similarity of tags. As each theme emerged, a theme title was recorded to group the Post-Its. In the case that a statement related to more than one theme, the statement was duplicated and positioned in each relevant theme cluster. This was done for all statements relating

TABLE 6.2
Example of Transcribed Statements on Tagged Digital Post-It Notes Arranged Thematically Based on Statement Content and Tags

Theme	Description
Curiosity	Being perceptive and open to exploring new ideas; exhibiting knowledge-seeking behaviour to expand beyond one's immediate knowledge
Creativity	Engaging in exploratory and innovative practices; exhibiting a willingness to create new ways to optimise situations and outcomes.
Critical Thinking	Being perceptive of situations and having the capacity to apply analytic and holistic thinking
Problem Solving	Ability to perceive situations and effectively and innovatively overcome problems identified
Synthesis	Ability to apply holistic thinking to perceive situations and bring together multiple strands of knowledge, which might facilitate identification of opportunities or new ways to approach problems
Interpersonal Skills	Skills to assist communication and facilitate collaborative and synergistic working relationships
Academic Skills	Skills essential for academic tasks, including writing, research methods, system knowledge and knowledge of research tools
Opportunistic	Demonstrating an open mind, and a willingness and self-efficacy to take on new opportunities that are presented
Kaizen	A desire for constant personal and professional development and optimisation
Vision	Demonstrating a strategic and purposeful approach to choices made; adopting a mindset conducive to knowledge seeking and innovation
Evolutionary	A willingness to adapt as requirements and interests change; adopting a mindset of personal and professional development

to each of the three focus areas. Figure 6.2 shows how statements with tags were organised into themes. Eleven themes emerged from the analysis. These are listed in Table 6.2, including a short description based on the main tags and content associated with the theme. In Section 6.6, each of the themes and associated content is discussed.

6.6 RESULTS

Results are presented in three sections corresponding to the three focus areas of interviews, as described in Section 6.5.3: (1) design experience; (2) academic experience; (3) the transition from design to academia.

6.6.1 Design Experience

Table 6.3 summarises the four themes and associated tags that emerged from participant responses to questions about design experience. The importance of high-level, broadly applicable skills was expressed in participant responses, as opposed

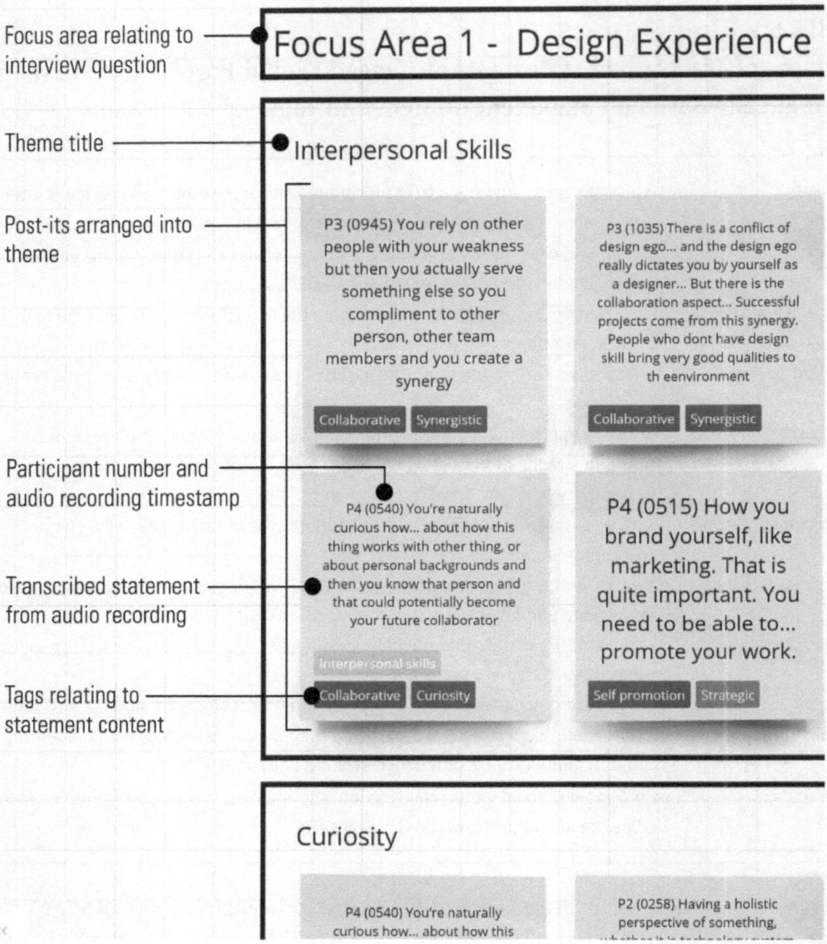

FIGURE 6.2 Affinity diagramming showing tagged post-its arranged in themes

to specific technical skills when discussing previous design experience and essential design skills. Summing up this trend was a response from P6, who stated:

> I don't think that skill is a particular skill … say this software or that software. It's about your mentality and how quickly you understand the gaps and needs and how quickly you can respond to that using any design.

Curiosity was a prevailing theme used to explain participants' approach to their work. P6 stated:

> I'm a very curious person and as soon as I get comfortable doing something, it means that I've nearly learnt most of the things that could be learnt in that area. So, I always want to find out answers for unknowns"

TABLE 6.3
Design Experience Themes and Concepts Summary

Theme	Tagged Concepts
Curiosity	Curious
	Knowledge seeking
	Perceptive
Critical Thinking	Perceptive
	Analytic
	Problem solving
	Holistic thinking
Interpersonal Skills	Synergistic
	Collaborative
	Perceptive
Creativity	Self-efficacy
	Exploratory
	Optimise
	Innovative
	Synthesis
	Creativity

The participant here identifies both their desire to expand on what they already know and recognises the importance of developing skills and capabilities. In this way, curiosity was also said to be important for developing personal networks and creating the opportunities that come with them. P4 stated: "You're naturally curious … about how this thing works with other things, or about personal backgrounds and then you know that person and that could potentially become your future collaborator."

This desire to explore the new also extended to creativity and the application of new approaches to tasks. P3 highlighted the importance of creativity in how they approach their work and how it facilitates taking on new and unfamiliar tasks. They stated: "you need that creativity, you need to have that realisation of, hey, I can do that." Similarly, P6 identified creativity as a core approach to their work. Whatever the task faced with, even if considered something that is not creative, they stated they always try to approach it in a creative way. They "always try to engage with new methods and technologies … always trying to make things more creative and innovative to keep up with tech."

Building on the importance of designers being curious and creative, P1 discussed the importance of critical thinking – to observe critically and to understand how things work. They explained one of their important traits as a designer as "being an observer. Having … a critical thought … filter, almost like a critical analysis filter wherever you go." The sentiment behind this statement was not to settle for the norm, but to critically assess opportunities for improvement and innovation. This also highlights some important qualities of critical thinking that emerged, which were being perceptive and adopting holistic thinking. P2 explains this in their ability of "Being able to adopt a holistic perspective of something. Being able to interpret the ambiguous and then being able to communicate that in a way that is appropriate for your audience." They recognise the importance of transforming observations and learnings into practical or creative outcomes.

Interpersonal skills were commonly recognised as important due to the emphasis on teamwork in design fields. For example, P3 discussed the importance of diverse team skills and the ability to interface with others: "You rely on other people with your weakness, but then you actually serve something else. So, you compliment another person, other team members and you create a synergy." This ability to perceive the strengths and weaknesses of one's self as well as others enables a synergistic approach to collaboration.

6.6.2 Academic Experience

Table 6.4 outlines the four themes that emerged from interviewees' responses about their entry into academia. They predominantly refer to their motivations and experiences.

When asked about their move to academia, a common theme identified as a catalyst by interviewees was a curiosity and desire to explore and discover more than what was currently on offer. An underlying vision of personal interest and purpose

TABLE 6.4
Academic Experience Themes and Concepts Summary

Theme	Tagged Concepts
Curiosity	Curiosity
	Novelty seeking
	Knowledge seeking
	Exploratory
Opportunistic	Novelty seeking
	Opportunistic
	Self-efficacy
	Responsive
	Personal development
	Professional development
Kaizen	Perceptive
	Reflective
	Personal development
	Professional development
	Optimisation
Vision	Purposeful
	Strategic
	Synthesis
	Innovative
	Knowledge seeking

was also expressed through a desire to seek and develop deeper knowledge of topics of interest. P3 explained that:

> Why I ended up in a PhD is … I think curiosity first ... If I don't like the design, I question things and why the design doesn't really work … and those questions you cannot find answers in industry. It just doesn't let you to think about that … And I think that's what I need … I wanted the discovery, the innovation, somehow finding solutions to our problems.

Similarly, P2 stated that

> I came through my undergrad and I felt like I had kind of skimmed the surface of everything and that I didn't get … a real in-depth understanding or experience in one thing. I thought a PhD would give me that.

A PhD was also viewed as a way to explore novel ideas that were of personal interest and not previously available to them. P5 stated:

> I was very interested in the social robotics field, but I was not very keen at looking at it from an engineering perspective. Because when I was working on those projects, the focus was more on the technology side. Like how can technology be designed or what technologies can be used? So, the focus was more on developing these technologies rather than understanding the human side of it.

In each of these statements, an underlying notion of personal development and knowledge-seeking behaviour can be seen. Participants identified the restrictions they felt in their previous experiences and saw academia and a PhD as a way to break away from that to explore their interests at a deeper and more purposeful level. After starting their PhDs, the academic work experience of interviewees was predominantly described as opportunistic, containing numerous short contracts. P1 stated that "I have had now one year of lecturing and then some research assistant contracts … I was tutoring and did have the odd contract like assisting someone with data analysis." A similar experience was explained by P2, who viewed these short projects as opportunities for development:

> *lots of little research contracts. So, just taking things as they come, kind of a little bit ad hoc. You know, just trying to get as much experience, kind of build a reputation for … a skillset and reliability and a good work ethic.*

This statement reflects both the desire for professional and personal development in the sense that the participant discussed both a skillset and their personal qualities that contribute to a reputation.

Kaizen emerged as a theme based on participants' discussion of incremental development and optimisation of their activities. The term Kaizen is a Japanese philosophy of continual improvement ("Kaizen," 2020). P1 stated: "I'm always … working out how I can streamline something or improve something. Even the most basic of tasks … I'll just think was it the best way? How efficient was that? What could be improved?" In addition to optimising activities, the process of development for participants can be viewed in a similar way as their move to academia. The act of continuous improvement served as an opportunity to innovate and be creative, therefore pushing beyond their current experience.

Vision was demonstrated by participants through identifying their interests and how these fit into their professional future. For example, P2 realised during their undergraduate design degree that their interest was not in developing technical design skills, such as computer-aided design. They stated: "I made a conscious decision very early on in my undergrad to push my career in a research direction …" P7 explains their strategic approach to develop professionally as a designer and academic, and drawing these together in their practice:

> I always wanted to become an academic. I actually have a very linear progression towards academia but I always in parallel did some practical work as well. I worked as a freelancer for architecture firms and did some artworks along the way. So, it was never a separate thing.

This synthesis of professional design and academia to drive professional development was prevalent across participants. Participants often identified areas of personal interest emerging in their design practice and explained its role in their academic pursuits.

6.6.3 Transition from Design to Academia

Interviewees were asked to discuss their perspectives on transitioning from design to academia, focussing on skills required for academic tasks and the transferability of their design skills. A summary of themes that emerged and associated concepts is shown in Table 6.5.

Participants discussed the importance and challenge of developing skills specific for academic work when transitioning to academic tasks. Academic writing was identified as a major challenge for most participants. P3 stated:

> The research skills and my writing. It took me two years to write a research problem to a PhD proposal … There were billions of ideas, billions of things I wanted to do but there is also this aspect. Being a researcher, being a PhD student … having that mentality is a different thing. So, putting in that two pages is a very difficult thing for a designer, and I heard this from other people too.

It was not only the difficulty of developing writing skills that was expressed but also the shift in the type of writing required for academia compared to industry. P7 identified that "even if it is a reflective type of writing, it's not the same that is happening in industry." This overall shift in skills from design to academia was described by P4 to be from visual to textual modalities.

> Have you seen the book called Designers Don't Read? I totally agree. Like when I was [a] designer there were books we used as a reference, but a reference for the graphics …

TABLE 6.5
Transition from Design to Academia Themes and Concepts Summary

Theme	Tagged Concepts
Academic skills	Research tools Academic writing Research methods System knowledge Analytic
Synthesis	Holistic thinking Perceptive Critical thinking Synthesis
Problem solving	Creativity Innovative Optimisation Critical thinking Perceptive
Evolutionary	Personal development Professional development Opportunistic Evolutionary

> In academia, you have to read. You know, like how do they tell a story? How do they make sense of their own storytelling? You know like why this picture is important, in what context? Everything has to become meaningful.

Here, the shift from visual to textual included a change in the meaning and function of these modalities as well as just type. In academia, it is essential for text and images to convey meaning, whereas in design it was suggested that they could serve functional purposes without having a depth of meaning.

Other areas of skill and knowledge development identified were research tools and methods. P5 stated: "I learnt to use different tools like Observer; Noldus Observer was a big thing and I got my post-doc because I knew Observer … And exposure to Atlas.ti, some of the analysis tools." P2 highlighted an area that could be improved as "Planning a project. We could use more education of that – planning and management. And maybe even a sharing of methods and approaches between disciplines and academics." In both these cases it was recognised that there are a lot of different options and ways of doing things. Being able to know which tools and approaches to learn and which not to learn was identified as important.

A broader area of development required was knowledge of the general structure of the academic environment and system. P2 stated: "You know, understanding the processes involved, the way the whole sort of system works in an educational system … There is quite a big leap there." P8 reflected the value of learning this system-level knowledge through their participation in an academic excellence program, which outlined various requirements and goals for what to achieve as an early career academic. They stated: "as an early career academic that course has been really helpful."

When discussing the shift in skills after transitioning from design to academia, the ability to synthesise information and engage problems in innovative and creative ways was highlighted. P3 explained:

> I can immediately open, create the [problem] frame bigger and make it smaller too. So, I think it's an important skill to have. Designers … they see the gaps, you know, better than the other people, better than other traditional researchers … We see the problem, and once we get those skills all together, we can easily play with that [problem] boundary.

Here, synthesis was discussed in terms of framing problems, narrowing down information, accessing information across boundaries and identifying gaps. Synthesis was also spoken about at a broader level through the integration of design and research. P7 stated:

> I like making things, and that's something that differentiates me, maybe. I like to test my ideas in ways that are more tangible, and I guess it does have an impact … I exhibit those works in creative mediums … cultural and art venues. And that kind of becomes a visual outcome or impact or engagement with the community.

The synthesis skills developed initially as a designer and transferred into academia serve as a potential means of differentiation and broader impact of research outcomes.

Problem solving through innovative and creative thinking was discussed, which expanded on the importance of creativity and curiosity expressed by interviewees earlier in the context of design. Interviewees expressed their determination to look at tasks in creative ways to analyse, innovate, streamline and bring new approaches to tasks. P6 stated: "We [designers] always look at things differently. We are looking at how we can create something out of nothing." P6 continued:

> As a designer, I need to find ways of improving things or doing or creating things. So, one of those being teaching in higher education in this field. In each country in each context the understanding of design was different. The need for the skillset required for a designer was different. So, I had to basically learn what I don't know and then learn the thing I didn't know.

In the final question of the interview, interviewees were asked to explain how they define themselves as an academic with a design background. A common approach was to describe their current position description or provide a summary of their current research and teaching interests or activities. Several interviewees had difficulty addressing this question and devising a concrete definition. After beginning to describe their research philosophy P2 conceded: "I don't know, I'll have to come back to that one. I'm still working on that." Similarly, P8 stated:

> I'm so new to this that I haven't found what my new description is. When I ran a business, we had an elevator pitch that we would use when we meet people. I probably haven't honed what my elevator pitch is when I meet people now.

Despite the difficulty devising a definition for themselves, it was acknowledged that their identity as a design academic was an evolutionary process – something that could evolve as research opportunities and interests change. P4 stated:

> I know there is some people, they'll kind of you know, brand themselves using very interesting … like a digital ethnographer. That's very, that's a bright name. Really nice and I'm jealous of that … Ok, I might have something like environmental gerontologist, but then, no that's my current research interest and I do think it's like design and your research interest evolve, and your research interest might have different iterations, you know, kind of slowly progress and venture to a new area … so still searching for my design identity.

6.7 DISCUSSION

Discipline boundaries serve as a useful point of reference when considering the capabilities important for design. An interaction designer designs interactive digital products, an industrial designer conceives product concepts to be manufactured, an architect designs structures in our built environment, so on and so forth. Designers from each discipline require specific skills to achieve appropriate outcomes. However, complimenting specific skills required for activities of designing and researching is a set of generalisable skills that can be applied to many tasks and situations. The importance of these generalisable skills alongside the specialist skills developed in a domain has been emphasised in this research, not only for the activity of designing but also the successful adaptation to academic activities. It is this adaptation, the approach to transferring existing skills and developing new skills for design academia, that this chapter has been principally concerned with. In addition to skill, adapting to new activities and situations requires an individual to possess personal traits that provide the mindset and willingness to adapt (Alessandra and O'Connor, 1996; Dweck, 2006; Savickas and Porfeli, 2012). Such personal traits were discussed by the participants as they reflected on their experiences as designers and design academics. Personal traits were attributed to leading them to academia, informing their practice and driving their evolution as design academics.

The Skills Map in Figure 6.3 illustrates the skills and traits that emerged from participant interviews as important for design, academia and transitioning to design academia as a new design academic. These are framed by the prevailing concepts of adaptability – capacity to adapt to the left of the map, and willingness to adapt to the right of the map (Alessandra and O'Connor, 1996; Savickas and Porfeli, 2012). Individual skills and traits are arranged in cells. They can be thought of as being on a continuum, where it is possible for a concept to reflect both capacity and willingness

FIGURE 6.3 Skills map

to adapt; many do this. The darker cells represent themes, whereas the lighter cells represent concepts associated with the themes they are adjacent to. Placement of themes and concepts was informed by the themes and tagged concepts that were reported in the results section. Just as in the results section, concepts can be related to multiple themes and thus, the Skills Map represents relationships among various skills and traits.

The skills and traits comprising the Skills Map are discussed in the following sections. For the purpose of this discussion, the Skills Map is divided into four groups. Each group is comprised of related themes and concepts that can be discussed together due to their similarity and complementary nature. Grouped cells are indicated by black outlines in Figures 6.4 through 6.7 in the following sections.

6.7.1 Group One: Curiosity

Group One reflects curiosity and an openness to explore the new and unfamiliar. It comprises the themes of Curiosity, Creativity and Problem Solving (Figure 6.4). Curiosity was often used by participants to describe themselves, and it was identified as an important influence in their career journeys spanning both design and

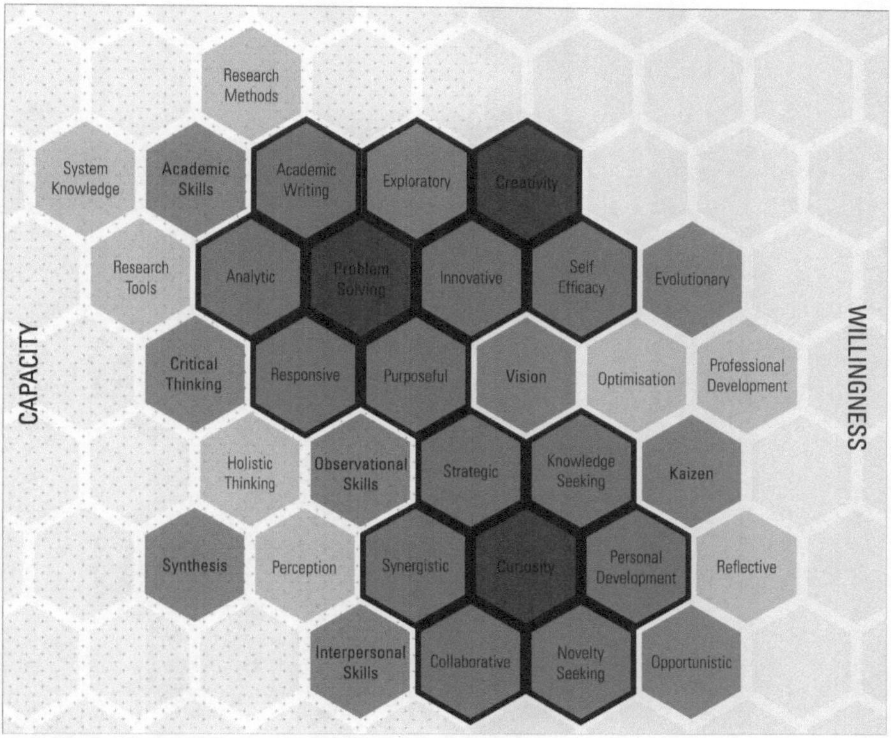

FIGURE 6.4 Group One: Curiosity

academia. Curiosity is an important aspect of intrinsic motivation, which is defined as "the doing of an activity for its inherent satisfactions rather than some separable consequence" (Ryan and Deci, 2000, p. 56). The link between curiosity and intrinsic motivation has been explained by Berleyne's theory (1960, cited in Oudeyer et al., 2016), which proposed that information gain, novelty or complexity provides an intrinsic reward to the brain. This intrinsic motivation aspect of curiosity was reflected in participants' desire to explore areas of knowledge further to develop a deeper understanding.

In vocational contexts, Schweizer's (2006) Novelty Generation Model identifies the trait of curiosity to be a precursor to the behaviour of novelty seeking, which is then followed by behaviours of creativity and innovative performance. This is consistent with the indicators of curiosity reported by Savickas and Porfeli (2012) in the Career Adapt-Abilities Scale, namely seeking different ways of doing things. Participant responses demonstrated this relationship of curiosity informing creative and innovative problem solving through their perspectives on developing and implementing work activities. Whether research or teaching, participants conveyed an openness to, and satisfaction from, exploring and discovering new approaches to common tasks they performed. For example, making learning experiences more engaging and creatively applying new research methods.

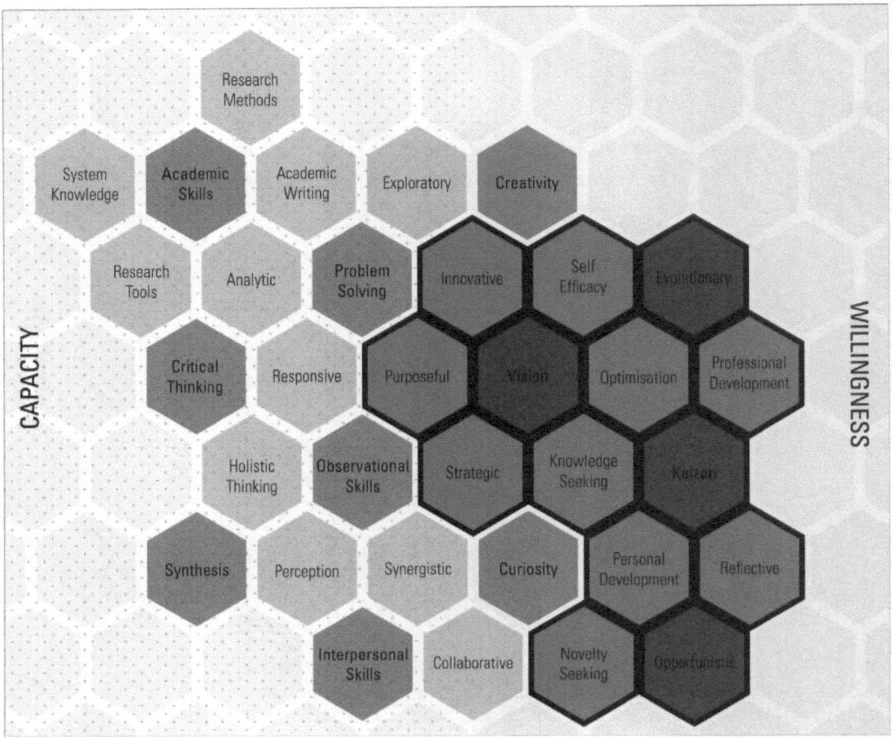

FIGURE 6.5 Group Two: Development

6.7.2 Group Two: Development

Group Two reflects an attitude of personal and professional development and comprises themes of Vision, Kaizen, Opportunistic and Evolutionary (Figure 6.5). Due to participants discussing the attainment of developmental outcomes, this group is associated with extrinsic motivation (Ryan and Deci, 2000). Vision, like curiosity, was an overarching concept discussed by participants and was a key factor in pursuing academia. Participants displayed a desire to engage bigger and more complex problems, and to become more adept in an area of knowledge and skill than they were able to in industry. The motivating factors and areas of interest that sparked this move were clearly articulated by participants. Once in academia, participants' approach to work and progression was Opportunistic. A common experience conveyed was accepting small opportunities to develop skill and form a reputation to foster more regular work opportunities. This duality of having strategic direction and being open and prepared for opportunities has also been highlighted by Miller (2021a) in her chapter on developing resilience in this book. It is about recognising what drives us and where we are best suited so that we can amplify our strengths and move closer to where we want to be.

Displaying this type of purpose and motivation is an important element for effective self-directed learners (Bouchard, 1996b; Clardy, 2000). For this reason, we also

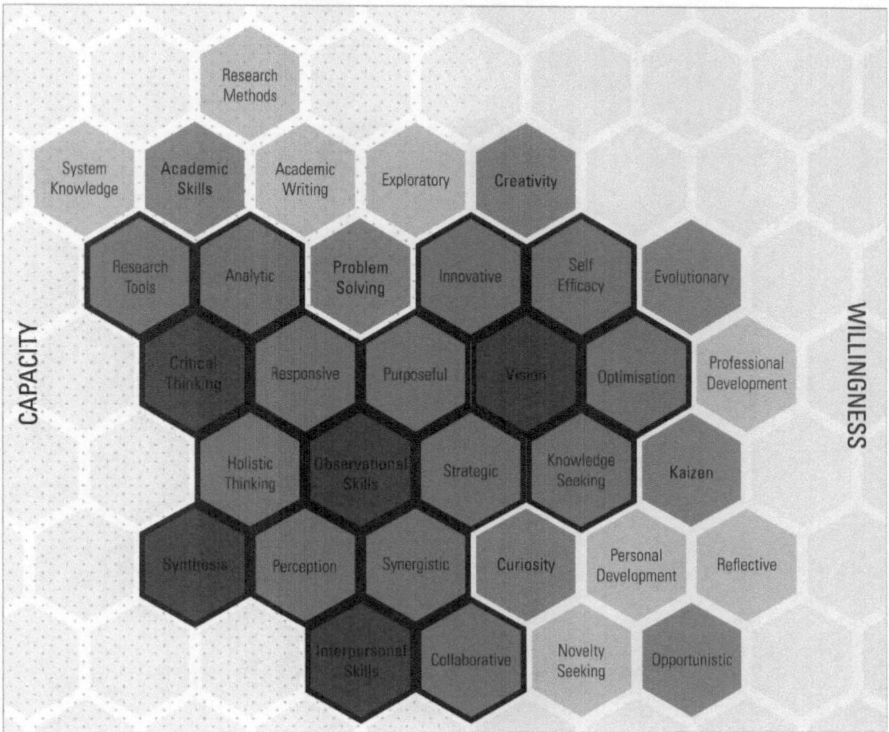

FIGURE 6.6 Group Three: Synthesis

associate the term Kaizen (continual improvement). Participants' desire to develop and grow was related to both personal benefits and to the benefit of others. For example, for their own benefit, participants aimed to develop skills and understand their strengths so that they could better contribute to their work and create opportunities. For the benefit of others, participants discussed the continual improvement of teaching and learning experiences for students.

The final theme in Group Two, Evolutionary, reflects participants' overall developmental mindset. Participants expressed self-efficacy as confidence grew in their ability to adapt to varying opportunities. The development of such a self-efficacious, and indeed opportunistic, mindset might be in part an environmental reaction to the short-term temporary teaching and research contracts that are often what is available to new academics (Wardale et al., 2019). In addition to being a response to the academic environment, the combination of being opportunistic and self-efficacious is important for self-directed learning. These traits support what is known as the "autodidactic leap," where learners take actions or make decisions which put them in positions requiring them to learn on their own, and often under pressure (Bouchard, 1996b, 1996a). For example, coming into a project without the full skills but believing they can be developed on the job was viewed as an opportunity for personal and professional development.

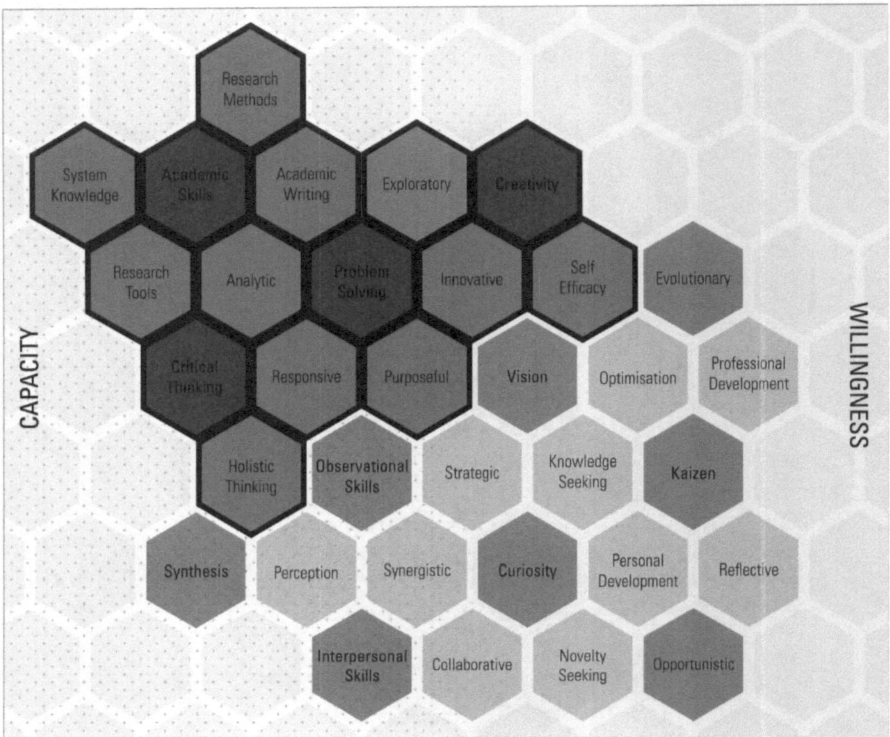

FIGURE 6.7 Group Four: Problem solving

6.7.3 Group Three: Synthesis

Group Three reflects the capabilities of problem synthesis and understanding. It comprises Interpersonal Skills, Observational Skills, Problem Synthesis, Critical Thinking and Vision (Figure 6.6). These capabilities share a relationship to designers' exploration knowledge, which is used to understand, form, refine and find requirements and solutions (Smithers, 2002). The importance placed on these skills is not surprising due to their emphasis in the activity of designing and designerly thinking (Dorst, 2011). A characteristic of designing is that it often deals with open, complex problems, which do not have a clearly defined problem, requirement or approach. Rather, this information must be actively constructed by the designer, informed by the types of characteristics outlined by participants in their interview responses; capabilities such as observing, communicating, and thinking critically (Smithers, 2002). In design thinking literature, this process is referred to as framing – the standpoint from which a problem will be addressed (Dorst, 2011).

6.7.4 Group Four: Problem Solving

Group Four reflects capabilities for problem solving, and thus is more outcome-focussed than group three. Group Four comprises Critical Thinking, Academic

Skills, Creativity and Problem Solving (Figure 6.7). These skills position an individual to execute effective problem solving and outcome generation in a variety of situations. Designers have adopted professional practices incorporating varied ways of thinking to achieve value and outcomes in response to complex problems. In the design thinking literature, designerly approaches to understanding and problem solving have been recognised to be of value and transferrable to domains outside of design (Dorst, 2011; Johansson-Sköldberg et al., 2013). They have been of particular interest to facilitate innovation in domains ranging from medicine to finance (Johansson-Sköldberg et al., 2013). In interviews, participants cited how their design experience helped them develop creative and innovative teaching practices or approaches to research.

Synthesis and problem solving in Groups Three and Four reflect the integration of understanding and exploring situations, and then creating outcomes. This is not a linear process, but instead is iterative and integrated (Smithers, 2002). When framing problems, designers move back and forth between analysis and creation, and detail and abstraction (Dorst, 2015). This designerly approach to exploring complex design problems is transferrable to research problems, which are also complex in scope and context. Research often requires multidisciplinary collaboration with fellow researchers, industry and a range of other stakeholders to converge on effective solutions. Designers' exploration knowledge, as described, allows for consideration of a breadth of factors to determine what is important and instructive. It also facilitates evaluation, reflection and self-regulation to produce satisfactory outcomes (Smithers, 1998, 2002). This mindset was ever-present in interview responses – being open to collaboration and optimising activity and outcomes to achieve better and new ways of doing things. These capabilities embedded into a mindset are important for adaptability to new contexts and activities, as adapting requires engagement in sensemaking and problem solving in new and unfamiliar contexts (Savickas and Porfeli, 2012).

6.7.5 Mindset of a New Design Academic

Participants in this research expressed with conviction their curiosity and ambition to explore new areas, their vision and openness to understand new situations, and self-efficacy to successfully execute and generate outcomes. Despite this, a candid uncertainty was present about how some participants described or positioned themselves and their identity as design academics. This uncertainty manifested in several ways. Participants attempted to consolidate their range of experience and current areas of interest. The result of this was a twofold recognition that developing and forming an understanding of personal strengths was in progress, and of the perceived inevitability of change and the opportunities it brings. This is something both of the authors relate to in their experience as early career academics.

Elsewhere in this book, Williams and Satherly (2021) similarly describe the challenges of establishing an identity as a designer in academia. They write from the perspective of experienced practitioners and academics embarking on a PhD while also performing a full-time academic role. Such a substantial change and the resulting uncertainty it causes pose significant challenges in terms of how people think and go about their work. In the context of this chapter regarding early career academics, however, we perceive this type of uncertainty not as a weakness. Rather, it is another trait that punctuates a collective mindset of early career academics. The mindset adopted by participants simultaneously recognises underlying capabilities as well as an openness to develop and take on emergent opportunities. For a new design academic, having this mindset with an attitude and capacity to adapt to change might be more important than having a stubborn and unwavering idea of where they belong. Curiosity, which we have identified as a key trait of this mindset, plays a central role in this proposition. Savickas and Porfeli (2012) explain that curiosity enables a person to explore multiple possible selves and alternate scenarios to develop and grow as a person. It facilitates the possibility of change and confidence to actualise choices. In this way, curiosity is linked to confidence and gives a person agency.

6.8 CONCLUSION

Academic environments are diverse, ever-changing and can be difficult to find a place in for new academics. We have found a place in this environment as early career academics due to a mindset that facilitated the development of an adaptable skillset and a curious attitude to seek, engage with and develop opportunities that emerge. This mindset was also found in the early career design academics we interviewed during this research. Each recognised the capabilities they developed in design and the impact they have had in their transition to academia. Their collective identity was recognised as one of curiosity and constant development, as well as an openness to the inevitability of change and the need to be adaptable.

Underlying this mindset was the strong value placed on adaptable skills. Critical thinking, interpersonal skills, holistic thinking, creativity and innovation, among others, were identified as essential skills for academic tasks that are transferrable from design experience. These allow for understanding and adaptation to various situations and their requirements, as well as facilitating the creative application of skill to produce innovative outcomes. Although it was also recognised that specific academic skills are required, and are indeed a significant challenge for new design academics, access to broadly applicable skills was seen as a core part of what it is to be a designer and design academic; they reflect strongly the mindset that fosters both the capabilities and attitude to be able to adapt.

Representing the outcomes of this research in a Skills Map has provided a stimulus for new design academics to reflect on their own experiences and consider the vision for their identity moving forward. While the Skills Map is not intended to be exhaustive in representing a universal experience or be prescriptive about skills and characteristics that must be developed, it is intended to be a tool for reflection. It represents a shared experience among several new design academics, and we hope it provides food for thought for designers entering academia or those who have recently entered academia and are looking for areas of development.

REFERENCES

Alessandra, T., & O'Connor, M.J. (1996). *The Platinum Rule.* Warner Books, New York, NY.

Benaquisto, L. (2008). Codes and coding. In L.M. Given (Ed.), *The Sage Encyclopedia of Qualitative Research* (pp. 86–89). SAGE Publications, Inc., Los Angeles, CA.

Bouchard, P. (1996a). *A Study of Self-Directed Professionals of High Attainment.* Paul Bouchard, GIRAT, Department of Education, Rm. https://eric.ed.gov/?id=ED436654

Bouchard, P. (1996b). Towards an etiological model of self-directed professional development. *18.*

Buchanan, R. (1992). Wicked problems in design thinking. *Design Issues, 8*(2), 5. https://doi.org/10.2307/1511637

Clardy, A. (2000). Learning on their own: Vocationally oriented self-directed learning projects. *Human Resource Development Quarterly, 11*(2), 105–125. doi: 10.1002/1532-1096(200022)11:2<105::AID-HRDQ2>3.0.CO;2-5

Cross, N. (1982). Designerly ways of knowing. *Design Studies, 3*(4), 221–227. doi: 10.1016/0142-694X(82)90040-0

Dalton, B., Simmons, T., & Triggs, T. (2017). Knowledge exchange through the design PHD. In L. Vaughan (Ed.), *Practice Based Design Research* (pp. 65–76). Bloomsbury Academic, New York, NY. www.bloomsburydesignlibrary.com/encyclopedia-chapter?docid=b-9781474267830&tocid=b-9781474267830-chapter7&pdfid=9781474267830.ch-007.pdf

Donley, A. (2012). Qualitative methods. In L. Grauerholz (Ed.) *Research Methods* (pp. 39–52). Infobase Publishing.

Donovan, J., Gomez, R., Swann, L., Niesel, C., Toth, B., & Poole, J. (2016). *Cotton Rapid Customisation Feasibility Study.* Cotton Research and Development Corporation. www.crdc.com.au/sites/default/files/pdf/3D%20printing.pdf

Dorst, K. (2011). The core of 'design thinking' and its application. *Design Studies, 32*(6), 521–532. doi: 10.1016/j.destud.2011.07.006

Dorst, K. (2015). The frame creation model. In K. Dorst (Ed.) *Frame Innovation: Create New Thinking by Design* (pp. 73–98). The MIT Press, Cambridge, MA.

Dweck, C. (2006). *Mindset: How You Can Fulfil Your Potential.* Random House.

Ehn, P., & Ullmark, P. (2017). Educating the reflective design researcher. In L. Vaughan (Ed.), *Practice Based Design Research* (pp. 77–86). Bloomsbury Academic, New York, NY. www.bloomsburydesignlibrary.com/encyclopedia-chapter?docid=b-9781474267830&tocid=b-9781474267830-chapter8&pdfid=9781474267830.ch-008.pdf

Gulari, M., & Fairburn, S. (2013, September 6). Is specialist designer an oxymoron? The value of specialisation in the design field.

Hamilton, G., Swann, L., Kutty, S., Nayak, R., Hearn, G., Donovan, J., Rittenbruch, M., Polson, D., & Hellens, R. (2018). *Detecting Opportunities and Challenges for Australian Rural Industries: Final Report.* Queensland University of Technology and Agrifutures Australia.

Hamilton, G., Swann, L., Pandy, V., Moyle, C., Opie, J., & Dawson, A. (2019). *Horizon Scanning—Opportunities for New Technologies and Industries* [Final Report]. Queensland University of Technology and Agrifutures Australia.

Hamtiaux, A., Houssemand, C., & Vrignaud, P. (2013). Individual and career adaptability: Comparing models and measures. *Journal of Vocational Behavior, 83*(2), 130–141. doi: 10.1016/j.jvb.2013.03.006

Holtzblatt, K., Wendell, J.B., & Wood, S. (2004). Building an Affinity Diagram. In K. Holtzblatt, J. B. Wendell & S. Wood (Ed.) *Rapid Contextual Design: A How-to Guide to Key Techniques for User-Centered Design* (pp. 159–180). Elsevier Science & Technology, San Francisco, CA. www.sciencedirect.com/science/article/pii/B9780123540515500094

Johansson-Sköldberg, U., Woodilla, J., & Çetinkaya, M. (2013). Design thinking: Past, present and possible futures. *Creativity and Innovation Management, 22*(2), 121–146. doi: 10.1111/caim.12023

Kaizen. (2020). In *OED Online*. Oxford University Press. www.oed.com/view/Entry/241008

Kolko, J. (2010a). *Exposing the Magic of Design: A Practitioner's Guide to the Methods and Theory of Synthesis*. Oxford University Press, New York, NY. www.oxfordscholarship.com/view/10.1093/acprof:oso/9780199744336.001.0001/acprof-9780199744336

Kolko, J. (2010b). Abductive thinking and sensemaking: The drivers of design synthesis. *Design Issues, 26*(1), 15–28. doi: 10.1162/desi.2010.26.1.15

Kolko, J. (2011). *The Conflicting Rhetoric of Design Education*. http://interactions.acm.org/archive/view/july-august-2011/the-conflicting-rhetoric-of-design-education1

Miles, I., Saritas, O., & Sokolov, A. (2016). Intelligence: Environmental and horizon scanning. In I. Miles, O. Saritas, & A. Sokolov (Eds.), *Foresight for Science, Technology and Innovation* (pp. 63–93). Springer International Publishing. doi: 10.1007/978-3-319-32574-3_5

Miller, E. (2021a). How to "Dare Greatly" in Academia. In A. Blackler & E. Miller (Eds.), *How to Be a Design Academic*. CRC Press, 177–188.

Miller, E. (2021b). A non-design in a design school. In A. Blackler & E. Miller (Eds.), *How to Be a Design Academic*. CRC Press, 45–55.

Morris, W., & Morris, M. (1988). *Morris Dictionary of Word and Phrase Origins* (2nd ed.). HarperCollins, New York, NY.

Oudeyer, P.-Y., Gottlieb, J., & Lopes, M. (2016). Intrinsic motivation, curiosity and learning: Theory and applications in educational technologies. *Progress in Brain Research, 229*, 257–284.

Ployhart, R.E., & Bliese, P.D. (2006, April 10). *Individual Adaptability (I-ADAPT) Theory: Conceptualizing the Antecedents, Consequences, and Measurement of Individual Differences in Adaptability* [Book-part]. Understanding Adaptability: A Prerequisite for Effective Performance within Complex Environments. https://doi.org/10.1016/S1479-3601(05)06001-7

Pulakos, E.D., Arad, S., Donovan, M.A., & Plamondon, K.E. (2000). Adaptability in the workplace: Development of a taxonomy of adaptive performance. *Journal of Applied Psychology, 85*(4), 612–624. doi: 10.1037/0021-9010.85.4.612

Ryan, R.M., & Deci, E.L. (2000). Intrinsic and extrinsic motivations: Classic definitions and new directions. *Contemporary Educational Psychology, 25*(1), 54–67. doi: 10.1006/ceps.1999.1020

Savickas, M.L., & Porfeli, E.J. (2012). Career Adapt-Abilities Scale: Construction, reliability, and measurement equivalence across 13 countries. *Journal of Vocational Behavior, 80*(3), 661–673. doi: 10.1016/j.jvb.2012.01.011

Schön, D.A. (1983). *The Reflective Practitioner: How Designers Think in Action*. Basic Books, New York, NY.

Schweizer, T.S. (2006). The psychology of novelty-seeking, creativity and innovation: Neurocognitive aspects within a work-psychological perspective. *Creativity and Innovation Management, 15*(2), 164–172. doi: 10.1111/j.1467-8691.2006.00383.x

Smithers, T. (1998). Towards a knowledge level theory of design process. In J.S. Gero & F. Sudweeks (Eds.), *Artificial Intelligence in Design '98* (pp. 3–21). Springer Netherlands, Dordrecht. doi: 10.1007/978-94-011-5121-4_1

Smithers, T. (2002). Synthesis in designing. In J.S. Gero (Ed.), *Artificial Intelligence in Design '02* (pp. 3–24). Springer Netherlands, Dordrecht. doi: 10.1007/978-94-017-0795-4_1

Szostak, R., Gnoli, C., & López-Huertas, M. (2016). The importance of interdisciplinary research and teaching. In R. Szostak, C. Gnoli, & M. López-Huertas (Eds.),

Interdisciplinary Knowledge Organization (pp. 1–33). Springer International Publishing, Switzerland. doi: 10.1007/978-3-319-30148-8_1

Wardale, D., Richardson, J., & Suseno, Y. (2019). Casual academics aren't going anywhere, so what can universities do to ensure learning isn't affected? *The Conversation*. http://theconversation.com/casual-academics-arent-going-anywhere-so-what-can-universities-do-to-ensure-learning-isnt-affected-113567

Wilkinson, D., & Birmingham, P. (2003). *Using Research Instruments: A Guide for Researchers*. Routledge Falmer, London, UK.

Williams, T., & Satherly, S. (2021). Getting that PhD: How hard can it be? In A. Blackler & E. Miller (Eds.), *How to Be a Design Academic*. CRC Press, 59–82.

Section 3

Being an Effective Design Educator

In this section, the focus is on the activities and strategies that underpin becoming a successful design educator. Chapter 7 documents how to run an off-site design intensive, highlighting the value of deep immersion in the design process and how collaboration skills and peer support networks are enabled through intensive, off-campus place-based education. Chapters 8 and 9 focus on online teaching.

Chapter 8 explores the value of developing and using short, engaging videos in teaching; while it is important to include the voice, wisdom and perspective of practicing designers in teaching activities, scheduling such sessions can be logistically challenging. This chapter explores how, through recording concise, structured interviews with local designers, the authors created reusable content that they integrated into weekly classes and active classroom discussions.

Chapter 9 turns to the challenge of teaching online during the COVID-19 pandemic, where the authors describe their experiences as design academics coordinating, developing and teaching units in interaction design and visual communication. The scale and speed of the transformative effort – across learning, communication and management – was a significant challenge, which these authors treated as a design problem for online learning, in which they needed to rapidly develop new online or blended learning experiences. Combined, the chapters in this section provide invaluable practical insight into innovative ways to teach design.

7 A Guide to Design Intensives

Andrew Scott and Sheona Thomson

CONTENTS

7.1 DESIGNING INTENSITY

Conducting design intensives outside the studio promotes immersion in the design process, collaboration skills and the formation of peer support networks. Design studio is the pedagogical hub of design education. In the design studio, "learning emerges through action" (Zehner et al., 2009, p. 6) as students undertake projects typically spanning weeks, or, more frequently, months. Teaching staff, often including practitioners, guide students through both the design project and the design process. At its simplest, the core of the design process consists of iterative ideation. Since design in the real world is a collaborative practice, studio-based education typically engages students in the evaluation of one another's individual work through critique and reflection in the iteration of ideas.

As a contrast to longer projects, shorter tasks or brief immersive projects – sometimes called "intensives" – offer significant opportunities to accelerate students' skills development, particularly if they are working in design teams. These concentrated design activities have their genesis in the "*en charrette*" design challenges of the 19th century *Ecole des Beaux-Arts* (Roggema, 2014, p. 15) and are similar to the "design sprints" common to software and user experience design. Design intensives complement longer-term design projects through their brevity, focus and short feedback cycle, fostering student engagement, socialisation and self-confidence, especially early in a design degree (Scott et al., 2001).

In our experience, locating these intensive activities off campus further amplifies their impact. While the design studio is a rich learning environment, the value of conducting design intensives off campus harnesses the power of what Sobel calls "place-based education," involving the natural environment as "the integrating context for learning." (Sobel in France et al., 2019, p. 12). Place-based learning has a long tradition in pre-tertiary educational settings as well as community contexts. A recent study of the benefits of field-based learning in higher education, including place-based immersions and retreats, revealed positive impacts of these types of experiences on peer-to-peer and student–educator relationships as well as on motivation and learning, providing "ideal environments for fostering the basic psychological needs of autonomy, competence and especially relatedness" (Fedesco et al., 2020, p. 78).

In this chapter, drawing on our experience conducting team-based design intensive activities over many years, we'll guide you in the selection of appropriate formats and settings to meet your needs and gain the learning benefits of "immersive" intensives. We'll discuss why you should consider field-based or off-campus design intensives, how to organise them, what to do when things go wrong, and present a range of options that will help you choose the right approach for your learning context.

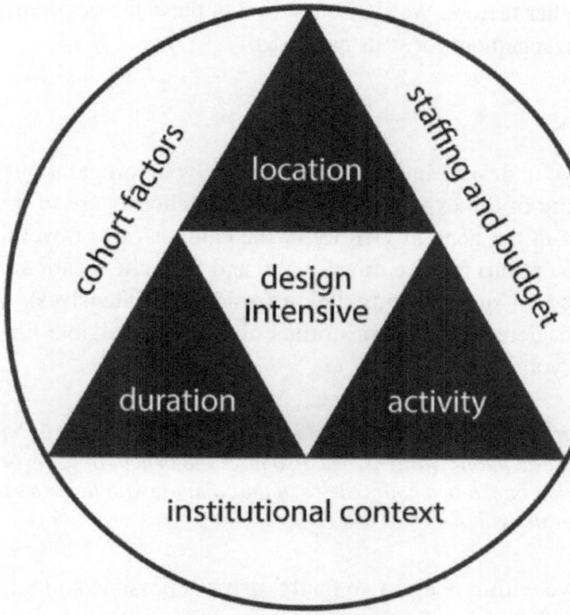

FIGURE 7.1 Three interrelated factors and associated influences affect the design intensive experience

Three interrelated factors will impact the nature of your design intensive experience – the activities you conduct, the duration of the intensive and the location where it is held – and it's difficult to consider them separately (Figure 7.1). Consideration of these three factors is shaped by your unique context, cohort characteristics and the resources you have to support activities.

In our experience, design intensives that are staged away from campus are most effective for the sense of immersion in novel experience, and this is the underpinning motivation for almost four decades of camping field trips at our institution. Taking students out of familiar surroundings and away from everyday distractions into remote, natural locations helps to create the conditions for students to tune into new ideas and new connections. Living closely together for several days encourages social interaction and bonding, and also fosters team collaboration and the creation of peer support networks. Citing Murphy (2001), Fedesco et al. (2020) reinforce the notion that sharing ordinary experiences "like eating meals together and sitting on a bus" increase students' sense of relatedness and connection. Quite simply, students get to know one another and many form lasting friendships that endure throughout their studies and beyond into professional practice.

Despite the advantages, the factors of accommodation, transportation, equipment logistics, available locations and costs can sometimes make off-campus design intensives difficult to organise. It's possible to mitigate some of these challenging location-related aspects by staging the intensive on, or close to, campus, while boosting the impact of the other factors. We'll discuss each of these factors in turn to help you identify the best arrangement for your purposes.

7.2 LOCATION

The location for your design intensive determines your preparations and greatly influences the nature of the experience. Broadly, the choices are to use on-campus facilities, locations in the general vicinity of the campus or to travel farther afield. Additionally, some options require travelling to and from the location each day (day intensives) or allow on-site accommodation (residential intensives). Your campus location, when considered against constraints of budget, available time and cohort size will influence your options.

Our campus is city-based and located at the edge of a thriving central business district adjacent to large public gardens. We also have the benefit of accessible natural environments within one to two hours' drive of the campus, with facilities and amenities to support our activities.

Staging an intensive within one or two hours' drive is possible so long as resources are available. Partial subsidisation by students can offset costs.

7.2.1 On-Campus Venues

If on-campus, a design intensive could be conducted within existing studio spaces, the special seminar or conference facilities that many universities have available. If spaces are available for the required duration this option may be attractive; however, the familiarity of the location and proximity to everyday "distractions" for students and staff may dilute the immersive potential of the activity. An on-campus design intensive has the benefit of drawing on local supports such as workshop spaces, computer labs and libraries. Tasks can be designed to be challenging and expectations for student output can be higher. While students will be interacting during working hours, the fact that everyone returns home in the evening may reduce opportunities for incidental social engagement unless these are built into the program.

Depending on your goals, the comparative ease of arranging a campus-based design intensive may offset the diminished sense of immersion. If you can schedule appropriate spaces for the required continuous time period, and if student work-in-progress can be left undisturbed overnight, conducting intensive projects on-campus is considerably easier than off-site options. Additionally, transportation needs are negligible. Students and staff simply make their way to campus as usual and project materials are likely to already be on site. You may not even need to arrange catering, although this is a compelling mechanism for bringing people together and minimising distractions and time away from the group.

We have conducted on-campus design intensives for small and large groups, sometimes in response to weather or scheduling problems (see Section 7.2.1). While they are less immersive experiences, they are still effective and useful. Additionally, they can be a good way to start building experience with this style of design teaching.

The key to success is creating a challenging, engaging design task framed within a constrained but viable time frame. The urgency of developing design solutions against the clock—with success!—is the vital ingredient. Time pressure pushes away mundane distractions when students rise to the challenge and student teams bond in adversity and success.

7.2.2 Near-Campus Venues

Leaving the campus ("escaping the studio") and travelling to nearby locations, such as a public park (with the relevant local government approvals), offers some of the benefits of remote locations without being as logistically challenging. Scheduling may be easier to arrange if the site is commonly available and there is no competition with other university users for campus space. The novelty of "escaping the studio" into a variety of other spaces adds a refreshing quality to the experience that students find memorable and interesting.

If the chosen site is close to campus, travel arrangements for students and staff will be routine. If the location is farther afield and not well served by public transport, it may be necessary to provide daily transportation to the site; in this case, cohort size will be a significant factor. The movement of project materials from campus to site and back requires consideration, and if it is not possible to leave project work-in-progress on location overnight this may constrain the scope of the project. Additionally, if the venue is public, such as a civic park, other people may be present, adding an additional layer of risk management in terms of student safety and public liability.

Consider the value of daytime catering to create a social focus around meal breaks. As it's likely everyone will return to their homes in the evenings, opportunities for incidental interactions between students should be maximised to increase the sense of immersion and novelty. Note that breaches in attendance can become an issue, either through late arrivals, early departures or unexplained absenteeism, impacting the students concerned and their teams. While skipping a class during a term project is rarely problematic, missing a day from a three-day design intensive is significantly counterproductive.

We have used a large public park adjacent to our inner-city campus on numerous occasions as the site for design intensives. Logistically uncomplicated, project materials can be delivered to the park using trolleys or vehicles. Temporary shade structures can be installed to increase the amenity of the venue, particularly in changeable weather. However, as student work-in-progress must be transported back to campus for storage, this constrains the complexity and scale of student installations.

7.2.3 Remote Venues

Travelling farther afield opens up exciting possibilities but greater challenges and increased costs, especially for large cohorts. While working in a natural setting may be difficult at times, a well-chosen site will offer significant benefits to students' imagination and inspiration. Everyday distractions are diminished or eliminated, increasing the sense of immersion. Living on site encourages serendipitous conversations between all participants – students and staff – and promotes deeper levels of socialisation. Some may struggle with the absence of streaming media, free Wi-Fi and the comforts of home, but most find the experience to be ultimately satisfying and memorable.

We hear of students forming lasting friendships that persist into professional life. It is likely that these peer support networks are an asset during their studies and contribute to students' enthusiasm and motivation for learning design. A study of the effects of field trips early in the first year of a course identified that students valued field trips most highly for increasing their motivation for the subject and enhancing their personal and social development (Larsen et al., in Fedesco et al., p. 67).

The farther afield you go, and the larger the cohort, the greater the challenge. Arranging and managing transportation of people and equipment, including accessibility constraints, accommodation, catering, health and safety and weather contingencies become significant tasks. For us, though, the benefits are compelling and have made mounting field trips to remote sites worth the considerable effort.

> Our experiences conducting remote design intensive field trips date from the mid-90s within the established tradition and experienced hands of the School of Design. Sites included a nearby island, a national park campsite, and, for 12 years, a lakeside camping ground.

7.2.4 Location Choices

Either of the three location ranges we have discussed can be made to work, as summarised in Table 7.1. On-campus intensives are effective and easier to conduct, especially if you have limited support and experience. Remote intensives are intrinsically rewarding, but more demanding, with large student cohorts. Build experience and support with small cohorts close to home before venturing farther.

7.2.5 Communications

Timely communication with staff and students becomes more important during an intensive. In principle, it's wise to cleave closely to the published schedule of activities and encourage participants to do likewise but, nonetheless, the exigencies of weather and other logistics sometimes dictate changes that need to be communicated efficiently. This can be more challenging in remote locations where telecommunications services are patchy or absent.

Email is poorly suited to distributing brief, time-sensitive notifications so we considered several popular messaging services for this purpose. Ultimately, we adopted the *Telegram* messaging service because it allows both one-way broadcast "channels," useful for sharing announcements with even very large groups (up to 200,000), as well as two-way "group" chats that allow for exchanges with students. Membership invitation links are easy to share via URLs online or QR codes printed on posters

TABLE 7.1
Comparing Location Advantages and Disadvantages

	Advantages	Disadvantages
On Campus	Relatively easy to arrange and affordable. No special travel arrangements necessary. Few logistics challenges (such as transportation and disposal of project materials).	Scheduling may be difficult. Familiar location with all of the usual distractions for students and staff. Residential intensives unlikely.
Near Campus	Potentially stimulating setting removed from routine daytime distractions. Scheduling may be easier than on-campus venues.	Arrange transportation for staff or students? No overnight storage of project materials or work-in-progress. Transportation for equipment will be necessary. Non-exclusive use of the site? Weather can be an important factor. Residential intensives are probably excluded.
Remote	Stimulating setting removed from routine distractions. Project sites continuously available. Residential settings encourage social bonds and networking. Scheduling may be easier than on-campus venues.	Considerable logistics challenges transporting people and equipment. Accessibility must always be a consideration for accommodating students with physical disabilities. Greater risk-management requirements. Some students may be unable to attend which may require a parallel campus program. Weather can create huge challenges.

displayed in the field. Other useful apps include *SafeZone* and *what3words*. Our institution permits off-campus sites to be temporarily included within the purview of *SafeZone*'s emergency contact features. Fortunately, we've not needed to put these features to the test, but it's comforting to have the assistance available.

What3words divides the surface of the Earth into three-metre squares and assigns three words to each. This makes it easy to share location coordinates on paper, which is useful for designating locations precisely to students. For example, the words "*sprinkler uttering weeklies*" translate directly to a specific location of importance on one of our remote sites. Other coordinate systems rely on URLs or long strings of digits which are less robust.

7.3 FORMAT

7.3.1 Duration

The duration of your design intensive sets the scope for what your students can achieve. While it's possible to conduct useful one-day intensives, they are not much different in character from an extended studio session. The possibilities posed by two-, three- and four-day sessions are more interesting. We have also conducted five-day remote design intensives but, at some point, the experience changes from an intense, energetic design activity into a more measured, sustained one.

Location has a bearing on duration – the more remote your venue the more time is spent travelling to and from it, which subtracts from the time available for design activities. The farther you travel, the longer you probably want to stay to make it worth your while. In practice, we avoid travelling for more than two hours from our central city campus because this consumes approximately half a day at the start and finish.

7.3.2 Two-Day Design Intensives

We consider two days to be just about the minimum amount of time for a design intensive, and even then only for on-campus or near-campus venues. To extend the useful time, you may run activities into the evening on one or both days.

7.3.3 Three-Day Design Intensives

Three-day design intensives held on, near or off campus allow reasonable time to conduct single projects. If the site is at some distance it can still be effective in a residential format because travel time only occurs at the start and end. Evening activities are also possible and even necessary to make the most of the experience.

7.3.4 Four-Day Design Intensives and Beyond

Two reasons to implement a longer design intensive are to conduct multiple projects or exercises in sequence and use sites that are farther away and thus take more time to reach. Depending on site-hire costs, this may only mean a moderate increase in cost compared to a three-day residential format and allows for more complex or numerous activities to be conducted.

7.4 TIMING

Scheduling a design intensive alongside regular classes can be difficult. To manage conflict, we suspend other first-year classes for the week of the intensive by agreement with our colleagues. While this is a long-standing practice in our school, it may be more difficult to introduce in other settings and require negotiation. Intensives can take place pre- or post-semester, during recess or on weekends. Mandatory student participation may be more difficult to request outside of semester but such timing may suit elective options. If there is no alternative, students will accept weekend scheduling but, in general, they prefer to avoid sessions that include Friday evenings and Saturday mornings if given the choice due to social, work and sporting commitments.

A two-day program is easier to accommodate within a conventional academic schedule. Deciding to run the intensive on a weekend often sidesteps this issue of conflicts completely with the benefit of greater room availability. However, weekends may be more difficult to schedule public spaces due to conflicts with public events. In either case, securing venues well in advance is essential.

7.4.1 Residential or Not?

Another important consideration is whether students and staff will commute to the site each day or reside there for the duration. The former is, obviously, little different from the routine university experience for most people, certainly at Australian universities where on-campus residencies are uncommon. At day's end everyone returns home to their families, housemates, pets, hobbies, entertainments and social commitments, reducing the intensity of the design intensive experience.

In contrast, on-site residential design intensives have a different character. Students spend the whole time in close proximity to each other and away from the quotidian elements of life. Thus, the design intensive becomes an exceptional experience, although not necessarily a comfortable one. Sharing extended time with people you don't know well is challenging, but it is also enriching. Outside of project-based scheduling, students share mealtimes and other living activities, often conversing long into the night, forming friendships and networks. Although difficult to measure, the benefits of this concentrated interaction are obvious upon return – if nothing else, students are considerably more talkative in class! In their review of field-based learning in higher education, Fedesco et al., (2020) cite an earlier study by Murphy (2001) into the social interactions of backpackers, which found that sharing the ordinary aspects of life helped to build connection and relatedness (Murphy in Fedesco et al., p. 78).

While principally focussed on benefits to students, residential intensives are also valuable for staff. Temporary respite from the relentless press of email, phone calls and meetings opens up time to discuss the topics that drew you to academia in the first place, and many collaborative research plans have been hatched around the campfire at the end of the working day. Essentially, residential design intensives provide students and staff with extended periods of undiluted time for design activities. Even allowing for sufficient unprogrammed time (to encourage bonding and networking), it's easier to conduct activities into the evening. Moving around a site after dark, especially far from the lights of the city, can be an "illuminating" experience.

7.5 ACTIVITIES

Design intensives enable projects and activities that would be less feasible if broken up into weekly classes. For example, in our case, large structures can be designed and constructed in days and then dismantled at the end of the design intensive. Conversely, the short burst of design intensives is less suited to complex projects that require extended time for research, reflection and the maturation of solutions. In such cases, consider using a design intensive as a concentrated research or concept-generation stage to initiate a longer project. We sometimes employ this "embedded design intensive" approach during the early ideation stages of a conventional studio project.

Our design intensive projects have, for the most part, revolved around physical installations constructed by students working in teams. There are several things we try to achieve in these tasks. Bear in mind that our first-year students are only weeks into their course – goals for more experienced students will differ:

- Challenge: challenging but achievable tasks that allow for diverse solutions and which give students a sense of satisfaction on completion. Ideally students have a sense of pride in their achievement that builds their confidence in their abilities as designers
- Achievable timeframe: we must be realistic about what students can achieve in the timespan of the design intensive
- Process focus: we want students to understand and apply design as a process. We encourage iterative exploration of multiple ideas on the path to developing a design solution. Tasks should challenge the problem-solving abilities of students, both practically and aesthetically
- Cater to disciplinary interests: while we aim to set projects that are relevant to students from seven design disciplines, this is difficult as first-year students often have a very limited conception of their disciplines. Some are quick to dismiss design tasks that fall outside their expectations

7.5.1 Single-Project Design Intensives

Design intensives may be structured around a single large project if time is limited or you wish to address a complex brief. Over the span of two to three days, students can be guided through a design process, perhaps culminating in the assessment of a physical design. Even after subtracting time for "set up" and "pack up" (shown in white in Table 7.2), this yields a continuous block of time that affords intensive immersion in the design task, a key feature of the intensive format.

Our 2019 design intensive was conducted in two three-day sessions of 350 students (Sunday to Tuesday, Wednesday to Friday). In each session, students worked in multi-disciplinary teams of seven. The three-day span of this project focussed explicitly on community formation. Students were asked to build effective communities even as they got to know each other.

Synopsis of design intensive brief: 2019
Based on the premise of prototyping ideas for a prospective event – the Festival of Circular Design – students designed and built a series of 'pop-up micro-communities'.

TABLE 7.2
Design Intensive Activity Structure, 2019 *Design Intensive Brief, 2019*

design → construct → inhabit	Day 1
evaluate → redesign → construct → inhabit	Day 2
present → assess → dismantle	Day 3

FIGURE 7.2 Example of a community structure built at the site, 2019

> These were hypothetical places for designers to work and live together during the festival. In their teams, students were invited to design, construct, inhabit and evaluate a prototype 'pop-up' micro-community that encourages cross-disciplinary design activity. Students simultaneously assumed the roles of designers and stakeholders, negotiating within and across teams to realise a prototypical micro-community.

The structured task of making and then remaking their communities also implicitly reinforced the iterative nature of design. This was an intentional strategy to counter the novice's impulse to accept the first idea that occurs to them. We were concerned that conflating student camping arrangements with working teams and communities of teams might demand too much of new acquaintances. In practice, though, it was effective due to the preparation we'd done in classroom workshops in the weeks preceding the design intensive (see Section 7.5). Most students arrived on site having already arranged their teams, and communities formed rapidly with minimal guidance from tutors. Figure 7.2 shows an example of one of the community structures viewed in the evening during the "inhabit" phase at the end of Day 2 (Table 7.2). Students and staff spent the second evening touring the installations and enjoying the variety of interpretations of the brief.

7.6 MULTI-PROJECT DESIGN INTENSIVES

It's also possible to conduct more than one project during a design intensive. Indeed, the majority of our first-year intensives have been organised as two back-to-back projects across four-day residential design intensives.

This approach allows teams to develop both design insight and group cohesion across two projects. Students apply what they've learnt from the first project, and then apply formative and summative feedback to the second, with opportunities to improve approaches to time management, negotiation, ideation and design resolution. Multiple projects also increase the variety of possible design briefs to accommodate diverse disciplines. To illustrate paired projects, we'll discuss two long-running examples used at a lakeside camping ground during four-day residential design intensives, *Elevate* and *Lightness*. *Elevate* was a short project, commenced on the afternoon of Day 1 and concluded at midday of Day 2:

> Your design team has been commissioned to design and build a prototype structure for use by communities to indicate locations in rural settings for community or agricultural projects. Design and construct a stable structure to raise a fabric marker ('flag') as high as possible. Your structure must also support an additional unspecified payload (GPS unit, signal lights, etc.) which may weigh as much as 2 kg. The raising mechanism must be easy for anyone to operate so consider ease of use and simple written instructions. Constraints include:
>
> - Use only the provided materials (of which you'll learn more at the start of the project)
> - Your flag will be elevated starting from a position completely on the ground by staff using your supplied instruction diagram
> - Your structure and flag must be robust and stable during the test period
> - Support the supplied 2 kg payload as high above the ground as possible, though it may be at a different height to your flag
> - Your structure must stand independent of existing structures or people
> - Your structure must rest on the ground without penetrating it with pegs or other components
> - Your structure must be erected at the specified location which may be changed on account of weather

This project challenged students to jointly consider a tangible physical goal (a tower with a flag-raising mechanism) and aesthetic demands (a well-considered appearance), with structural performance and aesthetic resolution considered equally. Students found the competitive nature of the task enjoyable and *Elevate* ran in various permutations for 7 years. Paired with *Elevate* was *Lightness*. In deliberate contrast to the goal-oriented *Elevate*, *Lightness* focussed on the nature of "site" and the expression of less tangible themes:

> Explore and express the qualities of physical 'lightness' through a temporary artefact that is thoughtfully embedded in a given context. This means translating an idea into a built form. This isn't a one-way process though because the form needs to be interpreted back into a reasonable approximation of the original idea by other people viewing and possibly interacting with your artefact.
>
> We'll give you specific materials to construct this artefact, and a general location for its installation. We'd like you to construct an object that will:
>
> - Express the qualities of physical 'lightness'
> - Employ some level of abstraction and avoid overly obvious literal interpretations of the theme

- Use the materials provided
- Be thoughtfully situated in the available landscape without damaging it
- Be easily disassembled so that the materials may be reused or recycled

As our most enduring project, *Lightness* originated in the first-year orientation field trips of the 1990s. It encouraged a wide variety of expressive outcomes, while challenging students to interpret less objective evaluation criteria (Scott et al., 2001).

7.7 ADJUSTING FOR CONTEXT

The location of design intensives has bearing on the project's potential. Due to weather conditions, we've occasionally had to relocate design intensive activities from a remote venue to a location on- or near-campus, forcing adjustments to the project briefs. The *Amplify* project is a case in point. This project ran on remote camping design intensives with ample breadth of site choice for students (see Column 1, Table 7.3). In relocating, once to on-campus facilities (mostly inside studios) and subsequently to a near-campus location (a large urban park), adjustments to the brief accounted for more limited site possibilities (see Column 2, Table 7.3). The use of keywords, chosen at random, substituted for the variety normally offered by working at a remote venue.

This comparison illustrates that, although working outdoors on a large project site offers a more stimulating design intensive setting for students, projects can still be adapted to function effectively in studios and other familiar locations.

7.8 BUILDING COLLABORATION

Design intensives are a great vehicle for teaching students about collaboration (Crowther et al., 2016). When students know each other well from several semesters of studying together, forming teams may be relatively straightforward. The process is more challenging at the very beginning of a course. Some strategies can make this easier.

- Team sizes of between five and seven people are effective on design intensives. Smaller groupings may still work, particularly for simple tasks, but larger teams are problematic and more likely to fragment
- With large first-year cohorts, enrolments can be volatile, with some students withdrawing in the first weeks of semester. It is best to form teams immediately prior to the design intensive, when membership is likely to remain stable. Once teams are formed, we resist requests to change teams without a compelling reason
- It's important to do groundwork prior to forming groups, introducing the theory of team collaboration, group dynamics and communication strategies. Theory is supported by in-class practical activities to help students understand that they need to work at working together

There are two broad approaches to creating student teams, allocated and elective formation. These approaches have advantages and disadvantages, and we discuss these with students as an aspect of collaboration theory (Sosa and Albarran, 2008).

TABLE 7.3
Example of Changing the Design Brief to Suit the Context – Differences in Bold

Remote Intensive Brief	On-Campus Intensive Brief
Design a structure that expresses and "amplifies" the aspects of **a site of your choice**.	Design a structure that expresses and "amplifies" **a specific theme**.
Your task is to explore the project area and choose a site within its boundaries that you find interesting. **Analyse the quality or qualities that make the site interesting** to you and design an installation that "amplifies" these qualities so that others can easily appreciate them.	Your group will **choose two keyword envelopes with a random word inside each. Analyse these keywords** and design an installation that "amplifies" either or both of these words. Take care to choose interpretations that others can easily appreciate and understand. You may also ask to choose a new random word, but you must give up one of your existing words before you receive a replacement.
	Explore the project area and choose a site within its boundaries that meets your needs.
Your installation will be assessed on its practical function (robust and accessible for the audience), aesthetic function (well resolved aesthetically) and symbolic function (amplifies site qualities clearly and interestingly).	Your installation will be assessed on its practical function (robust and accessible for the audience), aesthetic function (well resolved aesthetically) and symbolic function (amplifies site qualities clearly and interestingly).
Choose a site that you find interesting.	Choose a site.
Document the site in your Design Journal.	Document the site in your Design Journal.
Analyse the site's qualities and characteristics.	**Analyse the keywords for meanings and interpretations.**
Devise an installation that amplifies the essence of the site and communicates it to the audience.	Design an installation that amplifies the essence of the keyword and communicates it to the audience.
Build the installation on the site.	Build the installation on the site.
Mange the use of the installation.	Mange the use of the installation.
Dismantle the installation.	Dismantle the installation.

7.8.1 Allocated Formation

Allocation of students to teams is a straightforward way of ensuring discipline diversity. For example, if it's important for students to work in mixed-discipline teams, they can be allocated on the basis of their disciplines.

We've used random allocation for most design intensives. Since most students will be strangers in this case, it's important to implement structured "icebreaker" activities to promote sociability. Anticipate friction between members and prepare academic staff to intervene early to help teams work through such problems. In rare cases individual students can be moved to other teams, but this may be disruptive unless it happens very early in the design intensive.

7.8.2 Elective Formation

Elective formation allows students to self-nominate teams. If students don't know one another, this approach may be difficult. Additionally, self-nominated teams tend towards homogeneity. Counter this by creating opportunities in classes for students to meet and widen their network.

To promote students' exposure to different design fields, we set a criterion that each team must contain representatives of a minimum of four out of the seven disciplines in the course. Students used a variety of strategies to connect with others to meet this requirement, including advertising team vacancies in online discussion groups or using the recommendations of friends. This approach has proven successful: teams establish quicker than the allocated teams because the preparation takes place before the intensive.

7.9 MANAGING TEAM ISSUES

While teamwork and collaboration are an accepted part of the workplace, students may view it as an unnecessary evil. Prior experience of unproductive team members, conflicting schedules and priorities and absenteeism can build negativity. The design intensive format aims to ameliorate these problematic factors. Since full participation is necessary for positive team collaboration, learning and project outcomes, absence should be penalised. This signals that students need to take their commitment to design activities seriously. Absenteeism is, of course, greatly reduced on residential design intensives.

As team assignments of weeks or months require members to meet regularly and work to an agreed timeline, scheduling conflicts create challenges. Design intensives sidestep this issue by asking students to be fully present for a limited duration. The intensity and transparency of the experience largely reduces "social loafing," or the tendency of individuals to put in less effort for a group outcome than they would if they were working alone (Latane, Williams & Harkins, 1979). Karau and Hart (1998) suggest that self-evaluation of teamwork can eliminate social loafing; requiring that students reflect on their individual contribution to their team is valuable.

7.9.1 Building Effective Teaching Teams

Larger student cohorts usually call for larger teaching teams. Teaching teams may be composed of both academic staff – full-time academics and part-time tutors – and technical support staff who provide equipment, materials and advice on their usage.

7.9.2 Academic Staff

A clear benefit of short, intensive teaching models is that highly experienced design practitioners who are normally too busy to commit to weekly classes may be available to teach. We've built a committed team of designers who tutor on design intensives for just one week every year, with some travelling from interstate or even overseas to contribute.

7.9.3 Technical Support

Technical support is a vital component of design intensives, especially for large cohorts (150–350 students) and remote locations. For these, we rely on our workshop technicians for technical support, most heavily for transportation, distribution and collection of design project materials and tools. This is significant task when dealing with large cohorts due to the sheer quantity of materials and tools in use. In remote settings we've found that approximately 1 technician per 100 students is useful to maintain reasonable workloads and appropriate levels of support.

7.9.4 Team Communications

By their nature, design intensives call for careful planning to take full advantage of the limited time available, which means it's important that the whole team is aware of the program. Coordinating large student groups with many tutors can be a challenge, especially when working in remote locations where communications may not be available. This means it's important to establish a teaching team that is well aware of the purpose, nature and sequence of activities that will take place on the design intensive. Briefing the team is an essential first step, but providing reference documents for them to use on site is also highly useful. We've found it helpful to provide all staff with a "staff book" – a bound set of documents for use in the field. This is often more reliable than relying on digital devices that may be difficult to keep charged or protected on field trips. Our staff books contain everything needed for tutors to manage students and run the design projects, such as the schedule of activities, student and team lists, site map, assessment documents, staff contact list and emergency contact details.

As mentioned above in *7.2.5 Communications*, communications are useful for team coordination. If telecommunications are present on site, it's very helpful to create private channels for staff to use. As mentioned above, *Telegram* is useful for this.

7.10 INTENSIVE PLANNING AND SUPPORT

A significant amount of planning is needed to ensure the smooth running of design intensives, especially if conducted away from campus. Firstly, it's important that you, as an academic conducting a design intensive, have the support and backing of management and administration. This will help not only in the planning and budgeting stages but also while the activity is taking place. Organise a liaison person in the office to communicate with in regard to logistics arrangements, student matters and emergency incidents. This will provide you with peace of mind and remote support if difficulties arise.

7.10.1 Health, Safety and Risk Management

"Mishandling an emergency incident, a minor illness, or mental health situation can be devastating to participants and to institutions" (Fedesco et al., 2020). Comprehensive risk management must underpin the design and delivery of off-campus intensives. Make full use your university's risk-management procedures and advisors to anticipate and manage the risks involved. The first step in assessing risk is to observe the site and identify likely risk factors. These might be classified under several broad headings.

7.10.2 Setting

Every setting presents unique situations. Consider features such as bodies of water, uneven terrain, etc. that students and staff might encounter within or in transit to the site. Control factors include briefings, signage and supervision protocols.

7.10.3 Flora and Fauna

Local flora and fauna must always be considered, especially within Australia where we have some dangerous creatures. Investigate the likelihood of encountering animals, birds and insects that present risks. Useful control measures include briefings to inform participants, preventative measures (such as insect repellents) and behavioural guidelines.

7.10.4 Transport

Transport to and from the site, either provided by the university or in personal vehicles, always presents a risk. Control factors may include instruction to observe road rules and licensing. Also consider fatigue, which may be a factor in safely travelling great distances. Allowing students to make use of personal vehicles may reduce reliance on hired transportation but brings with it additional challenges for managing arrival and departure, such as the risk of contraband and the need to control site access and parking.

7.10.5 Medical Conditions

It may not be safe for all students to attend a remote design intensive. Medical conditions, mobility issues, caring responsibilities and proximity to emergency services will colour the decision to limit access for some students. Without invading their privacy, require students to share relevant details of conditions that might affect their wellbeing while away from home. If the design intensive is a compulsory assessment activity it may be necessary to offer an on-campus alternative for students unable to safely participate in remote settings. For this reason, we always offer a parallel design intensive activity close to campus for those students unable to be away from home overnight.

7.10.6 First Aid

The greater the remove from campus infrastructure, the more important it is to have access to first aid expertise and equipment. Take well-stocked first aid kits even when operating near or on campus to treat the minor injuries that sometimes occur when students are building structures. In remote settings these become even more important. Ensure that there are always staff in attendance who have current first aid qualifications. Everyone on the teaching team should know who these people are so they can direct students in need. Identify the locations of medical services in the area, such as clinics and hospitals. Provide all staff with contact information for emergency services.

7.10.7 Conduct

The conduct of participants can have great bearing on both the success of the design intensive and the wellbeing of all involved. Beyond the conduct requirements placed upon students and staff by the university, it's useful to impose special requirements for the field trip design intensive. We require all participants to observe a "safety agreement" as a condition of attendance. Students who breach the agreement are required to leave the site and are consequently unable to complete the team project, affecting their academic outcome.

7.10.8 Planning the Program of Activities

Devising the activity program for your design intensive begins by allocating time for tasks. This will depend on your circumstances, context and goals. In can be difficult to anticipate the time required for supporting tasks off campus. When devising your design intensive program, first consider the time students will need for their design activities and also whether participants are staying on site (residential). You'll have significantly more time for activities in a residential format because travel is only a factor at the start and finish, although your site location will impose travel time. Daily activities can commence quite early and conclude well into the evening.

Tips for conducting activities with large groups in near-campus and remote settings:

- Addressing large groups is difficult outside so avoid this as much as possible
- Adhere to the published schedule of activities as closely as possible; changes are difficult to communicate and lead to confusion
- Coordinating activities across a large site can be difficult so brief tutors about activity goals and empower them to make decisions on the spot. If available, use mobile communications, as discussed in above in *7.9.4 Team Communications*, to aid the teaching team
- Designate clear locations for specific items and services such as project materials, first aid, water, etc.
- It may take longer for people to move from location to location than you expect

- Set-up and clean-up activities will be much faster if students are encouraged to assist
- Cultivate a supportive community environment and encourage students to look out for one another. This approach aids bonding and collaboration but also makes everyone safer
- Be aware that, for some young students, the design intensive may be the first time they have been away on their own. Encourage students to approach staff if they are uncomfortable or anxious and consider designating a particular person in a counselling role. Designate male and female staff members whom students may approach if they need assistance overnight
- If camping, be prepared to assist inexperienced campers with locating and setting up tents. Have spare tents, mattresses and sleeping bags on hand

7.10.9 Plans B, C and D

Conducting design intensives outside the studio puts you at the mercy of the weather and contingency plans should be prepared. Certain times of the year may offer more stable weather, depending on location, but inclement weather is always a possibility. This can be managed by having undercover options nearby. If built structures are not available on site then it may be necessary to hire temporary shelter structures. You should also have a plan for dealing with extreme weather or other conditions that would put participants at risk, such as cyclones, storms, bushfires or flooding (depending on your region). This is stressful to contemplate and much harder with very large cohorts but putting plans in place in advance can make it manageable. In the last three decades we have been forced to cancel remote field trips due to severe weather conditions twice. On another occasion, our campsite became unavailable at short notice due to a change of ownership. We were able to arrange on-campus or near-campus design intensives to allow the design intensive to take place, although sometimes we've needed to reschedule to successive weekends.

7.11 CONCLUSION

Intensive, immersive project-based learning experiences conducted early in a design course give students a practical insight into the design process and team collaboration while accelerating their socialisation with peers and teachers. In our experience, first-year design intensives establish a sense of belonging and camaraderie that develops into professional networks once formal studies are over. If these intensives are held in extraordinary off-campus locations, greater value may accrue due to the memorability of the experience, the novelty of the place and the extended period of shared living. However, on- or near-campus intensives can also achieve positive outcomes by taking students out of their ordinary schedules to work together on a focussed task.

The interrelated factors of setting, duration and activity inform the character of the design intensive and each can be adjusted to meet a range of learning outcomes and contexts. Cohort size and proximity to familiar campus supports and services affect the level of challenge in conducting intensives: large cohorts and remote settings demand more thorough planning.

Build an experienced teaching and support team by starting small and close to campus, expanding as you grow more confident. Well-designed and orchestrated intensive projects, with clear success conditions and tight feedback loops, promote a strong sense of accomplishment in teams of novice designers that carries forward into subsequent learning.

REFERENCES

Crowther, P., Scott, A., & Allen, T. (2016). Perceptions of collaboration amongst novice design students. In Tucker, R. (Ed.), *Collaboration and Student Engagement in Design Education* (pp. 126–144). IGI Global.

Fedesco, H., Cavin, D., & Henares, R. (2020). Field-based learning in higher education: Exploring the benefits and possibilities. *Journal of the Scholarship of Teaching and Learning, 20*(1), 65–88. doi: 10.14434/josotl.v20i1.24877

France, D., Mauchline, A., Whalley, B., Doolan, M.A., & Bilham, T. (2019). Authentic learning through place-based education. In Bilham, T., Hamshire, C., & Hartog, M. (Eds), *Reframing Space for Learning: Excellence and Innovation in University Teaching* (pp. 12–28). UCL Institute of Education Press.

Karau, S.J., & Hart J.W. (1998). Group cohesiveness and social loafing: Effects of a social interaction manipulation on individual motivation within groups. *Group Dynamics: Theory, Research, and Practice*, 2(3), 185–191.

Latané, B., Williams, K., & Harkins, S. (1979). Many hands make light the work: The causes and consequences of social loafing. *Journal of Personality and Social Psychology, 37*(6), 822–832.

Roggema, R. (2014). The design Charette. In Roggema, R. (Ed.), *The Design Charrette: Ways to Envision Sustainable Futures* (pp. 15–34). Springer.

Scott, A., Jerome, K., & Thomson, S. (2001). Orientating design education: Implementing an orientation field trip as an introduction to design education. *Proceedings of ICSID Educational Seminar 2001* (pp. 62–68). Korea Institute of Design Promotion.

Sosa, R., & Albarran, D. (2008). Supporting idea generation in design teams. *Proceedings of EDPE 08, Barcelona.* Retrieved from: www.designsociety.org/download publication/28074/supporting_idea_generation_in_design_teams

Zehner, R., Forsyth, G., Musgrave, E., Neale, D, de la Harpe, B., Peterson, F., & Frankham, N. (2009). *Studio Teaching Project (STP) Final Report.* Retrieved from the Universities Australia Learning and Teaching Repository website https://ltr.edu.au/resources/STP%20Report%20Vol%201.pdf

8 Using Video and Blended Learning

Sheona Thomson and Andrew Scott

CONTENTS

8.1 HOW TO SHARE PROFESSIONAL VIEWPOINTS WITH LARGE CLASSES

In 2018, as we were developing a new first-year "common" core unit in our undergraduate design course, we wanted to fix a longstanding problem. The unit required students to engage across disciplines and work in multi-disciplinary project teams, supported by learning about process, stakeholder perspectives collaboration, visualisation and the impact of design. From past experience, we knew that first-year students taking common units sometimes struggle to understand the relevance of multi-disciplinary curriculum. They are keen to start learning about their chosen

FIGURE 8.1 A video episode as viewed by our students on YouTube (Fashion designer, Gail Reid of Gail Sorronda, Brisbane, 2019)

design discipline and perceive broad topic content to be a waste of time and fees. To relate each discipline to the unit content, we decided to incorporate the representative perspectives of practising designers on the unit themes.

While this can normally be achieved through invited guest lectures, our case was complicated by a compressed five-week schedule of lectures, with multiple weekly deliveries required due to class size. Few design practitioners would be available to attend multiple sessions in a week. Also, with the shorter teaching term, accommodating the in-person "voices" of seven disciplines would be challenging. Our solution was to record concise, structured interviews with local designers, asking each of them the same set of questions related to the broad unit topics, creating reusable content to integrate into our weekly classes. Not only would these interviews give students access to the insights of practising designers from their chosen discipline, we hoped it would encourage students' broader appreciation of the various design disciplines, their differences and similarities.

This chapter explains how, in 3 months, we created 169 episodes of engaging video to share and explore the perspectives of practising designers with first-year students in active classroom discussions. Figure 8.1 shows an example of the video we created and delivered on YouTube, with this chapter illustrating how we activated classroom discussions and broadened discipline knowledge by creating compelling design videos on a shoestring budget.

8.2 THE IMPORTANCE OF INDUSTRY PERSPECTIVES IN THE FIRST-YEAR CLASSROOM

Incorporating professional perspectives into the classroom overlays the design theories and processes students are learning about with practical relevance and discipline specificity. Watching and listening to practising designers enthusiastically describe their processes, including who they collaborate with and how they communicate with stakeholders, helps students to develop their fledgling understanding of their chosen field. While expert perspectives can be presented in-person through invited lectures, there is a considerable workload involved with sourcing and scheduling guest presenters accompanied by the risk that a busy practitioner may become unavailable at short notice. Typically, only one guest would present in a session, which limits the number of perspectives that can be shared in a finite period of time.

8.2.1 The Benefits of Capturing Perspective on Video

Capturing the authentic perspectives of practising designers on video alleviates the need to find and schedule guest presenters. It is also possible to capture many more individual perspectives on video than could be reasonably scheduled in-person. Compared to the one-to-many lecture mode, video can generate the feeling of a more direct communication to the viewer. Evidence shows that connection to video is emotional and cognitive, positively affecting students' motivation and creating conditions that can enhance further learning (Cruse, 2007).

Video can be incorporated into classes in a variety of ways to meet formal and informal learning goals and outcomes. One such goal is professional socialisation (ASHE-ERIC, 2001). Students often commence their studies with an idealised version of their chosen profession. Over time, through a complex and dynamic process with university education at the core (see Figure 8.2), students acquire the values,

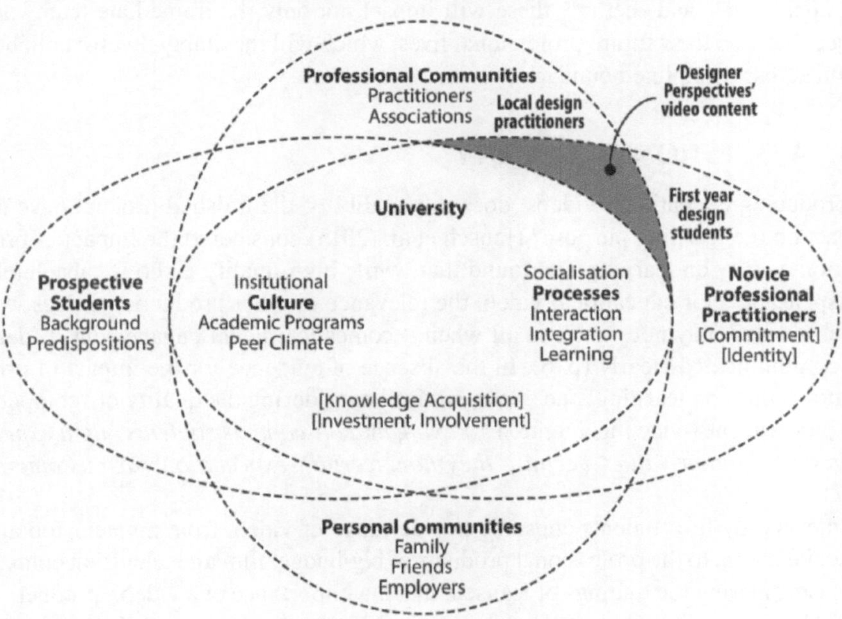

FIGURE 8.2 The Weidman, Twale, and Stein Graduate Socialisation Framework showing the dynamic nature of professional socialisation and the elements that promote professional identity with "Designer Perspectives" overlay

attitudes, skills and knowledge of their chosen discipline to become practising designers. A key influencing factor external to the university is the professional community that students will enter upon graduation. As educators, we can help students refine their expectations about a profession or discipline by creating and maintaining a diversity of connections between the university and the professional community. Arguably, this connection is most keenly felt by students when a practitioner brings relevance to what is being learned in the classroom. As we will discuss, video is an efficient and engaging way to bring the knowledge and experience of the profession into the first-year classroom.

Once captured, video content can be used to spark discussion, deepen or broaden understanding of ideas, and link classroom learning to the contexts of professional practice by giving students a feel for the real work of practising professionals. While students are learning the basics of process, visualisation and collaboration, the authentic advice and insights captured in the videos can help students see a path through their studies to a meaningful career.

Additionally, the video interviews let students explore fields beyond their chosen discipline. Our unit is one of four in the curriculum that prepares students to work across discipline boundaries on inter-disciplinary or trans-disciplinary problems and issues. For the major assessment task, students work together in mixed-discipline teams of five to seven applying a team design process to a common design challenge. Using the video content, we can prepare students for working collaboratively by exposing the mindsets of many different types of designers, not just those from their chosen discipline. Students can compare and contrast different discipline processes and approaches, and see how these will impact not only the immediate teamwork project but also their future professional lives, which will inevitably involve collaboration across discipline boundaries.

8.3 A QUESTION OF QUALITY

In producing educational videos, does the quality of the finished product have an impact on the learning purpose? Hansch et al. (2015) considered the impact of production quality on learning and found that, while high-quality audio is "absolutely indispensable" for educational video, the relevance of other production values was uncertain and may not matter at all when it comes to student engagement if ideas are communicated clearly (p. 6), In the absence of evidence for the impact of production values on learning, one expert advised considering the quality of "competitors" for students' viewing attention. "*Paying and on-campus students might expect higher production values because they have certain expectations as customers*" (p. 7).

In everyday life students consume a wide range of video, from amateur footage on social media to the professional products of big-budget film and television companies. Considering the findings of Hansch, that the importance of a video's production value depends on its context and audience, and noting that higher quality supported longevity of use (p. 7), we decided to achieve the best affordable quality on our limited budget, while ensuring that the message and meaning of the content was authentic and clear. We concentrated on recording the words of practising designers as directly as possible in the hope of delivering relevant and compelling content to students. Might seeing successful local designers and hearing their stories inspire our students to create their own successes?

We also decided on a short, episodic format that would be accessible to students and adaptable to the classroom. Considering evidence that "students learn better by observing dialogue than observing exposition" (Poquet et al., 2018, p. 155), which reinforces Mayer's (2006) "personalisation" principle for effective multimedia learning that "people learn better when words are presented in conversational style rather than formal style," we used a structured interview format.

8.4 STRUCTURED INTERVIEWS

The interview questions serve two purposes: (1) to address our unit learning goals and outcomes; and (2) be useful as a resource for other course units. We shared the interview questions and our reasons for asking them in advance with our designers as shown in Table 8.1.

Interviews were edited to produce three distinct viewing options. First, we produced a full-length interview for each designer, showing all of their answers in sequence. This allows students to delve further into the ethos of a designer that interests them, and is useable as a case study within other units. Secondly, we produced seven discrete episodes per designer, featuring their answers to each interview question to allow students to easily compare multiple responses to a single question in a short time. We discuss this approach and our results below in *Using Episodic Video in the Classroom*. The third output is a short teaser episode introducing each designer and sharing a brief piece of advice, introducing students and other viewers to the content. This episode is available online at URL https://youtu.be/l6I7T-DJXC4Q (21-Jul-19). Counting all 3 formats together, this amounts to 169 videos organised into 8 video playlists.

8.5 USING EPISODIC VIDEO IN THE CLASSROOM

While students may enjoy relevant video content in the classroom, their engagement may be passive, especially for extended viewing. Students are passively engaged when they are "oriented toward and receiving information from the instructional materials without overtly doing anything else related to learning" (Chi and Wylie, 2014). The ICAP framework of Chi and Wylie (2014), which proposes that students' engagement can be distinguished as interactive, constructive, active and passive, informs our learning design. The premise of the framework is that, as students move through the modes from passive to interactive, learning increases. Table 8.2 sets out examples of learning activities by modes of engagement.

To create the greatest impact when using video resources, learning design should encourage interactive student engagement both in the classroom and online. By using the short, episodic videos in the context of small group discussions, we sought to actively promote interaction in the classroom. Guided by our questioning, in-class

TABLE 8.1
The Interview Questions with Notes for Our Interviewees

	Question	Reasoning
1	How do you describe yourself as a designer?	The first question is to help students identify the type of design you do and how you think of yourself. Suggestion: relate how you describe your occupation to people at parties.
2	Just briefly, tell us about your design process, the steps you typically go through during a design project? How long does it take?	The question will help students grasp your design process. It might be helpful to outline only the most important stages in the design process to avoid too lengthy an answer. Suggestion: relate how you describe this to your clients.
3	Who do you collaborate with during a typical project?	We're interested in who you collaborate with during typical projects and what portion of this work you do personally. You might mention in-house team members, consultants, stakeholders and community groups and, of course, clients.
4	How do you visualise and communicate your design ideas?	We'd like students to have some insight into the tools and methods you use to develop and communicate design concepts, however humble or high tech they might be. For example, how important is sketching? How much do you prototype design ideas? How do you relate design ideas to clients or stakeholders?
5	How does your work have impact on people?	These videos will be used as the first in a series of units intended to help students understand how design makes a positive impact on the world, hence the name "Impact Lab." This might be a good place to share what motivates you in your design work.
6	What are the challenges and opportunities you see for the future in your discipline?	We'd like you to speculate on the opportunities and challenges that your discipline may face in the future. Here are some topic suggestions to get you thinking, but please discuss what matters to you: public or client perceptions of the discipline, the competitive landscape, government regulation, the changing nature of the discipline, sustainability, automation, etc.
7	What advice would you give to someone starting their design journey?	The last question is a chance to share a few words of advice with someone setting out to become a designer. What might you say to your younger self?

activities were designed for constructive (generating) and interactive (dialoguing) engagement.

As classroom discussion topics were related to a unit assessment task that students would complete individually, we also considered this relationship such that students could commence (and even complete) their individual task at the end of class, when discussions with their peers were fresh in memory. To maximise students' direct and choice-driven engagement with the video content, we avoided showing

TABLE 8.2

ICAP Framework by Chi and Wylie (2014), Examples of Learning Activities by Engagement Mode

	PASSIVE Receiving	ACTIVE Manipulating	CONSTRUCTIVE Generating	INTERACTIVE Dialoguing
LISTENING to a lecture	Listening without doing anything else but oriented towards instruction	Repeating or rehearsing; copying solution steps; taking verbatim notes	Reflecting out-loud; drawing concept maps; asking questions	Defending and arguing a position in dyads or a small group
READING a text	Reading entire text passages silently/aloud without doing anything else	Underlining or highlighting; summarising by copy-and-delete	Self-explaining; integrating across texts; Taking notes in one's own words	Asking and answering comprehension questions with a partner
OBSERVING a video	Watching the video without doing anything else	Manipulating the tape by pausing, playing, fast-forwarding, rewinding	Explaining concepts in the video; comparing and contrasting to prior knowledge or other materials	Debating with a peer about the justifications; discussing similarities and differences

Note: Reprinted from "The ICAP Framework: Linking Cognitive Engagement to Active Learning Outcomes" by Chi, M., and Wylie, R. 2014, Educational *Psychologist*, 49(4), p. 221.

episodes on the main lecture theatre screen and instead asked students to come to class with a mobile device and headphones or earbuds for private listening. The majority of students used laptops for this purpose, but some viewed the content on their smartphones.

For each classroom discussion, which centred around one of the interview questions, we created a YouTube playlist. The URLs were shared in three ways to make them easily accessible using a broad variety of devices:

1. A QR code on the main theatre screen, which allowed students to capture the URL by aiming their smartphone camera at the image
2. A shortened URL, which they could type directly into their browser
3. A link provided on Blackboard within the weekly learning page

We allowed students 10 minutes to individually select and view 2 or 3 episodes from the 21-episode playlist. As an example, Figure 8.3 shows the *Question 2 Design Process* playlist. While students could opt to watch videos related directly to their design discipline, for comparison and discussion purposes, we encouraged them to view at least one response from a designer from another discipline.

After viewing, students gathered in small groups of two or three to discuss what they had just seen in relation to the week's assessment task. For example, for the *Question 3 Design Process* topic, students were asked to contrast the differing duration of projects in design disciplines and also differences and similarities in how these designers work. The class activities were explicitly designed to involve students through constructive and interactive engagement modes, to compare, contrast, incorporate prior understanding and expectations into new understanding about design and what designers do.

As in-class discussions were also aligned to weekly reflective tasks, students were well prepared to complete their assessment. Additionally, in-class discussions were structured to facilitate student socialisation, networking and cross-disciplinary awareness in preparation for the team design project conducted later in the unit (discussed in the "Escaping the Studio" chapter).

8.6 WHY YOUTUBE?

We considered several options when choosing the delivery system for our video episodes. Our institution's bespoke video-delivery solution is integrated with our learning management system (LMS); however, early testing showed that embedding a large number of episodes into the unit site was labour intensive. Additionally, the workload involved in updating content was onerous, requiring manual replacement in the LMS. Instead, we turned to that backbone of internet culture, YouTube, as the delivery platform. YouTube offered several distinct advantages over other solutions.

1. Management: YouTube offers allows set of videos to put together in a "playlist." This allows grouping all of the responses to *Question 2 Design Process* into a single *Question 2* playlist, for example

FIGURE 8.3 Students chose episodes from the YouTube playlist assigned for that class activity and watched a selection on their personal device with headphones (mobile screen shown)

2. Flexibility: over time, we can easily add and replace videos in the playlists as needed. Directing students to playlists avoids the need to maintain embedded links to individual videos
3. Familiarity: students are familiar with YouTube. Once the playlist is accessed, it is easy to browse the episodes on any online device, desktop or mobile. This was effective in the classroom
4. Technical features: YouTube has evolved to offer a useful range of features that students found helpful. Despite inaccuracies, the automatically generated subtitles were used by a third of the class. Some students also found

it convenient to adjust the playback speed: faster to allow perusal of more content, and slower when transcribing quotes for their reflective writing. Additionally, YouTube provides us as channel owners access to "analytics," which give insights into how the channel episodes are being viewed.

8.6.1 Why Unlisted?

While YouTube makes it easy to share content publicly, we chose to publish via unlisted (private) URLs for the following reasons:

- We promised our interviewees that the recordings were for internal use only in the hope they'd be comfortable speaking candidly to students*
- We wanted to avoid prior exposure for incoming students so that the content would be "fresh" for them
- The videos were produced with funds dedicated to course development and enrichment. While there might be promotional value in making content public, this would undermine perceptions of exclusivity and value for enrolled students

To control content distribution when sharing unlisted URLs we request that people not share them publicly. Unlisted videos are not indexed by search engines, which reduces the extent of broad external exposure of the content.

8.7 TEACHING OBSERVATIONS

As we were planning this approach, we were concerned if lecture theatre Wi-Fi networks would be sufficient for simultaneous video streaming by up to 200 students. In practice, we did not experience any such problems. This may be due to our robust campus network or YouTube's ability to adjust video resolution to match the available bandwidth. More importantly, we were also unsure if all students would have the personal device and headset needed to avoid a cacophonous clash of noise from so many classroom viewers, but this too proved to be unfounded. We warned students well in advance and all of them had the necessary equipment, either laptops, smartphones or tablets. This video series was used in two iterations of our unit, *DYB Impact Lab 1*, with a 2019 enrolment of 743 in Semester 1 and 160 in Semester 2. Inspection of our YouTube channel "analytics" revealed the following data:

Total watch time in hours: 1,377.7
Total views: 44,176
Average watch time in hours: 476.1
Average episode views: 259.6
Most popular episode views: 872

* Interviewees reviewed "rough cuts" of the recordings and only one person asked to have some comments removed from their episodes.

Anonymous, routine unit surveys conducted during and at the end of semester revealed positive comments explicitly relating to the videos, though relatively few in number. Students appreciated the insights and advice from people working in the fields, although some wished for more content for their discipline.

> I am really enjoying this unit the videos are great in linking the subject to the real world.
>
> This unit has been insightful in viewing the world from a designer's perspective.
>
> I really enjoy the YouTube videos, getting advice from people working in the field is incredibly helpful, I really think the YouTube video series is a really good idea.
>
> I really value the videos provided that are being used to help us learn however, I wish there were more visual communication designers and I believe that adding more interviews with them would help future students due to the increasing popularity of the new course.

Overall, we are satisfied with the reach of this content and hope that it will remain useful for several years.

8.8 VIDEO PRODUCTION NOTES

In the following section, we'll briefly discuss our production process and the lessons we learnt undertaking this project. This may be useful if you're considering undertaking such a project, either to understand which tasks you'd be comfortable taking on and for which you'd hire professional assistance. At the outset, we were determined to stretch our new unit development budget of $6,000 (AUD) as far as possible for the production of video resources with a useful life of three to five years. We chose to hire graduate videographers but handle the producing, directing and editing roles ourselves.

8.8.1 PRODUCTION PERSONNEL

Neither of us, the authors, had any formal video production training, only an interest in video and years of informal experiments with consumer video editing software, such as iMovie.

We shared producing, directing and interviewing roles, often switching roles on a shoot. As producers, we followed up leads to arrange interviews with designers based on our discipline backgrounds. We contracted two recent QUT Film and Television graduates to undertake 60 hours of filming each, which consumed all of our budget. These young professionals brought their own equipment to the job, specifically professional-quality mirrorless cameras and lenses, lighting and audio equipment. Their expertise enabled us to capture high quality vision and audio, usually in the offices of our interview participants. Most interview shoots consumed at least an hour of travel time (including parking and moving equipment) plus as many as two hours for "bump-in," photography and "bump out." It's standard industry practice to pay crew for no less than half a day's work so we tried to shoot several interviews back-to-back whenever possible.

8.8.2 Production Schedule

Preparations for conducting the video interviews occupied approximately three months. Starting in December 2018, we canvased colleagues for recommendations of local designers to showcase and set an ambitious goal of interviewing 3 designers per discipline, potentially 21 interviews in total. When contacting prospective interviewees, we provided designers with an information sheet outlining shooting arrangements, personnel involved, equipment used, time required, questions to be asked and their choice of shooting at their offices or on campus. Responses to enquiries were very positive despite the Christmas holiday period, with most designers agreeing to be involved. Negotiations continued through January 2019 and overlapped filming in February.

Filming took two months. The first interview took place in early January. Twenty interviews had been recorded by mid-February, just under two weeks before classes began and the final recording took place on the last day of February. Post-production began in parallel with the recording. We prioritised editing episodes that would be required for the first week of classes and the remaining episodes were completed by the third week of classes in early March. Including initial contact with prospective interviewees, the total production time amounted to approximately less than four months.

8.8.3 Preparing to Shoot

To maximise authenticity we hoped to shoot the interviews in the designers' work environments. For many students, this would be their first glimpse of a designer's workplace. This wasn't always possible due to either how the designers work, the desire to avoid disruption to the workplace or the lack of available shooting locations. Ultimately, of the 21 designers interviewed, 15 were filmed at their place of work and the remaining 6 recorded on campus. It's surprising for the layperson how much equipment may be required to record a simple interview with one person. Transporting all of the necessary equipment and personnel is time-consuming and fatiguing, especially in the central city locations where many design firms are located. Figure 8.4 shows the typical arrangement we used, with minor variations, for the interviews.

We always attempted to place the interviewee (1) at least several metres in front of an interesting background, usually their office work environment. Some distance helps the videographer create depth-of-field separation, which creates a subtle blurring of the background (evident in Figure 8.1). The interviewer (2) was seated face-to-face with the subject. The interviewer was off-camera and their voice would be edited out of the final recordings (see Episode Format below).

The videographer (3) framed shots and, once recording commenced, could step back from the camera. Recordings ran continuously for up to 30 minutes, which was a technical limitation of the mirrorless DLSR cameras used for filming, but this was almost always sufficient. Two cameras were used for vision, a main camera (4) for a head-and-shoulders view and a second camera (5) for a wider shot. This dual-camera arrangement is not essential, but gives the video editor an easy way to disguise edits

FIGURE 8.4 Our typical video recording setup (at the offices of Anna O'Gorman, 2019)

by switching between angles while adding visual interest to what is otherwise a largely static recording. Both cameras recorded at high "4K" resolution, allowing the editor a variety of cropping options for the 1080p "HD" resolution used in the final video episodes.

While both cameras recorded audio using their built-in microphones, this audio was not used for production because of its poor quality. Instead, an overhead boom mic (6) or, in some cases, a lapel ("lavalier") microphone, was used for sound, recorded to a digital audio recorder. This was synchronised with the vision during editing. Finally, the subject was illuminated by the "key light" with diffuser (7), positioned to the left of the cameras, with an additional "backlight" (8) behind them (and out of frame). The key light models the interviewee's face clearly and the backlight provides definition and highlights to their outline. Many videographers use a third "fill light" to soften shadows but we used ambient light for this purpose.

8.8.4 Conducting the Interview

We realised that to produce natural recordings it was important to make the subject as comfortable as possible during filming. Recording an interview calls for behaviour that makes most people feel self-conscious – "Please sit here and try not to move too much. Look this way. Try not to look at the cameras." Prior to our first interview, we experimented with the subject addressing the unseen audience by looking directly into the camera lens. It became apparent that this feels very artificial, and it seemed likely that many people would find it difficult to avoid looking at the interviewer. Instead, we set out to create a relaxed, conversational atmosphere.

To establish rapport during the shoots each of us played the role of interviewer for those designers we had contacted, while the other person gave direction. Typically, the interviewer would chat with the subject while the equipment was being set up which gave ample time to discuss the purpose of the questions while building rapport. This could be too successful – there were times when the interviewer would become so absorbed in the "conversation" that they'd lose track of the questions or make conversational noises, "hmm," "yes" and so on. To counter this, we created a checklist to assist the smooth running of the interview. Ticking off the questions in turn helped the interviewer maintain the flow of the "conversation" without missing questions. Because we decided to remove the voice of the interviewer and replace their questions with titles, these utterances would interfere with the editing process. This required restraint from the interviewer to avoid speaking while the subject answered a question, which doesn't come easily. Because our interviewees were given the questions in advance, some of them tended to merge their responses to multiple questions, which created problems later while editing to make discrete question episodes.

Here, we discovered the importance of the director's role. It became the job of the person not asking the questions to carefully monitor the flow of the interview and keep track of what was said and who was speaking. At the conclusion of recording, we'd spend one minute in silence while we recorded "room tone" using the microphones. This could be used in production to subtract ambient background noise to make a cleaner voice track, which was necessary on several occasions.

8.8.5 Packing Up

Finally, we packed up equipment, thanked our guests and departed. While packing up, we'd transfer the two video recordings and the sound to an external hard drive so that editing could commence soon after. As a matter of good practice, our videographers insisted on duplicating essential files immediately to guard against catastrophic data loss. Interviews took between 60 and 105 minutes to conduct, not counting travel time. We allowed 45 minutes for setting up equipment, and with experience we reduced this to approximately 30 minutes. Shooting occupied 30 to 45 minutes and packing up about 15 minutes.

8.9 POST-PRODUCTION

Post-production is the process of creating finished video programs from the resources gathered during production. We chose to undertake post-production ourselves instead of paying others to produce the video episodes. There were four reasons for this:

1. A desire to extend the budget and produce as many video resources as possible

2. An interest in maintaining control over the content with a mind to producing further variations and edits at a later date
3. An interest in developing video production skills that we might apply to projects in future
4. We had access to hardware sufficient for editing

8.9.1 Episode Format

As mentioned above, we devised a short, episodic format for the videos. By using title cards to frame each question we removed the presence of the interviewer, which allowed us to share the interviewer role without concern for consistency, streamlined the interview recording process and maintained the focus exclusively on the designers being interviewed.

8.9.2 Hardware and Software

We lacked formal training in non-linear editing techniques although we'd created short amateur videos for private and informal teaching uses in the past. This project, however, demanded the use of more advanced features and techniques so we were attempting to master them while using them. This was time-consuming but ultimately satisfying. We relied on a variety of sources for information, including YouTube and lessons available on Lynda.com (now known as LinkedIn Learning).

We used available desktop computers for video editing. Editing 4K video demands ample RAM and plentiful storage; however, an expensive computer is not necessary. Post-production was undertaken on a comparatively out-of-date 2013 Apple iMac with 32 GB of RAM and 3 TB of internal storage, using 2 TB external hard drives for editing and file backup. Project files (21 interviews and 169 discrete video episodes) consume 2.1 TB of storage. The use of dual monitors was helpful. High-quality non-linear editing software is readily available – the Adobe Creative Cloud software suite includes *Adobe Premier* (Adobe, n.d.), and Blackmagic Design (n.p.) offers a free version of their *Da Vinci Resolve* editor – we adopted Apple *Final Cut Pro X* (Apple Inc., n.d.) for this project.

8.10 LESSONS LEARNT

8.10.1 Costs of Production: Time and Money

When undertaking projects like this it can be difficult to forecast how long it will take and therefore how much it will cost. Table 8.3 records approximate time in hours for each part of the process.

During our production period we shot between one and five interviews per week, dictated by the availability of our interviewees. It was always easier and faster to shoot on campus, but shooting on location produced more compelling content. Post-production commenced as soon as we had recorded the first videos and was easily the most time-consuming phase of the project. Due to inexperience, we required

TABLE 8.3
Approximate Time in Hours for Each Stage of the Project (Unless Otherwise Noted Tasks Show Time per Person)

Stage	Hours	Notes
Pre-Production		
Project planning, writing questions, budget, staffing (two people)	38	Undertaken over a span of approximately two months
Arranging interviews	1	Per interview
Pilot interview (two people)	2	Testing the format
Production		
Shooting on location (videographer, director and interviewer)	3	Including travel time
Shooting on campus (videographer, director and interviewer)	2	Less when interviews could be conducted back to back (rare)
Post-Production		
Designing the video style and format	20	Learning to use software
Designing titles and motion graphics	20	Learning to use software
Music	3	
Editing per 25-minute interview	6	Varied with interview duration and interviewee
Distribution on YouTube	4	Preparing playlists, uploading, titling, revising

some days of experimenting and learning to master unfamiliar techniques and new software. More experienced editors would likely be much more efficient. Editing speed improved considerably after the first few episodes were produced but time per interview varied. Some interviews were simply longer in duration and some interview subjects required significantly more work to edit. Some interviewees paused often whilst speaking or repeated themselves; editing these interviews for brevity was intensely time-consuming. In general, the "tighter" the edit, the more time it took to produce.

8.11 CONCLUSIONS

We conclude that educational and engaging video production for internal teaching purposes is within range of determined academics should time and modest resources permit. Students respond well to concise video resources, especially when they see them as relevant to their studies. They also don't require high production values to gain benefit from the material. Although it's too early to tell, we hope that students feel more connected to the local profession and that future work integrated learning opportunities will flow from the introduction to so many local designers. Should you choose to incorporate this approach into your own teaching practice, please consider the following recommendations:

Pre-Production

- Allow at least a month to arrange interviews, and more during holiday periods
- Use trained videographers if possible. Our graduate videographers were excellent to work with despite their limited professional experience
- Test and refine your interview and filming methodology by conducting test shoots and editing them prior to interviewing your first participants

Production

- Use the best cameras you can; however, even high-end smartphones can be effective
- Take care with lighting, both in the position of light sources and their colour temperature. For example, a mix of incandescent and fluorescent light sources may lead to unpleasant colour casts that are challenging to rectify in post-production. Use a colour chart to record a few seconds of video on each camera before or after filming to make post-production colour grading easier
- Use good-quality microphones to record participants clearly and minimise background noise while filming. Sound recording is harder than it seems. Background noises, such as a doors slamming, squeaking chairs, air conditioning hum and traffic might not be noticed during an interview but become distracting while listening to a recording. Camera microphones capture too much of this noise and it is difficult or time-consuming to remove it in post-production. Record 30 seconds of ambient audio without speech at the end of the interview to provide "room tone" for post-production sound clean-up

Post-Production

- Have confidence that you can master basic editing techniques but allow time to learn
- Be realistic about time needed for editing – as a rule of thumb, allow ten minutes editing for each minute of final video
- Use a computer with plenty of RAM and ample file storage. Ensure you have enough storage and make regular backups of all essential files
- Consider the use of YouTube for content distribution and use "unlisted" for content that you wish to keep in-house. YouTube is flexible and familiar to students

Learning Design

- Use your own video resources to directly address the learning needs of your students.
- Combine short, episodic video resources with classroom discussions and exercises to create engaging and interactive learning experiences

REFERENCES

21–7 (2019) YouTube video, added by Impact Lab 1 [Online]. Retrieved from https://youtu.be/l6I7TDJXC4Q

Adobe. (n.d.). Video editing that's always a cut above. [Online]. Retrieved from www.adobe.com/au/products/premiere.html

Apple Inc. (n.d.). The most powerful take ever on postproduction. Retrieved from www.apple.com/au/final-cut-pro/

Blackmagic Design. (n.d.). Da Vinci resolve 16 [Online]. Retrieved from www.blackmagicdesign.com/products/davinciresolve/edit

Chi, M., & Wylie, R. (2014). The ICAP framework: Linking cognitive engagement to active learning outcomes. *Educational Psychologist, 49*(4), 219–243.

Conceptualizing Socialization in Graduate and Professional Programs. (2001). *ASHE-ERIC Higher Education Report, 28*(3), 11. Retrieved from https://eric.ed.gov/?id=ED457710

Cruse, E. (2007). Using education video in the classroom: Theory, research and practice [Online]. Retrieved from http://www.libraryvideo.com/articles/article26.asp

Hansch, A., Hillers, L., McConachie, K., Newman, C., Schildhauer, C., & Schmidt, P. (2015). *Video and Online Learning: Critical Reflections and Findings from the Field.* (Technical Report #HIIG Discussion Paper Series No. 2015-02). Retrieved from https://papers.ssrn.com/sol3/papers.cfm?abstract_id=2577882

Mayer, R. (2006). Ten research-based principles of multimedia learning. In O'Neil, H., & Perez, R. (Eds.), *Web-Based Learning: Theory, Research, and Practice.* NJ: Lawrence Erlbaum and Associates, New York, NY.

Poquet, O., Lim, L., Mirrahi, N. and Dawson, S. (2018, March). Video and learning: A systematic review (2007–2017). In *LAK'18: International Conference on Learning Analytics and Knowledge* (pp. 151–160) New York: ACM. doi: 10.1145/3170358.3170376.

9 Reframing Learning via Technology
COVID-19 – Hacks and Reflections

Jennifer Seevinck, Nick Kelly, Heather McKinnon, Richard Evans, Brett Fyfield, Leo Rezayan, Jeremy Kerr, Anastasia Tyurina and Jane Turner

CONTENTS

9.1 GETTING OUR TEACHING ONLINE, IN TWO WEEKS

Suddenly, in March 2020, we had to get our teaching online within only one month of starting our academic year. We had already seen students start to disengage and display some anxiety from attending face-to-face classes, and we began to realise that, at some point very soon, we would go home and go online. Critical to understanding the challenge of teaching in the COVID-19 pandemic is to recognise the time constraints – we had effectively two weeks in which to make a plan, develop resources and communicate it to students so that we were ready for the first assignment submission. The scale and speed of the transformative effort – across learning, communication and management – was a significant challenge.

We, the nine authors, are design lecturers coordinating, developing and teaching units in *interaction design* and *visual communication*. This chapter is about the units we were teaching in design principles, theories and processes during Semester 1 of 2020, when the global health pandemic hit.

Prior to the COVID-19 outbreak and the associated safety precautions (social distancing, self-isolation and staying home), we would engage with our students in predominantly face-to-face formats: through either a one-hour lecture and two-hour tutorial, or a three-hour studio. Some of our units featured group assignments, in which students would work together through collaborative design and individually, where they benefit from social critique sessions. Some units required access to prototyping facilities in an on-campus workshop, but mostly students in our units worked to design content in, and for, the digital medium – our medium as interaction and visual designers. Students were surveyed for their thoughts about technology access – drawing on a questionnaire designed by fellow chapter author Andrew Scott. This feedback gave us confidence in students' technology access to continue study online, while at the same time the university was able to keep computer labs open for student access where needed.

The quality and nature of digitally augmented experiences is our core business – from educators and researchers in this space through to how we practise that specific experience design understanding to create online and blended experiences. This expertise informed how we understood and responded to the pandemic: as a design problem for online learning, in which we needed to rapidly develop new online or blended learning experiences.

In designing our response, we considered tools already used in teaching and research or available through our university: learning management and presentation software such as Blackboard (Blackboard Inc., 1997) and Articulate (Schwartz, 2002); communication software such as Discord (Discord Inc., 2015), Slack (Butterfield et al., 2013), Zoom video conferencing (Yuan, 2011) and Teams (Microsoft, 2017); and additional online learning tools such as Padlet image sharing and commenting (Goel, 2008). We implemented solutions informed by our understanding of course content and what it means to create for digital media – guided by a sense of "what suited" and what "might work" and, most importantly, what was "ready to hand" (Heidegger, 1962).

From the very start we knew that online teaching was not as simple as taking a face-to-face tutorial with 25 students and 1 staff member and putting it online for multiple hours of synchronous video interaction. Differences in context and activity are just some of the factors that impact the student experience – from their sense of autonomy, relatedness with others and motivation, through to their physical health and ability to focus (Martin, Kelly and Terry 2018). Other issues include adapting/recreating learning content to make it suitable for online (if that is even possible!), student technical difficulties, the cost of their data plans, management of anxiety (of both students and teaching staff), through to ongoing feedback and coordination efforts. The necessity for the rapid turnaround compounded the concerns and meant that our solutions were imperfect: the redesigned delivery was done in crisis; it was hasty, but, importantly, it was also thoughtful.

Our experiences, insights and reflections on this rapid adaptation of teaching during the global COVID-19 crisis are shared in this chapter, where each unit coordinator discusses their unique concerns and practical design solutions. In our university, "unit coordinators" are responsible for the design, organisation and delivery of unit content, as well as administration of students and teaching staff. Units run for 12 weeks, normally with weekly lectures and tutorials or studios.

We all faced the same challenges of switching to online teaching and engaging students to attend, interact and learn, while addressing anxiety and mitigating the impact of too much screen time. In teaching design practice, we were also concerned with how we might stimulate creative student experiences. Anastasia's augmented reality lessons brought creativity into students' personal physical environments through teaching technical skills in an open-ended way. Jen used the web camera and Padlet view of student work during critique sessions to generate new opportunities to leverage the perceptual processes of seeing – a key mechanism for creative interpretation.

Jane explicitly identified the role of technology in the learning experience to set up a Discord server for playful and social interaction in learning, addressing the transience of the digital medium and placelessness. Brett and Rich also recognised the social dimension of design to find a way to maintain the efficacy of group assessments, despite the inability of students to meet physically during lockdown. Jeremy's focus on contemporary visual design led him to draw on pop culture to frame the learning medium and experience. Heather reflected that delivering online offered increased flexibility, but weighed it against significantly more preparation for that online delivery. She also identified a loss of spontaneity, of teaching practice where you "read" your class experience and situation to determine how to continue and *reframe* the delivery as informed by the *situation talk-back* (Schön, 1983). The significance of situated learning was also raised by Leo, who laments the limited potential for learning by watching online – for those students who would previously sit at the back of the class, not actively participate but still be gaining something by "being there." Our experiences vary with our teaching values, cohorts and learning requirements.

9.2 THE CASE STUDIES

In these sections, each unit coordinator discusses their approach and reflects on its success.

9.2.1 Nick Kelly, a First-Year Interaction Design Unit (200 students, Equivalent 1-Hour Lecture, 2-Hour Tutorial)

Nick describes his team's approach when reshaping the tutorial experience for online and how they worked to address attendance, engagement and too much screen time. This semester-long unit teaches the foundational principles of interaction design with an initial focus on user experience research (half of the unit) and then on students' "first steps" in designing and representing interactions. For their first assignment, students conducted user experience (UX) research about an interactive display: a large, high-resolution projection screen space with 16 interactive touch screens below it called *The Cube* (www.thecube.qut.edu.au/). The space is used by approximately 10,000 high school students each year for learning experiences. In adapting for the crisis, we needed a way to give students the "experience of being a UX researcher" without being able to directly be in contact with the space that they were designing for (where typically they would interview users, explore the space and conduct observations over time).

We addressed this challenge by:

- Forming students into 3–4 person groups and then inviting them to interview one another about their (pre-crisis) experience in the space over Zoom
- By going into the space to film an in-depth speak-aloud tour while interacting with it
- Crowd-sourcing questions from the students, collating these and then conducting subject-matter expert interviews with the "owners" of the space, and sharing the answers from these interviews with students to provide a basis for their UX research
- Finding online resources (photos, videos) from prior events shown on *The Cube* that students could reference

Another challenge concerned the length of synchronous online interactions. We drew on online pedagogy literature (e.g. Anderson, 2008) and realised that a two-hour Zoom session of synchronous activity would not be feasible given students' ability to focus and look at a screen; instead, we split each tutorial in half. For example, 24 students in 1 tutorial became 2 half-tutorials of 12 students and ran for half as long, improving the tutor–student ratio. A disadvantage was that students now had half as much contact time. This was addressed by crafting highly scaffolded learning tasks to be done in small groups between tutorials. These tasks added up to about 1.5 hours weekly.

The biggest challenge in going online was the students who disengaged. Those who came along for the ride appear to have had an excellent experience; yet just under half chose not to come to online tutorials, instead addressing the assignments using asynchronous resources. Online learning seems to be more challenging for some students, and my sense is that some who may have had a patchy attendance in person ended up with zero attendance online due to a lack of engagement. For next semester, I'm trying to keep these disengaged students in mind and am thinking "how can I structure this unit so that students get invested in the subject early on?" Building an online rapport in the first few weeks and providing certainty around structures and expectations in the online learning will help with this.

My understanding from this experience is that students need certainty as a precondition for learning. They can handle really diverse approaches to pedagogy, but they need to feel trust in the system and feel cared for, as if there is a "hierarchy of needs" for learning to happen. Somewhere near the base of that pyramid is "having a clear picture of what the whole semester will look like." Without that, there is a pervasive fear and mistrust that makes it hard for learning to be the students' focus.

I was surprised by how hard it was to get students engaged in online learning. Getting students to show up for Zoom seemed to require many reminders – I got more attendance when I sent more reminders. Once they were there, encouraging them to turn on their cameras and microphones was another challenge. And then creating the space where people would actually talk freely with their opinions was another level on top of that. I didn't expect it to be that hard, but it became easier the more sessions we did.

9.2.2 Heather McKinnon, a Second-Year Design Studio (60 Students, Equivalent 3-Hour Studio)

The abruptness of the transition of my second-year design studio meant focus went towards transforming and simplifying three key elements of the studio (Figure 9.1): Sharing and Feedback, Content Delivery and Connection.

Dividing elements of the unit into three categories seemed to help communicate new unit setup and expectations to students. The flow of the new asynchronous studio week ended up as a three-step process for all students to follow every week:

- Each student to watch the content videos and complete assigned design activity
- Students upload completed work to Padlet for sharing and feedback
- Students connect with peers and tutors for discussion, feedback, questions and casual conversation. The variety of connection points for students with their teachers each week included Blackboard, Zoom, Slack, Padlet and email

Most students adapted to the new expectations relatively quickly over about a two-week period. The transition ended up working quite well for most students and the majority of them completed their work on time. Overall, once everyone had adapted to the transition, the course seemed to run adequately until the end of the semester. Nonetheless, I am not sure how enjoyable it was for everyone, myself included.

From a teaching perspective, while I thoroughly enjoyed not being locked into a rigid teaching timetable as it created a very flexible way of working, and it all ran quite smoothly, I found this way of teaching very challenging. Looking beyond the rushed way that it all occurred, challenges for this medium seemed to be the amount of time it takes to prepare, record and upload content, organise materials with enough guidance that students understand, organise the teaching team and then

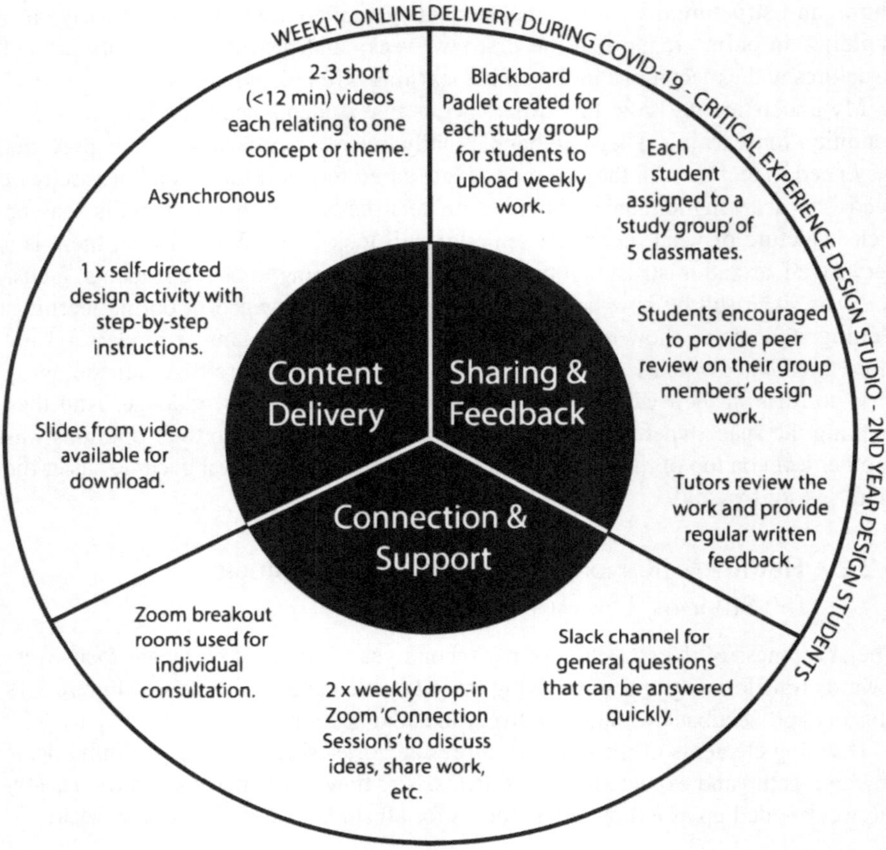

FIGURE 9.1 Three categories of consideration for transition to online teaching

be sure everyone is on track through regular feedback. However, the most significant challenge and perhaps the biggest lesson from this for me personally was the lack of opportunities for spontaneity and enjoyment within my online version of teaching. In contrast to in-person teaching, this very ordered and pre-prepared online medium left very little room to be spontaneous, to gauge the understanding of students or to expand on a topic just because someone asked a question that triggered a thought or idea. I now realise that these spontaneous moments are a significant part of my teaching experience. Moving forward, it is this element of spontaneity and enjoyment that I will work very hard to try and integrate more into my online teaching delivery. My current ideas for this include:

- Integrating live Zoom sessions to chat about one particular design example or case study. Both teachers and students could take turns bringing design examples and running the session
- Split the cohort into sides and organise a live debate against each other

- Integrate boutique online "events" into the course. In lieu of excursions and guest lectures we could all tour an online exhibit at MOMA together, watch a documentary whilst live chatting or even chat about design over a meal together

Perhaps consideration into integrating opportunities for spontaneity and enjoyment into the course could help increase the enjoyment of the online teaching and learning experience for everyone involved.

9.2.3 Brett Fyfield and Richard Evans, Third-Year Design Studio (15 Students, Equivalent 2.5 Hour Studio)

As the COVID-19 crisis unfolded, students were preparing to submit their first assessment item which required them to explore the theme of privacy by design through augmented reality (AR). We quickly recognised that our approaches to teaching and research within Interaction Design put us in a good position to transition to online design practice and teaching. From prior work (Czaplinski et al., 2017; Fyfield and Czaplinski, 2017), we learned that keeping students engaged through social connection would be crucial to maintaining studio morale over the coming months. The transition to online presented us with an opportunity to share something central to design practice – the ability to thrive on uncertainty and apply design-based approaches to problem-solving (Dorst, 2018).

Our approach acknowledged the difficulties students were facing through an intentional and critical questioning of the problems posed by the assessment and tasks for critical reflection in each virtual studio. We made a conscious decision to retain the group work in the second assessment item despite obvious challenges for supporting group work. We explicitly re-wrote the assessment brief to include a requirement for augmented reality design prototypes that maintain social connections during a time of physical distancing. In the first virtual studio, after an extended two week mid-semester break, we asked the students what they wanted, anticipating that they might ask for asynchronous academic–student interaction, and were surprised to learn that they wanted more time in the virtual studio for feedback in their groups.

We spaced out the delivery of the subject/unit across the week with 20-minute pre-recorded videos and set readings, followed by two-hour collaborative sessions in the virtual studio. The pre-recorded videos act to connect theory with design practice and prepare the students for conversation about the broader social, technological and ethical considerations for AR in the virtual studio – online using video conferencing software. The students come together in the early part of each online synchronous virtual studio to share challenges, successes and plan for the week ahead. This design "stand-up" reassures them of their progress and allows us to adjust our delivery and address concerns.

The next 45 minutes was a discussion of readings and a demonstration of methods and techniques. After a short break, we placed the students into virtual "breakout rooms" within their teams to practise methods and techniques. During this time, one of us would circulate amongst the breakouts to encourage, clarify and provide feedback on group work. After about 45 minutes of group work the students came back into the main room to share their concepts and prototypes. Students used this opportunity to gather peer feedback and, in some cases, conduct their user interaction analyses and evaluations in-situ. We believe this pacing with group work in the studio was why we saw the same level of attendance in online studio sessions that we would expect in the physical studio. We acknowledge and appreciate the challenging circumstances that students faced in their personal lives and are grateful for the way they conducted themselves in the studio.

When we consider the possibility of social distancing restrictions in place for future studio units, there are many things we have learned from this experience. First, in facing the challenge of being prepared for both physical and online delivery, we would feel confident that supporting group work online is not only possible, but preferable. Our rationale is that globally distributed teams are the norm, especially in design, and now more than ever in other fields. Our feeling is that students can rely on one another to complete group work when that work is part of a shared challenge, and that it is a mistake to discard group work because it is "too hard" to coordinate or assess online. We had the benefit of meeting our students face-to-face in the studio early in the semester and believe that making this human connection at the outset was essential to the success of the unit. We are unsure what the future may bring, but will be investigating ways of maintaining social distancing in physical spaces through scheduling times for smaller groups of students. Our aim here would be to establish the social contract in an intimate studio setting before sustaining the delivery online.

9.2.4 Leo Rezayan, Second-Year Design Unit, (75 Students, Equivalent 1-Hour Lecture and 2-Hour Tutorial)

In this studio unit an advanced tangible media design project is produced, necessitating theoretical knowledge and an understanding of the processes that underpin the tangible, as well as the embodied, ways in which people interact. With a focus on design-by-making a key component is the studio opportunity for students to interact with staff and peers collaboratively and engage with processes of design. The unit relies on lectures and group round-table discussions, where students are required to present their work during the semester. With COVID-19 necessitating a switch to online teaching, the main challenges were how to transfer a one-hour lecture and two-hour tutorial each week.

My tutors and I consulted the students and concluded that shorter tutorial sessions might be needed – like Nick's unit, we went from a 2-hour face-to-face tutorial session with a ratio of 1 teaching staff to 24 students, to shorter sessions – 4 online 30-minute tutorial sessions via Zoom with the ratio of 1 staff member to 6 students. These were used for 1-on-1 design feedback with the tutor – each student in that 30-minute session had a chance to talk and show their work, and the rest of the group (5 others) and staff would comment and give feedback.

Halfway through the semester, we sought feedback from students during our online tutorial sessions on how we could improve these experiences considering the online delivery method. Based on this, we planned for online and offline support for the second half of Semester 1; consisting of Padlet for offline and an additional 30-minute weekly session – using Zoom – for questions and feedback. In the week leading up to the first assignment submission we had very good attendance. Students used their webcams to show their prototypes or shared pre-recorded videos demonstrating these, exemplifying students' engagement online, seeking peer review and teaching staff feedback.

We found that it can be beneficial to have a few touch points during an online session – such as where the tutor opens the session by briefing all the students for a few minutes. The online session could then be broken down into smaller groups for doing the activities, giving feedback and peer reviews or discussion. This requires one to have the right tool (online platform) to interact with the class as a whole and to be able to break down into the smaller groups. We found that having a minimum of 5 to 8 minutes per student to provide more detailed feedback, along with a smaller group of 4 or 5 students per 30-minute session seemed to work well.

In general, I felt that online zoom sessions worked well for this unit. Having said that, challenges with engagement remain. An online session, when compared to a face-to-face teaching and learning environment, is missing engagement from just being in the classroom and being exposed to what is happening during the session – when students can hear one another's ideas and just simply see what is happening during the session without necessarily focussing on every single detail. As face-to-face teaching and learning environments exist in the physical world they take up real space, and, as a result, offer non-fragmented visibility. This is when students and teaching staff talk, present and share ideas and others can follow, creating a situation where seeing is understood as being seen (Robertson, 1997). Interacting in the physical world is observable, comprehensible and promotes communicative and performative tasks as well as collaboration. Furthermore, in a face-to-face session with a larger group of students, the more active and energetic students impact the experience of the session for everyone by asking questions, giving feedback and engaging with the content.

9.2.5 Jeremy Kerr, First-Year Design Unit (400 Students, 1-Hour Lecture and 2-Hour Tutorial) and Third-Year Design Unit (30 Students, 1-Hour Lecture and 2-Hour Tutorial)

These two visual design units share a traditional on-campus delivery format where theory, principles and case studies are delivered through in-person lectures, with follow-up tutorials allowing students to apply this learning in portfolio-based assessments. They also learn new software skills – Adobe Illustrator, Photoshop and InDesign in the first-year unit and After Effects in the third-year unit – through demonstrations by teaching staff and hands-on activities tailored to develop portfolio pieces. Central to the transformation to online delivery was "playful" experience design and approaching learning as "edutainment" (Jarvin, 2015; Němec and Trna, 2007). These approaches were already embedded within the teaching ethos of the units, in initiatives such as youth-oriented and pop-culture driven assessment items and gamification across student feedback sessions. In the case of the latter, in the first-year unit, this involved using an "emoji scale" (from "smiley" face to the "poo" icon) to encourage a less formal and intimidating introduction to peer feedback. In moving to online delivery, however, this allowed for further playful approaches to be explored beyond a conventional university format.

For lectures, both units moved to a video-driven mode and these were conceived as a series of upbeat "micro-lectures," approximately half the standard lecture hour, to adjust for screen-based attention spans (Subramanian, 2017). The delivery of this recorded content was integral to both units as a large proportion of students were international exchange students and had returned to their home countries and different time zones. The format allowed for asynchronous learning, whereby education could occur when it best suited students and when individuals were most receptive to tuition.

Within this "bite-sized" lecture format, the presentation was approached more informally and was less "lecturely" in tone. This saw new approaches introduced, such as video sequences of staff re-enacting popular film title sequences to demonstrate design strategies, software lessons based around distinctly humorous visual designs and theory presented in a more conversational way, not unlike popular "YouTubers." While licenced video software tutorials (such as those offered by LinkedIn Learning [Weinman, 1995]) were available for use, self-produced videos were developed as these allowed for customised entry points for the students and learning that aligned with the overall high-spirited approach. Such strategies proved successful, as evidenced by high student engagement rates, and in feedback emails where students made specific note to enjoying and "having fun" with these units (see Figure 9.2).

Building upon the popularity of such approaches, a series of further "playful" initiatives was devised throughout the semester, including a co-designed music playlist using Spotify streaming software (Ek and Lorentzon, 2006), curated by the students themselves. This publicly available playlist was not only used within "drop-in" online design workshops occurring through Zoom classes, but for students to play and feel "connected" with their cohort while working independently on assessments. While conventional classes did not exist, this served to facilitate a sense of a shared community amongst students.

In redesigning the units, students were also given an increased freedom in assessment options, allowing for opportunities to design within any area of popular culture that resonated with them, or to respond directly to their experience of the COVID-19 situation. In so doing, they blurred the line between a standard university unit and personal interest or enjoyment. For example, the poignant video images created by students Jennifer Olid and Mun Sam pictured in Figure 9.2 represent their personal responses to the COVID-19 situation (inset top and bottom, respectively). Screenshots from the micro-lectures and other self-produced video content, created to support playful online learning experiences, are also pictured.

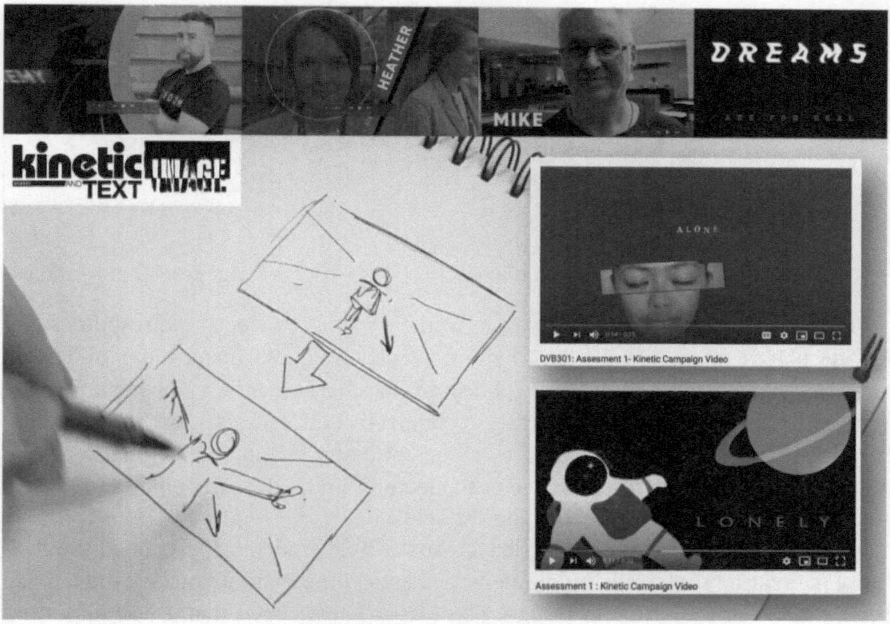

FIGURE 9.2 Top and drawing close up show video content created for the third year unit. Inset YouTube videos are two student works exploring the COVID-19 pandemic experience

9.2.6 Anastasia Tyurina, Second-Year Studio Unit (90 Students, 1-Hour Lecture and 2-Hour Tutorial)

In turning to online teaching, I worked to adapt existing methods, prioritising student creativity by scaffolding their learning of new complex techniques, e.g. design for Augmented Reality (AR), in a number of ways. I focussed on the processes of making – creating my own examples for the student weekly tasks, I illustrated the expected level of skills at this stage of the course and provided a clearer context for the final assessment. I also developed step-by-step software use demonstrations (Adobe CC range and AR apps) to guide students in completing their weekly tasks. I kept the technology-heavy demonstrations pre-recorded as this allowed students to prepare questions before logging in to the next studio. It also allowed me to schedule a follow-up discussion and to scope ideas for future demonstrations.

Weekly tasks were developed to encourage creativity as well. Some of the tasks were to create a collage poster, GIF animation and AR work. Figure 9.3 shows an example of the task where a black and white photograph of the studio interior expands to show colour and moving elements through the addition of AR technology. In this task, the use of AR enabled students to see their environment differently and visually experience the moving elements and colours in the still black and white image, provoking a creative interpretation of the image. The animation in Figure 9.3 can be seen using the QR code provided. It needs to be scanned to obtain the link for installing the EyeJack app (free, 56 MB), which enables activation of the animation triggered by the image.

FIGURE 9.3 Photograph of the studio interior used as a trigger image for AR animation

Online scaffolding was developed to frame content, concepts and resources and provide places to share work and receive feedback. These include use of Padlet, guest lectures and TED Talks, YouTube and LinkedIn Learning videos that allowed students to learn from different creative perspectives. Lectures were pre-recorded, though I still see purpose in the lecture content being live as the "regular learning atmosphere" created by the presence of the lecturer and fellow students can increase student engagement (Wright, 2014, pp. 120–132).

Common themes in student feedback showed that students appreciated the flexibility in creating something more interactive and a little bit more niche, which made them excited for the future of design. Students acknowledged that it was important to be aware of where technology is leading design and how they can follow it to stay "in the loop." The most-enjoyed aspect of the assessment was creating content with the purpose of making a change or sparking a thought.

9.2.7 Jen Seevinck, Third-Year Studio Unit (170 Students, Equivalent 1-Hour Lecture and 2-Hour Tutorial)

Reframing to see a situation or problem in a different way is a key design process, enabling new insights and generating multiple possible solutions. In design, such reflective practice (Schön, 1983) techniques are often taught "apprentice style" – where an expert models design *moves* for the design situation, drawing on her professional expertise in problem-setting, and the student watches and learns. As creative design is wicked, varied with multiple possible solutions, this is not a process of rules for execution "… where ends are fixed and clear" (Schön, 1983, p. 41); rather, it is a practice that must be developed. Moving this practice to the online space was a key challenge in this unit, particularly because students were learning to design reflectively – with data.

In prior years, the new data design concepts would be explained in lectures and applied in tutorials. To facilitate learning during the COVID-19 lockdown I created lectures, segmented into shortened videos between 5 to 20 minutes each, and integrated them into a web app using Articulate (see Figure 9.4, left [Seevinck, 2020]). This maximised student accessibility and agency because they could engage with the content on different devices and contexts (mobile, desktop, at home or work, etc.), review and pause content to cover sub-topics, tracking their progress throughout. This interactive learning design actively involved the students in their learning.

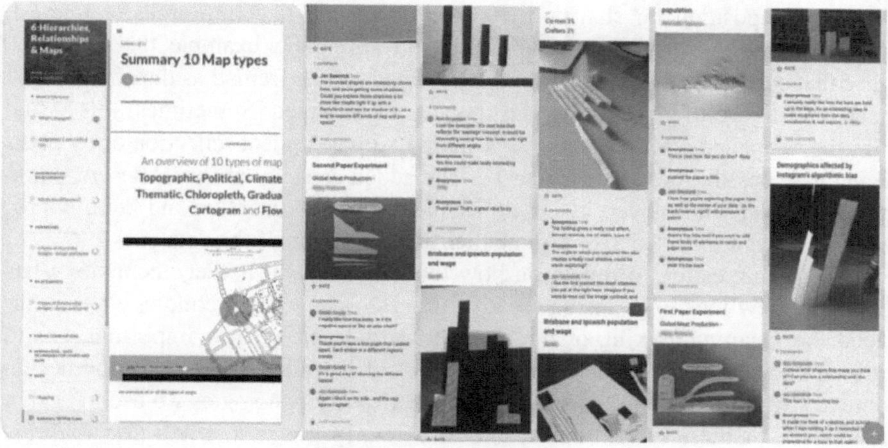

FIGURE 9.4 "Online lectures" created in a web app format (left, Seevinck 2020); Student work and feedback using photo sharing app (right, Towers et al., 2020)

I made these available two days ahead of the timetabled face-to-face lecture – rationalising that this flexibility of access could address students' changing situations (e.g. work shifts) and any technical problems. Importantly, this reliable and flexible delivery also worked to stabilise the learning environment in the uncertain COVID-19 context, to reduce the feeling of helplessness many of us were experiencing. Providing choice in how to engage while ensuring robustness in the delivery (e.g. through regular, early release) arguably facilitates agency and security in the face of crisis and, by extension, mitigates anxiety.

Tutorials focussed on critique and reframing for data design – looking at designs and data, learning to read data and seeing data in new, creative ways, to create visual data-based designs. For these sessions we used a combination of video- and chat-based interaction, where in-class activities and examples were shared in "screen share" mode while a tutor explained and gave directions. Students engaged with these activities for 10–20 minutes, in small virtual groups ("breakout" rooms in Zoom). During a paper-based workshop where students explored data materially and conceptually, they generated different sculptural forms of their data visualisations rapidly to ideate.

Paper explorations were posted to a shared Padlet image space (Figure 9.4, right [Towers et al., 2020]), which supported comments and discussion from peers and the tutor. This has its drawbacks – it is a slower and less granular visual feedback than face-to-face interaction can allow because of the difficulty of gesturing to specific formal elements or manipulating design directly to reframe and prompt ideation. Despite these challenges, the Zoom- and Padlet-based interactions were still fairly effective: student cameras, used for photographing their activities and posting to Padlet, became themselves effective for ideation.

One surprising development was how the camera for posting to Padlet could offer new creative interpretations. The camera can facilitate design thinking: as students frame a composition for sharing they are prioritising, selecting areas of interest, "*visually thinking*" through this compositional framing. For example, the "negative" space that surrounds the paper figure they created can be framed as the composition focus, and figure-to-ground relationships can be reversed. Sometimes, this visual thinking happened after the photo was shared – upon student reflection or by peers seeing new possibilities in their images. The added capacity of visual design thinking could invert and subvert assumptions, prompt further ideation and literally provide new ways for students to reframe their interpretation of the data.

I find it interesting to reflect on how this (camera and Padlet) technology has facilitated new meanings in the creative process – how the technology has influenced the meaning-making or message. It reminds me of how important it is to critically engage with media for experience – of learning or whatever the design context. I will continue to engage critically with the technologies we use to design and teach – as these inform our design thinking. In this unit, I will also fold this into future unit delivery to keep the tutorials face-to-face but highlight the smart phone camera as a mechanism for reframing, to increase the potential for "seeing the data differently."

9.2.8 Jane Turner, Fourth-Year Studio (30 Students, 5-Hour Studio) and Second-Year Unit (100 Students, 2-Hour Studio)

Reflections here are in the context of two design studios: second-year game designers and fourth-year interaction and visual designers. These units represent two distinct cohorts of students. One is a group of final-year interaction and visual design students undertaking an independent capstone project, and the other is a larger group of science and engineering faculty students enrolled in a games and interactive environments degree.

The move online was interesting because while the students are, by the very nature of their courses and units, all engaged in using software for various purposes (making games, creating images, programming interaction), finding themselves using it as a communications medium, being embodied by software rather than using it to embody ideas is a very different experience. Software, as Johnson observed back in the early days of network-mediated communications (Johnson, 1999), does not merely permit communication, it becomes an active participant contributing to the culture of the communication. Moving the game design students into a fully online environment was relatively direct. The workshop content was delivered via the already established unit Blackboard site's real-time video conferencing tool Ultra (broadcast with live chat available), and workshop activities were enacted via a Discord server with group and whole cohort channels.

Discord was already familiar to many of the game design students as a communications channel they might use outside of university. It seems to work well as a virtual design discussion environment, although the flatness of the channel hierarchy can be annoying when there are 40 different channels lighting up at the same time and groups are calling out for the teaching team to pop by and visit! At first sight, Discord looks very similar to other communications software that allows both synchronous chat and asynchronous messaging such as Slack. However, Discord has some legacy features as a result of its roots in Internet Relay Chat (IRC) and its early uptake by game players. It has a "mood" that fosters playfulness in the form of shared memes and jokes. Small bots can be added that enact environment chores ranging from basic greetings to playing derivations of Nintendo's *Pokémon* and *Animal Crossing* franchises.

Its architecture, or the culture of its interface, is predicated towards creating community and a sense of participant ownership and playfulness. For example, the usual changeable individual profile icons can be supplemented with a direct announcement of games currently being played or apps being used and individual "tag lines" and emojis. Discord boasts an extraordinary range of available emoticons that are often explored as visual utterances in their own right.

Moving the fourth-year students into a fully online environment was smaller scale as they were a cohort of 30 students, each working on an individual project, and subsequently less dramatic but also somehow less alive. Again, briefings were delivered via Blackboard Ultra, but instead of holding discussions in a chat server, briefings were followed by Zoom group-based discussions or "virtual desk crits" to discuss work in progress. Zoom provides an apparently ideal replacement of the physical environment with its easy access and synchronous video and voice communications, and yet the playfulness and sense of participant ownership visible in the second-year Discord server communications between teaching staff and students disappears in Zoom, replaced with a more structured and more corporate mood.

What seems to be missing in the use of Zoom is more than the "liveness" (Auslander, 2008) and presence of the physical space. The placelessness of the conference room is replicated in Zoom and what is removed is the opportunity of place and place-making. Place is more than "space"; it allows accumulation of meaning and meaning-making (Dourish, 2006; Tuan, 1977). Activities that create place leave their mark and have persistence in the place. Zoom, being designed as videotelephony and conferencing software, is a dedicated point for communication and, as such, has no persistence as a space.

A chat server like Discord, however, offers a sense of participant ownership and playfulness, as well as some interesting place-making opportunities and potentials, as a result of its combination of synchronous and asynchronous communication. Discord has a persistent existence after a fashion. Student experience doesn't disappear, but traces remain that evidence process. Students will pop back to visit via a server, often if they find something that they think I might be interested in, sometimes just to say hello when they see I am online. It would seem that some software design, not unlike its counterpart in the physical world of architectural design, really can make a difference to experience.

9.3 DISCUSSION AND CONCLUSION

We have, as necessitated by COVID-19, and in rapid turnaround, replaced face-to-face, in-person learning with online learning. It was not an ideal situation for reframing the learning experience – but it has been thoughtful and very much the embodiment of Schön's "reflection-in-action" (1983). We relied heavily on technology in various ways: synchronous and asynchronous, passive and active; playfully, flexibly, reliably and creatively. We all engaged in exploring the "ready to hand" as Heidegger (1962) calls the practical and useful. We used what we had and drew on our expertise to adapt the ready-to-hand technologies and reframe the situation. Reflecting on this delivery has revealed some significant insights moving us forward as we travel into the next semester, still socially distant and reliant on online learning. These insights include:

- Virtual attendance is a challenge – getting students to "show up" to online classes. Sending multiple reminders has been found to be an effective mechanism. Reducing the length of the online class time and focussing topics on those necessitating interaction is another approach taken – and something which may also be making attendance more likely. Initial whole-group synchronous contact at the start of a session has also been advocated as it can pull the group together and get them on track
- Students who do not come to tutorials miss out on situated and social learning, such as when watching someone else go through design *moves* (Schön 1983). Since technical issues can also be constraining virtual attendance, educators have felt an onus to create offline catch up content. This is, however, a lesser experience, and risks becoming prescriptive/rule-based thinking rather than modelling design thinking. It also adds significantly to staff workload
- The interactive learning experience – studio or tutorial – has changed significantly. Video interaction is more intense and focussed than face-to-face, and it is possible to get through content more quickly. The converse of this is quicker fatigue and, with limited physical cues, less situated meaning-making and reduced spontaneity
- Lectures have been largely unchanged since medieval times and critiqued for their reliance on passive learning. Changing lectures to an online format creates an opportunity for more active learning. We expect that online lecture delivery that is active or episodic will stay a preferred approach for many educators moving forward
- Social distancing and lockdown affected student assignments and capacity for group work. Many staff switched to individual assessment. Fyfield and Evans, however, kept their group work, making the point that online collaboration is a growing norm globally
- The crisis experience created anxiety that impeded learning. It may be possible to create a playful experience alongside creative reflection on COVID-19, effecting light-hearted relief and, potentially, personal catharsis. Connectedness and mechanisms for spontaneity that are social can also address this, as can group work, ultimately benefitting well-being. Flexible delivery with active learning can afford agency

Overall, as we have found in our reflections, during crisis, there is a need for certainty to create the necessary conditions for learning – and certainty in how learning will occur is a significant aspect here. Our responses here were peculiar to, and leveraged, our discipline – designers necessarily engage with unpredictable and wicked problems of rapidly reframing for the unknown. We used a "design thinking mindset" in learning design – applying it to this crisis situation to improve outcomes for our students.

Design thinking must be critical in nature – and as we work and reflect we turn this lens on the technologies. These media influence the nature of the user's experience, the possible connections and quality of the meaning-making between people. As experienced designers in online learning we necessarily engage critically and experiment with technology and media, yet this is not without its challenges and dangers – as Willis (2006) attests: what we design and put out into the world has the potential to design us back. Creating online learning environments is far more than effective delivery, as the example of the students using Discord demonstrates, it is also about creating opportunities for "being human" through creativity and place-making.

During this time, staff discussions covered topics ranging from reducing cognitive load for students through to the need for technical support for the online learning environments. Other challenging issues include limiting access to the design fabrication workshop into prescribed timelines so as to manage the resource while adhering to new social distancing requirements, or limiting the range of online teaching platforms to pre-set designs or endorsed tools to reduce the complexity of technical support.

Reducing access to set times and set tools is, however, not without its drawbacks – design iteration happens at different cycles and tools inform thinking. The focus in the face of the pandemic has been on the pragmatic and the functional. Rationalising pressures are not uncommon in times of stress – typically budgetary, but also in a global health crisis like we experience here. The reach for simplicity is understandable – it can afford gains in mass deployment and a sense of control to manage anxiety within the uncertain situation. Fortunately, as designers, we have capacity and training to respond to uncertain situations and so-called "wicked" problems thoughtfully and through robust discussions, and we continue to do so.

Moving forward, despite some restrictions being lifted during next semester's delivery, we remain mostly online. Social distancing impacts the face-to-face collaboration, yet, as described, collaborative online interaction is increasingly common in design. Further, we believe that certainty and stability are critical to the students being able to learn and these factors point towards staying online. Social distancing to prevent disease transmission impacts student access to the fabrication workshop given its size and capacity; however, a more organic access model has been implemented that combines prescribed times with delegating activities to technical staff (from students), and prioritising certain units for booking slots by students throughout semester. This more complex solution better suits students' ongoing design iteration – a fundamental tenet of design practice.

The cognitive load of student users, as they may become overwhelmed with technologies, designs and platforms – while also needing to experience diverse modes in order to innovate in design – remains an important consideration. We have discussed that to *scaffold but not limit* educator decision making is a valuable way forward. For example, providing "sensible defaults"; that is, reasoned and contextually appropriate, can assist educators in selecting learning technologies and platforms, but should not limit experimentation across this emergent, transdisciplinary and critical space. Ongoing experimentation in digital media – beyond hegemonic tools and the values they impose – is extremely valuable.

Teaching in the time of COVID-19 has been challenging. Our direction and lessons come from design, which integrally involves thoughtful experimentation, speculation and critique. We conclude that, even under crisis, these should not be sacrificed to rationalising pressures. "*You may not control all the events that happen to you, but you can decide not to be reduced by them*" (Angelou, 2008). Despite the challenges of COVID-19 we value critical and thoughtful design with technology; not only is this key to our discipline, it is also who we are. Experience design is a broad space and the design of its technologies, interactions and communications will be limited only by the imaginations of our students. Experience design is what we, as design academics, do and teach as we lead the designers of tomorrow to define our collective futures.

REFERENCES

Anderson, T. (Ed.). (2008). *The Theory and Practice of Online Learning*. 2nd ed. Athabasca University Press.

Angelou, M. (2008). *Letter to My Daughter*. 1st ed. Random House.

Auslander, P. (2008). *Liveness: Performance in a Mediatized Culture*. 2nd ed. Routledge.

Blackboard Inc. (1997). *Blackboard Learn*. www.blackboard.com/en-apac

Butterfield, S., Costello, E., Henderson, C., & Mourachov, S. (2013). *Slack*. Slack Technologies. https://slack.com/intl/en-au/

Czaplinski, I., Sillence, M., Parsons, S., de Laat, M., Devine, C., Phillips, M., Fyfield, B., & Boman, M. (2017). What about me? Staff perspectives on the implementation of Intensive Mode of Delivery in an undergraduate science program. In R. Walker & S. Bedford (Eds.), *Research and Development in Higher Education: Curriculum Transformation—Volume 40: Refereed Papers from the 40th HERDSA Annual International Conference* (pp. 106–116). Higher Education Research and Development Society of Australasia. https://eprints.qut.edu.au/110954/

Discord Inc. (2015). *Discord*. https://discord.com/

Dorst, K. (2018). *Notes on Design: How Creative Practice Works* (Main edition). BIS Publishers.

Dourish, P. (2006). Re-space-ing place: 'Place' and 'space' ten years on. *Proceedings of the 2006 20th Anniversary Conference on Computer Supported Cooperative Work – CSCW '06*, 299. doi: 10.1145/1180875.1180921

Ek, D., & Lorentzon, M. (2006). *Spotify*. NYSE:SPOT. www.spotify.com/au/

Fyfield, B., & Czaplinski, I. (2017). By design: Facing the academic challenges of implementing technology enhanced learning in higher education and the example of a third year biology unit. In H. Partridge (Ed.), *Me, Us, It! Proceedings ASCILITE2017: 34th International Conference on Innovation, Practice and Research in the Use of Educational Technologies in Tertiary Education* (pp. 245–254). ascilite – Australasian Society for Computers in Learning in Tertiary Education. http://2017conference.ascilite.org/wp-content/uploads/2017/11/ASCILITE-2017-Proceeding.pdf

Goel, N. (2008). *Padlet*. https://padlet.com/

Heidegger, M. (1962). *Being and Time* (J. Macquarrie & E. Robinson, Trans.). Harper & Row.
Jarvin, L. (2015). Edutainment, Games, and the Future of Education in a Digital World. *New Directions for Child and Adolescent Development, 2015*(147), 33–40. doi: 10.1002/cad.20082
Johnson, S.A. (1999). *Interface Culture: How New Technology Transforms the Way We Create & Communicate* (Reprint, Subsequent edition). Basic Books.
Martin, N., Kelly, N., & Terry, P. (2018). A framework for self-determination in massive open online courses: Design for autonomy, competence, and relatedness. *Australasian Journal of Educational Technology, 34*(2), Article 2. doi:10.14742/ajet.3722
Microsoft. (2017). *Microsoft Teams.* Microsoft. www.microsoft.com
Němec, J., & Trna, J. (2007). Edutainment or entertainment. Education possibilities of didactic games in science education. *The Evolution of Children Play, 24th ICCP World Play Conference*, 55–64. /paper/Edutainment-or-Entertainment.-Education-of-Didactic-N%C4%9Bmec-Trna/a6370a469f23819207b809f335004cb94f85efbe
Robertson, T. (1997). Cooperative work and lived cognition. A taxonomy of embodied actions. *Proceedings of the Fifth European Conference on Computer-Supported Cooperative Work*, 205.
Schön, D.A. (1983). *The Reflective Practitioner How Professionals Think in Action* (1995th ed.). Temple Smith.
Schwartz, A. (2002). *Articulate Rise.* Articulate. https://articulate.com/360/rise
Seevinck, J. (2020). *Sample of Online Teaching Resource Created for Design Students* [Figshare]. doi: 10.6084/m9.figshare.12782972.v1
Subramanian, K.R. (2017). Product promotion in an era of shrinking attention span. *International Journal of Engineering and Management Research, 7*(2), 85–91.
Towers, R., Rodriguez, M. V., Roberts, A., & Gowty, S. (2020). *Sample of Online Studio Environment Created for Visual Design Student Critique and Feedback* [Figshare]. doi: 10.6084/m9.figshare.12782990.v1
Tuan, Y. (1977). *Space and Place: The Perspective of Experience.* University of Minnesota Press.
Weinman, L. (1995). *LinkedIn Learning with Lynda.com Content.* Microsoft. www.linkedin.com/learning/
Willis, A.-M. (2006). Ontological designing. *Design Philosophy Papers, 4*(2), 69–92. doi: 10.2752/144871306X13966268131514
Wright, R.D. (2014). *Student-Teacher Interaction in Online Learning Environments.* IGI Global.
Yuan, E. (2011). *Zoom.* Zoom Video Communications. https://zoom.us/

Part 2

Becoming a Leader

Part 2 is for those who are interested in leading in academia in the design fields, developing their own careers and those of others. It is intended to guide readers in growing as academics and in leading, mentoring and developing others. This is important because design has the potential to have a huge impact on the world, but designers are not traditionally seen as potential leaders. Therefore, opportunities exist to fill this gap, and, to grasp them, design academics need to strategically focus on leading their own careers, as well as leading various teams of their colleagues, students, industry partners and academic peers.

Part 2 starts with an internal focus on academics leading themselves in order to make the most of their time, energy and talent. It builds to a more external focus on developing a reputation and profile by leading others, both close to home and around the world, and then finally to leading a whole group or school.

Part 2 contains contributions from a diverse range of experienced design academics in various formal and informal leadership roles. Focussed on providing practical, real-world advice, the case studies, examples and ideas included are intended to guide readers in developing a leadership portfolio and, if desired, undertaking a substantial role as an academic design leader. This book will enable readers to move from design academic to academic leader in their field, whether they wish to be a global research leader, Head of a School or Faculty, or leader of a group of teachers or researchers, peers or students.

Section 4

Leading Yourself

This section focuses on the skills and attributes it takes to "lead yourself" as a design academic. Chapter 10, drawing on Brene Brown's concept of vulnerability, focuses on the importance of developing the courage and resilience to "dare greatly" in academia. As well as highlighting the importance of embracing a growth mindset, this chapter documents several practical strategies for developing resilience, tackling imposter syndrome and fostering work–life balance. Chapter 11 continues this focus on intentional career development and management, documenting how mid- and senior-career academics manage their time. Combined, while acknowledging that many design academics feel overburdened, tired, isolated and excessively busy, this section focuses on the tangible steps we can take to wrestle back some control over our work week and our careers.

10 How to "Dare Greatly" in Academia

Evonne Miller

CONTENTS

10.1 DEVELOPING THE RESILIENCE TO "DARE GREATLY"

In this chapter, I reflect on some of the lessons I have learnt and the practices I have adopted in the hope that they will help in building your academic career – and developing the confidence and resilience needed to "*dare greatly*" and lead in academia. I have separated this chapter into three parts, which I will address in turn: (1) developing resilience; (2) maintaining work–life balance; and (3) resilience and growth as an academic leader.

Absolutely, without a doubt, the one factor that predicts (sustained) success in academia is resilience. Others have talked about the challenging process of securing that elusive first academic job, a topic beyond the scope of this chapter (see Kelsky, 2015; Hume, 2010). While I focus on how you might better weather the "ups and downs" of academic life as a more senior academic, resilience is also critical in the job-hunt. We need resilience to find the energy to deal with and overcome the complaints, the rejections (I am thinking about you, Reviewer 2!), the dispiriting moments and our own missteps, as well as the negative and, at times, toxic personalities of our colleagues and the disappointments that are part and parcel of academic life. At the same time, as Clark and Sousa (2018) write in their honest, practical and inspiring book, *How to be a Happy Academic*, many academics feel overburdened, tired, isolated and excessively busy, struggling to learn the ropes and meet "sky-high expectations." So, what can we do?

We can learn much from the work of Angela Duckworth (2016), who has written extensively about grit, arguing that it is not innate talent that predicts whether we achieve our goals, but our perseverance and passion; as well as the work of Carol Dweck (2006), whose concept of growth mindset reframes failure as an accepted part of the learning process. We must work to embrace, as Swan and Rezayan (2021) describe in detail earlier in this book, the resilient growth mindset of "always learning."

Failing is inevitable in academia but, as Clark and Sousa (2018) remind us, wasting this failure is not. We need to purposively embrace Kolb's (2015) experiential learning cycle, which offers a model of learning something new, experimenting with the concrete experience of this new input, evaluating and then integrating or rejecting this new information – an iterative process of reflection, observation and active experimentation that enables us to grow from the next new piece of information. We must experiment, learn and grow – not stagnate. And, we must learn, as vulnerability and leadership researcher Brene Brown (2015) explains so eloquently, to *dare greatly*. The title of her best-selling book, *Daring Greatly* draws on the famous Theodore Roosevelt quote which argues that it is not the critic who counts or the person who points out the mistakes, but the credit must go to the person who is trying, failing, picking themselves up and knows that, if they fail, then they will do so while "daring greatly."

10.1.1 Strategically Identify and Amplify Your Strengths

If we are to develop the resilience needed to "dare greatly" as academics, then we must be thoughtful and strategic in our actions. We should start by doing an honest self-assessment about what our university or discipline/field expects of us, and where our unique skills, interests and talents lie. As the chapters in this book illustrate, design academics have crafted very different careers for themselves, with varying levels of teaching, research and engagement. It is up to you to decide what type of academic you wish to be, and to identify your unique contribution – ideally ensuring it aligns with your university's priorities.

Here, I am inspired by management guru Peter Drucker, who argues that we must do an honest self-assessment. There are, he notes, many areas in which each of us has no talent, skill or aptitude, and so we need to (1) honestly assess our talents; (2) figure out which activities best suit our interests, skills, personality and temperament; and, finally, (3) play to our strengths.

> One should waste as little effort as possible on improving areas of low competence … Do not try to change yourself – you are unlikely to succeed. But work hard to improve the way you perform. And try not to take on work you cannot perform or will only perform poorly … Successful careers are not planned. They develop when people are prepared for opportunities because they know their strengths, their method of work, and their values. Knowing where one belongs can transform an ordinary person – hardworking and competent but otherwise mediocre – into an outstanding performer (pp. 101, 104–106, Drucker, 2005).

Clark and Sousa's (2018) success pyramid provides a valuable framework and process to help us think through our career identity, values, and the critical indicators, priorities, goals and tasks. Drawing on Duckworth's (2016) concept of grit, Clark and Sousa (2018) remind us that academic success, whether in research, teaching or engagement, is "achieved far more by good habits than talent or skills," and by "the disciplined pursuit of what really matters" (p. 13 and p. 70). Similarly, Karen Kelsky, in her influential book, *The Professor Is In*, shares a reader's story that illustrates the importance of learning to focus on what matters, and not being distracted by easier but less significant activities. In the excerpt below, one academic describes doing a lot of "busy work," but putting off and not doing the harder, big-ticket items – in this example, high-quality research outputs. I know that this story resonates with me, as I am sure it does with many of us, illustrating how we may have a tendency to pick the lower risk and easier path; we must, however, develop the resilience and wisdom to work on what really matters.

> During my years as a tenure track assistant professor, I went about publishing and doing research the way I do the grocery shopping: concentrating on the sale items like conferences, book reviews, on-line collaborations, i.e., all things that seemed 'affordable.' As a result I stock up on unnecessary items and find myself too tired to focus on the important things, those items that do not go on sale, but that are the building block of a good kitchen: articles and books. Although I have managed to publish quite a bit, I have squandered a lot of time and energy, because I did not have a clearly elaborated research plan. One high-risk, high-cost item—a book proposal successfully written and pitched to the leading press in your field, or a journal manuscript dragged through the excruciating review process of the top journal in your area—is worth ten cheap alternatives such as book reviews (Kelsky, 2015, p. 83).

10.1.2 Developing Support Systems and Sharing Failures

Part of the reason we often aim lower than we should is because we fear the critical critique that is part of academic life. Whether it is the brutal feedback from Reviewer 2, grant rejections or tough evaluations from students, as academics, we are relentlessly exposed to critique and criticism which challenges our identity and worth (Gabriel, 2010). To succeed in our chosen profession, therefore, each of us must learn to develop resilience. And to be resilient, we need the support of our colleagues. We need to know that they too have been rejected from that journal, that they too hate Reviewer 2, and how they have managed to pick themselves back up. Clark and Sousa (2018) specifically encourage us to share more, to tell our colleagues about both our accomplishments and our failures. Luckily, academics are becoming much more open about the many bumps in their careers – and how they have navigated them.

The "*CV of Failures*" is one such initiative. Explaining that academics too often "construct a narrative of success that renders our setbacks invisible both to ourselves and to others," so successful careers "seem to be a constant, streamlined series of triumphs" (p. 467), Stefan (2010) encouraged academics to keep – and share – an alternative CV of their failures. While I don't currently keep a formal record of unsuccessful grant applications, I would estimate a 10–20% success rate – or lower.

And it really is a lottery. I remember one year, a grant of mine was in the top 90%, but not funded. *Very, very, very close.* I resubmitted the same grant the following year: bottom 50%, not funded. One *very* successful colleague told me last year he has applied, unsuccessfully, for 17 Australian Research Council (ARC) Discovery grants since 2005. That is resilience. I do vividly remember one ARC reviewer, early in my career, that called me a "shallow, dangerous theorist" – referring to a proposal about sustainability and the legal system (an unfair assessment, I thought!).

We need to be more open about these knockbacks and the strategies we use to deal with them. I still vividly remember one professor who taught me years ago in graduate school sharing how she, and all students in her lab, posted journal article reviews publicly on their lab door – this public sharing helped take "the sting" out of negative comments. I don't do that yet, but, in the spirit of openness, here are some recent comments I have received: "*I have several serious concerns with this article which I will outline below*"; "*This is where the article is at its weakest*"; "*this paper provides an OK*

rationalisation for the study and for using qualitative research"; "*there are a number of typos (extra words/missing words)*" and, most cuttingly: "*all of these issues come from a core problem clearly present throughout the manuscript. Basically, to be blunt, the author does not really know the scholarship dealing with X, and the issues I have with this paper are as follows …*"

That last comment really stung, as it was in a field I know well – and I have over 100 peer-reviewed publications. I wondered if perhaps I had not cited that reviewer's work and, indeed, they did suggest I read and reference a book I was not familiar with. You may be surprised to know that I addressed those issues in a strong revise and resubmit, and successfully got that paper published. No matter how tough the review, I will always reply to a revise and resubmit decision. That's resilience in action. Several years ago, I submitted a paper to a journal, and got a *very* tough rejection. Last year, they announced a special issue on that topic. I resubmitted the *same paper* (with some minor modifications) that had been rejected in 2016, and it was accepted – with revisions. The point I am trying to make is that you are very likely to receive this message from journal editors: "*I am sorry to have to deliver the unpleasant editorial news that I must decline to accept this paper for publication.*" To succeed in academia, you need to develop and build the *personal* resilience to keep going in light of constant rejection.

We also must develop a network of supportive peers. In her chapter, Blackler (2021) discusses her experience setting up regular "*Shut Up and Write*" sessions and *Promotion Club* – a safe, supportive place where academics in our school intending to apply for promotion to Associate Professor or Professor can share insight and knowledge. Often, academics treat the promotion process in a very confidential manner; Promotion Club disrupts that script, with an overwhelming positive impact and an 80% success rate so far. The quality of applications improves as people openly share ideas about structure and narrative, and if a person is unsuccessful in their application, they do not suffer alone: their sadness is acknowledged, constructively. Interpersonal relationships have developed and deepened, and these personal, emotional connections benefit everyone. Successful applicants are celebrated;

unsuccessful applicants are encouraged to seek feedback about how to improve, and the team supports them to move forward, drawing on their own knowledge and networks. People see others fail, pick themselves up, learn, go on and improve. It is inspiring, and helps each of us build the resilience we need in academia.

10.2 FINDING AND MAINTAINING THAT ELUSIVE WORK–LIFE BALANCE

Sustaining work–life balance is a challenge for many academics – but critical for crafting resilient, happy lives. As I drafted this chapter in November 2019, a Twitter war on academic work–life balance broke out. Prominent UK academic, historian Mary Beard, tweeted that she worked 100 hours a week. While many replied that they did as well, many others disputed the claim, asking how that is mathematically possible and querying the sustainability of such working patterns. In response, Professor Beard explained she worked from 6am to 10pm every day, skipping lunch and having a dinner break with her husband. She lamented a desire to relax and drink more port in the evenings, but felt an obligation to share her knowledge with others; for example, responding personally to school children emailing her for advice on projects.

I appreciate both views. There is no doubt that, at certain times during our career, academics will need to devote long hours to the job – whether it is developing a new course or unit, submitting a research grant or meeting a book manuscript deadline. Academics are usually personally passionate about the topics they teach and research, and, at times, the boundaries between work, life and leisure are extremely blurred. However, I am acutely aware of how important it is for us "to consciously develop, retain and defend a life outside of work" (Ellison, 2019, p. 105).

We all need some major interests outside work because, as Drucker (2005) explains, no one "can expect to live very long without experiencing a serious setback in his or her life or work, whether it is tragedies in one's personal life or being passed over for promotion" (p. 109). Whether it is alpaca farming, yoga, mountain biking, sewing or voluntarily coaching at the local soccer club, each of us needs to find and pursue our personal passions and meaningful leisure activities outside of work. Academic and social anthropologist Peter Worsley (2008), in his memoir *An Academic Skating on Thin Ice*, explains how buying and maintaining a second-hand narrowboat, and subsequently exploring the over 2,000 miles of English canals, helped to maintain his "mental equilibrium during the bad times" (p. 172).

While many academics do as billionaire investor Ray Dalio (2017) advises in his book *Principles*, making their passion and their work the same thing, research highlights that non-work passions helps people develop a *sustainable career*. There is a whole field of research exploring what a sustainable career looks like, with De Vos, Van der Heijden and Akkermans (2019) defining it is the integration of career, personal passions and family, where an individual purposively protects their well-being and productivity. Engaging in leisure can help us "switch off," and develop valuable new social networks, perspectives and value outside of our work identity. Maintaining a strong non-work identity is also particularly valuable later in life as retirement beckons. Many of us may have observed older family members and colleagues view retirement with anticipation and/or fear. While it may feel decades away, each of us must at some time in our lives ensure we have thought about and prepared for our post-work life and identity. Retirement, as Ellison explains, is:

> a different kind of threat, that of exchanging constant occupation with a vacuum. This ending can trigger grieving rather than feeling rewarded for years of dedication in the field. Whilst carrying out important work, there may have been no time, perceived need or energy for anything outside of this. But at the end of a career, is one's contribution remembered? (Ellison, 2019, p. 106).

10.2.1 Why Rest, Relaxation and Rejuvenation Matter

As well as making time for non-work activities and leisure, we must also make time for rest. In her thoughtful book documenting innovative leadership behaviours and practices in a diverse array of global multi-ethnic settings, from humanitarian aid to sport, medicine and prisons, Ellison (2019) notes that – if we are to develop and maintain the resilience needed to cope with both crises and the "daily grind," we *must* prioritise rest and relaxation. Elsewhere in this book, Chamorro-Koc and Caldwell (2021) have reflected on different experiences of "*running the academic marathon*," hinting at the importance of timing and taking breaks. Ellison describes one particularly memorable example in her book, of how Phil the Prison Superintendent has slowly learnt to better manage the stress of prison riots, which may involve blame, fear, unprocessed emotions and deaths. Realising that he "becomes snappy, tense and short-fused with people in the days straight after an incident" (p. 71), Phil has learnt to monitor and respond to "mental and bodily signs indicating to him that he needs to prioritise his personal wellbeing" (p. 71), taking a few much-needed days off work post-riots to reduce his stress by walking his dog and taking his children to school.

Everybody's stress pressure points and reduction mechanisms will be different. And while academics are not dealing with prison riots like Phil, we can be inspired by his example and proactively monitor our stress levels and appropriate responses, taking conscious steps to ensure we don't burnout. Many years ago now, I remember reading the reflections of a senior executive from a consulting company: he said consultants would come to him to quit, citing stress and over-work. Checking, he would find that, invariably, they would have accrued months of unused annual leave. His argument was that, instead of getting to the point where quitting was the only option, as knowledge workers, we need to try to implement our own boundaries – and take some of that leave now, rather than simply accruing it. Anticipating peak workload times, such as marking and grant deadlines, academics must plan to take leave afterwards and/or before they know a busy period is coming up.

Too many academics have weeks of unused leave, often lamenting it is too difficult to take large blocks of time off. Personally, I have found that simply planning to take one day and a quarter and creating an unexpected long weekend has surprising benefits, with that brief time of relaxation and respite enabling me to "physically and emotionally refuel" (Ellison, 2019, p. 73). Over the past few years, I have also built in a weekly routine for relaxation: Wednesday night is "my time" – I joined a gym with a steam room and sauna that also offers restorative Yin Yoga. Enjoying that, after a group workout, is now my Wednesday-night ritual: 4.30pm bootcamp, 5.30pm steam room, 6.30pm Yin Yoga, followed by dinner alone or, sometimes, a massage or facial. The steam room has become my reflection place, where I have practised my promotion interview answers and frequently pondered how the day has gone.

That's my personal routine – and to sustain ourselves, in the constant busy-ness that is contemporary academia, each of us needs to find our own way to build rest, recreation and relaxation time into our daily lives. Whether it is binge-watching *Breaking Bad*, a weekend camping in a national park, running a marathon or booking into a one-day art class, sustaining ourselves requires consciously and proactively planning times to self-nurture, as, during a crisis – such as the COVID-19 pandemic – (and I would also argue, just the normal work–life juggling act that is contemporary academia) "there may not be time or mental 'bandwidth' to do this" (p. 73, Ellison, 2019).

I must acknowledge, I am nowhere near perfect – I am frequently rushing, late to yoga class or cancelling at the last minute. But I am trying. I am putting steps in place to build in opportunities for exercise and rest. I am also purposely "*boxing myself in*" by making commitments that force me to leave the office on time. For example, I book in for, and endeavour to attend, a 4.30pm group boot camp, two to three times a week; sometimes I cancel, but, for the most part, I get to class and ensure my mental and physical health is more important than shorter-term work deadlines and dramas. No one else can do this for you, because all academics:

> feel their work is vital, which it often is. They may not notice a depletion in their energy, compensating for it with snacking on sugar-loaded foods, drinking more alcohol, not exercising or, conversely, becoming hyperactive. Some people find they cannot rest. They cannot calm down. They are addicted to busy-ness. In this regard, leaders must develop a habit of gently monitoring those around them. Spotting people who deny their level of exhaustion, or who may be using work as a source of self-worth, identity or as an avoidance strategy for something going on at home, can be a useful informal barometer of psychological health (Ellison, 2019, p. 73).

10.2.2 Mental Health

A detailed discussion of mental health in academia is beyond the scope of this chapter, but we know it is problem: a recent survey of British academics found 43% exhibited symptoms of at least a mild mental disorder (Gorczynski, Hill, and Rathod, 2017), and many workplaces – including my own – are introducing mental health first aid training.

Part of looking after our own mental health is building in boundaries and routines, and fighting the implicit and explicit expectations and narratives of what a "successful academic" looks like. As Brene Brown explains, we need to let go of and challenge any assumptions that exhaustion, busy-ness and extreme productivity are status symbols of achievement – rather, each of us needs to model appropriate boundaries; for example, by shutting off email and prioritising a life outside the office.

Of course, this is not easy – and no one is perfect. There will be times when we fail, and, in fact, I have seen many debates about the challenges for people who are working part-time and flexible hours. For example, if you work late at night so you can pursue a hobby or watch your children's soccer (i.e., it suits your work–life pattern), some have suggested adding words such as these to your email signature, explaining your non-standard email time: "*I work flexibly, and email outside standard office hours. I do not expect you to.*" This is something I do. In 2017, France actually introduced a new law that established workers' "right to disconnect"; to reduce burnout and protect private time. Companies with more than 50 employees must establish hours when staff should not send or answer emails. I wouldn't have thought it mattered, but now that I reflect on it, I used to have a boss who would send emails at all hours of the night and weekends. And you would feel that you should reply; I have noticed a difference now I no longer work closely with them, and so am personally *working towards* trying to only send emails during business hours.

At a personal level, I also know that engaging in some time-out behaviours (exercise, yoga, dinner alone) recharges my batteries, reduces my stress and the likelihood of burnout, while maximising my productivity. When we are stressed-out, tired and running on empty, tasks take longer and we make mistakes. In many ways, therefore, self-care – whatever that looks like for you – is a critical step in developing a sustainable career as a design academic. We must, as Drucker, notes learn to play the long game and "learn to develop ourselves. We will have to place ourselves where we can make the greatest contribution. And we will have to stay mentally alert and engaged during a 50-year working life" (2005, p. 100).

10.3 RESILIENCE AND GROWTH AS AN ACADEMIC LEADER

While other chapters on this book focus specifically on leading a design school (Scharoun and Muratovski, 2021) and research leadership (Foth and Rittenbruch, 2021;Blackler, 2021), in closing, I also want to briefly reflect on the importance of setting boundaries, of holding feedback "lightly," and growing as an academic leader. All of us will have, at various times, thrived and suffered under different academic leaders, for, as Goffee and Jones (2000) note, many people sadly "don't have the self-knowledge or the authenticity necessary for leadership" (p. 5).

If you are appointed into a formal leadership role, I strongly advise you to seek out training and engage in relevant reading. In their wonderful 2000 Harvard Business Review article *Why Should Anyone Be Led by You?* (and subsequent 2015 book), Goffee and Jones (2000; 2015) identify that, alongside energy, skills and strategic vision, leaders who "excel at inspiring people—in capturing hearts, minds, and souls" (p. 2) share four unexpected qualities: they (1) *selectively show their weaknesses* (and by showing that they are imperfect, just like you, they highlight their humanity, authenticity and approachability. Goffee and Jones note that the shared flaw should not be a critical flaw (for example, not being able to read spreadsheets if you are a financial advisor), but a flaw that emphasises your humanity; (2) *rely heavily on intuition*; (3) *manage with tough empathy*, giving people what they need, not necessarily what they want; and (4) *reveal their differences* and dare to be different, building "on what's unique about themselves" (p. 1). After all, as Goffee and Jones so astutely note "the challenge facing prospective leaders is for them to *be themselves, but with more skill*" (2000, p. 7).

Leadership is not easy. Ellison has called for higher ethics in leadership, arguing for "inward honesty and outward consistency," for "revealing bad practice and amplifying good leadership," and for "greater diversity of thinking" (p. 150). I agree. But no one is perfect, and we often learn the most through our failures. I have held different research leadership roles for nearly a decade, and it is my failures – when I send an email rather than setting up a meeting, when I let my hot head momentarily get the better of me, when I made a flippant remark in humour rather than pausing to listen – that have taught me the most. I am continuing to learn, grow, develop and change as an academic leader. And just as we assign our students books and journal articles to read in order to grow, I am building my own knowledge through reading time-management and leadership books – a practice I highly recommend. Currently, I am reading *Leadership on the Line*, which has really resonated as it talked about separating yourself from issues, something I struggle with:

> Whether you like it or not, you will embody issues in the eyes of other people, and sometimes they will attack you when they see you running with the ball down the field … when you lead, people don't love you or hate you. Mostly they don't even know you. They love or hate the positions you represent. Indeed, we all know how quickly idealization turns into contempt when suddenly you disappoint someone (Heifetz and Linsky, 2017, pp. 196–198).

Instead of making up stories in our heads about something that has happened, and thinking and talking about an issue with our friends and family members for hours, we must learn to calmly confront what scares us because, as Brene Brown so wisely notes, "it saves a tremendous amount of time and mental capacity to just turn around, face and confront whatever is at our heels head-on" (Brown, 2018, p. 297). This, developing the ability to have brave conversations, to name an issue and discuss it, is a skill I am slowly developing. If, like me, you find it difficult to have tough conversations, I recommend experimenting with the DEAR MAN strategy from Dialectical Behavioral Therapy (Linehan, 2014), an evidence-based cognitive-behavioural psychotherapy approach that emphases mindfulness, interpersonal effectiveness, acceptance and emotional regulation. The DEAR MAN acronym is designed to help you healthily assert yourself to others, with each letter representing a different step to take in a conversation:

D: Describe the situation (be objective, non-emotional and stick to the facts, like a good reporter would)

E: Express how you feel about the situation (remember to use "I" sentences, don't blame and say "you," which puts the other person on the defensive)

A: Assert yourself (ask directly for what you want. This can feel uncomfortable, but remember people are not mind-readers)

R: Reinforce your request (make sure the other person knows why they should grant your request)
M: Mindful (don't be distracted, stay focussed on this moment and this specific situation)
A: Appear Confident (regardless of how you feel inside, present yourself confidently)
N: Negotiate (remember, this is a conversation about resolving the problem together)

I have found that the DEAR MAN structure is quite helpful in thinking about how to approach a difficult conversation or resolve a conflict, and it also seems to enable me to assert myself and my preferences more calmly. DEAR MAN makes having hard conversations much easier, and I highly recommend experimenting with this approach. Additionally, most universities have free therapy available for staff, and seeking help from a professional is very often a positive experience.

10.4 FINAL REFLECTIONS ON RESILIENCE, RELAXATION AND GROWTH

In closing, this chapter has reflected on my experiences in academia – my continual battle to maintain an appropriate work–life balance, the importance of resilience and a learning, growth mindset, what I am learning from contemporary leadership literature, as well as some practical tips for managing stress, aiming for a work–life balance (or integration) and approaching difficult conversations. The challenge of academia, as others have described, is that it is at least three jobs in one – being an educator, researcher and administrator. To survive and thrive, each of us must identify our unique strengths, and being passionate, resilient and brave enough to simply keep trying. Like Brown (2018), I too want to look back on my life and say "I contributed more than I criticized" (p. 259) and I entered the arena and risked getting a little dirt and blood on my face.

However, sustaining our career passion requires resilience and a focus on rest and rejuvenation. We must preserve our unique "life juice—your creativity and daring, your curiosity and eagerness to question, your compassion and love for people—can seep away daily as you get beat up, put down, or silenced" (Heifetz and Linsky, 2017, p. 225). And so, I end this chapter with a challenge: I challenge you to pick something from one of the chapters in this book to experiment with – whether it is setting up a promotion club in your own department, reading a leadership book that intrigues you, prioritising your health and wellness by "boxing yourself in" to a weekly art class next semester, or trialling the DEAR MAN acronym, please take one purposeful step towards a more balanced, resilient and happier academic life.

REFERENCES

Blackler, A. (2021). Research leadership: Local and global. In A. Blackler & E. Miller (Eds), *How to Be a Design Academic.* Taylor & Francis, 211–226.

Brown, B. (2015). *Daring Greatly: How the Courage to Be Vulnerable Transforms the Way We Live, Love, Parent, and Lead.* Avery.

Brown, B. (2018). *Dare to Lead: Brave Work. Tough Conversations. Whole Hearts*. Vermilion.
Chamorro-Koc, M. & Caldwell, G. (2021). Running the academic marathon: Planning and executing as planned. In A. Blackler & E. Miller (Eds), *How to Be a Design Academic*. Taylor & Francis, 189–207.
Clark, A. & Sousa, B. (2018). *How to Be a Happy Academic: A Guide to Being Effective in Research, Writing and Teaching*. Sage.
Dalio, R. (2017). *Principles*. Simon & Schuster.
De Vos, A., Van der Heijden, B. & Akkermans, J. (2019). Sustainable careers: Towards a conceptual model. *Journal of Vocational Behavior*. Advance online publication.
Drucker, P. (2005). Managing oneself. *Harvard Business Review*, January, 100–109.
Duckworth, A. (2016). *Grit: The Power of Passion and Perseverance*. Scribner/Simon & Schuster.
Dweck, C. (2006). *Mindset: The New Psychology of Success*. Random House.
Ellison, R. (2019). *Global Leadership and Coaching: Flourishing under Intense Pressure at Work*. Routledge.
Foth, M. & Rittenbruch, M. (2021). Building a research group. In T. Blackler, & E. Miller (Eds), *How to Be a Design Academic*. Taylor & Francis, 289–306.
Gabriel, Y. (2010). Organizational studies: A space for ideas, identities and agonies. *Organizational Studies*, *31*(6), 757–775.
Goffee, R. & Jones, G. (2000). Why should anyone be led by you? *Harvard Business Review*, September-October, 63–67.
Goffee, R. & Jones, G. (2015). *Why Should Anyone Be Led By You? What It Takes to Be an Authentic Leader*. Harvard Business Review Press.
Gorczynski, P.F., Hill, D. & Rathod, S. (2017). Examining the construct validity of the transtheoretical model to structure workplace physical activity interventions to improve mental health in academic staff. *EMS Community Medicine Journal*, *(1)*, 1–4.
Heifetz, R. & Linsky, M. (2017). *Leadership on the Line: Staying Alive Through the Dangers of Change*. Harvard Business School Press.
Hume, K. (2010). *Surviving Your Academic Job Hunt: Advice for Humanities PhDs*. Palgrave Macmillan.
Kelsky, K. (2015). *The Professor is in: The Essential Guide to Turning Your PhD into a Job*. Three River Press.
Kolb, D.A. (2015). *Experiential Learning: Experience as a Source of Learning and Development*. Pearson.
Linehan, M. (2014). *DBT Skills Training Manual*, 2nd ed. Footprint.
Scharoun, L. & Muratovski, G. (2021). Leadership of a design school. In A. Blackler & E. Miller (Eds), *How to Be a Design Academic*. Taylor & Francis, 307–326.
Stefan, M. (2010). A CV of failures. *Nature*, *468*, 467.
Swan, L. & Rezayan, L. (2021). Exploring the mindset of a new academic. In A. Blackler & E. Miller (Eds), *How to Be a Design Academic*. Taylor & Francis, 83–111.
Worsley, P. (2008). *An Academic Skating on Thin Ice*. Berghahn Books.

11 Running the Academic Marathon

Planning and Executing as Planned

Marianella Chamorro-Koc and Glenda Amayo Caldwell

CONTENTS

11.1 INTRODUCTION

What comes to mind when you hear the following phrase? 'Academic work is, by its very nature never finished' (Morrish, 2019, p. 5). For us, this phrase resonates with our everyday lives as Design academics. It brings to mind those casual hallway conversations about the perennial long night hours of work to meet deadlines; the juggle between parenting duties and trying to fit it all in the day; the writing of papers en route to a conference; the fake holidays or the days where we take leave but work from home. It also brings to mind that, for most of us in the Design disciplines, we feel fortunate to do this job that allows us to work on an area of knowledge that we love. However, as indicated in a recent global university survey (The World Universities Ranking, 2018) and literature on academic labour (Hall and Bowles, 2016; Morrish, 2019), the darker side to the busyness of academic work – including academics feeling overwhelmed with multi-faceted performance expectations and struggling with mental health – has been particularly noted in recent years.

We – the authors – are academics in Industrial Design and Architecture at Queensland University of Technology. We have young families and we push ourselves to "do it all" without compromising our family lives. As mid-career academics, we have experienced the effects of "the anxiety machine": a term coined by Halls and Bowles (2016) in reference to the university as a system driven by productivity and income metrics to reflect academic performance. In this time of higher performance expectations, we view our careers as *Running the Academic Marathon* – a permanent form of engagement with our academic work that requires us to be mindful and effective with how we work, a topic further explored in this book in Miller's (2021) chapter on resilience. The present chapter, however, is about *Planning and Executing as Planned*. It shows some of the strategies that work for us to make our workload viable while running the academic marathon; it also demonstrates that mapping our academic journeys can provide valuable perspective to our career planning. In the following sections, we introduce the current context of academic work and our concept of running the academic marathon. We then discuss our methodology and findings of the unique paths and workload experiences at early career academic, mid-career academic and professorial stages. Finally, we conclude by providing seven strategies for "getting things done" in the midst of ever-increasing performance expectations and metrics.

11.1.1 The Context of Academic Work: Running the Academic Marathon

In a relatively short period, working in academia has shifted from a focus on traditional scholarly work to a focus on performance parameters where project management skills are paramount to manage rapidly increasing academic and administrative responsibilities. Higher education is a competitive market and, in Australia and globally, government cuts to university funding have resulted in disruptive change (Mukerjee, 2014) to our priorities and the work we do as academics. Some of these changes are driven by digital innovations, which impact our teaching and research activities (McCluskey and Winter, 2012); others are driven by attitudes about the demographic composition of the academic workforce (Aiston and Jung, 2015) and of leadership positions (Lewis, 2018). All of this translates into our academic work becoming increasingly pressured by higher accountability and a focus on key performance indicators as a measure of our excellence (Deem et al., 2007; O'Connor and O'Hagan, 2016). In our view, and as articulated within several chapters of this book, such changes are deeply transforming the ways in which academics plan their careers and how they feel and perform in their roles.

Literature on academic performance highlights that academics feel a disconnection between career planning and meaningful work (Lewis, 2018; Lieff, 2009), and that the complexities around meeting performance metrics lead to occupational stress (Gillespie et al., 2001; Gorczynski, Hill and Rathod, 2017; Winefield and Jarrett, 2001). In addition, like Rafnsdóttir and Heijstra (2013), we believe that this sense of under-accomplishment is exacerbated by limitations experienced in dedicating time to other important components of both academic work (such as supervising and mentoring colleagues) and personal life (such as being a spouse, a parent, a carer, a sibling, a human). It is in this context that we construct the metaphor of *Running the Academic Marathon* to describe the academic career, for which an efficient process of *Planning and Executing as Planned* is critical to maintain good performance, reduce stress and lead impactful careers.

A marathon is a long physical, mental and emotional effort that requires training, practising and focussing – just like our academic careers. We like using the marathon metaphor to differentiate it from running a "sprint": an endeavour that consumes all our energy, focus and strength to achieve a goal in the shortest possible time. In academia, we work on multiple tasks in parallel – strategise, prioritise and perform in teaching, research and service – which require multiple types of sprints to succeed in short-term goals or projects. As academic work is a continuous endeavour, the results of those multiple sprints form part of our running of the academic marathon. The results of our sprints give us an idea of our overall performance every year, usually at the time of our annual performance review.

In sports, running marathons requires stamina, which is the strength and energy needed to exert oneself for an extended period of time. Building stamina requires exercising, planning, keeping a healthy lifestyle, consistency and minding recovery times (see Miller, 2021). We can compare the building of stamina in sports to maintaining good performance in academia. This is an important relationship to consider as studies demonstrate that, with higher performance expectations, academics tend to work without rest and keep pushing themselves to accomplish tasks at the cost of their own wellbeing (Gillespie et al., 2001; Gorczynski et al., 2017). Gillespie et al.'s study across 19 Australian universities shows that approximately 25% of those sampled were experiencing high levels of stress or anxiety (2001). Since then, figures indicate an increased rate of poor mental health in academia, as shown in a study across UK universities, which identified that over 45% of academics suffer from mild mental health issues (Gorczynski et al., 2017). Stressful workloads lead to mental burnout, and, similar to performing in sports, building stamina in order to run the academic marathon while doing meaningful work is critical for the wellbeing of academics and their potential to meet career goals.

A growing number of studies in academic productivity (Chase et al., 2013; Clark and Sousa, 2018; Pencavel, 2014) demonstrate the importance of understanding how much and how we work in order to avoid experiences such as burnout. For example, a 2018 study by Koens et al. revealed that, although Dutch academics spend most of their time doing research, it is not on "their" research. The higher the position, the more prominent this was – Dutch professors, for instance, would work a 55-hour week and spend 17% of their time on research, but only 9 hours and 20 minutes on their own projects.

Similarly, Robinson's compilations of over 150 years of research in productivity (2009; 2012) summarises that when people spend more than 50 hours a week working, they take more time to accomplish work tasks than if they were to work a normal 40-hour week. As trends on performance indicators continue to increase academic workloads, we tend to take on more work rather than strategising around the tasks that are most meaningful to us. This shows that *Planning and Executing as Planned* is an important mindset in order to achieve the goals that that are meaningful to our academic careers. While there is a vast number of tools available for time management, these are not specific to the needs of academics at different stages of an academic career.

Running the academic marathon is a complex task that is different for academics across their different levels of experience, their varying workloads, interests and according to their work styles. The academic career is one of progression that is signalled by three distinct stages: early career academic, mid-career academic and professor. There are critical milestones at each stage (Hall, 2017). For example, for the early career academic, the transition between completing a PhD and finding a permanent academic position can be one of the most challenging periods of their career. For mid-career academics, securing research income through prestigious national grants is a critical achievement. The following section outlines our methodology for understanding the patterns of work at each of the three stages of an academic career in Design. From these patterns we produce seven strategies for *Planning and Executing as Planned* to maintain stamina, work effectively and achieve meaningful goals when running the academic marathon.

11.2 METHODOLOGY

As Design academics, our teaching and research work is driven by Human-Centred Design. This approach understands people's perspectives, needs and experiences as the centre of design innovations. To understand the academic marathon, we therefore adopted a Human-Centred Design perspective with a focus on how Design academics plan and execute their work tasks. We drew from our own academic experiences as mid-career academics and captured insights from colleagues at the early career academic and professorial stages. With QUT Ethics Approval, we firstly conducted a survey to gather insights from early career colleagues. The survey consisted of 16 questions about the composition and yearly distribution of their workload, the way they organise it, and compared their emotional perception of their academic workload and the significance of the tasks that comprise it. We then used the same survey to interrogate our own experiences as current mid-career academics to reflect upon our past experiences as early career academics. From the understanding gained of how Design academics navigate their work, we generated academic journey maps for one of our early career academic colleagues, from one author's past experience of being an early career academic, and for both authors in their current experience of being mid-career academics.

These maps visualise patterns of the early career and mid-career academic workloads, and help to identify pain points in the academic year. We laid out the maps based on the academic calendar. In Australia, the academic calendar starts with a pre-semester time from December to late February; in some courses, this period constitutes summer semester. Semester One typically begins in the last week of February, with 13 weeks of teaching and 2 weeks of assessment. A mid-year break occurs for approximately 4–6 weeks from mid-June to mid-July. Semester Two begins in late July and ends in late October, with final exams completed by mid-November. We also laid out the maps with reference to the type of tasks that the workload comprises; in Australia, the workload typically includes teaching, research, Higher-Degree Research student supervision, leadership and committee participation, writing and conference travel.

With reference to both the academic calendar and the typical academic workload, we traced each individual academic experience by identifying the moments where meaningful activities were achievable (high points), and the moments where the clashing of many activities impeded progress towards personal goals (low points). We converted these responses into lines that we traced with different colours per task and mapped across key segments of the academic calendar. We then distinguished the meaningful tasks from the routine ones.

To gather insights from Design academics in the professorial stage, we interviewed (through audio recording) three professors and asked: "In your role, how do you Plan and Execute as Planned?" We present several of the answers in a table and discuss the recurring themes. Both the maps and the discussion of the interviews show that the academic marathon academics at all three stages are running. Our examination of experiences of each stage is essential for gaining a broad understanding of how Design academics work in order to contextualise the strategies we present for *Planning and Execute as Planned.* We purposefully scoped this study to only focus on how academics work and not on how academics navigate their careers, which is an important issue in career planning and the subject of other chapters in this book (see, for example, Evans et al., 2021; Miller, 2021; Scharoun and Muratovski, 2021).

11.3 EXPERIENCES OF RUNNING THE ACADEMIC MARATHON

The following sections present academic journey maps from three Design academics at early and mid-career stages.

11.3.1 Early Career Academic Journey Maps

Figure 11.1 reveals the journey map of an early career academic who, after completing their PhD and doing a post-doctorate and sessional teaching for three years, transitioned into an ongoing lecturing position. The map indicates that during the last year teaching tasks and workload take up most of their time. Pain points are identified during the teaching semester where meaningful teaching and research tasks

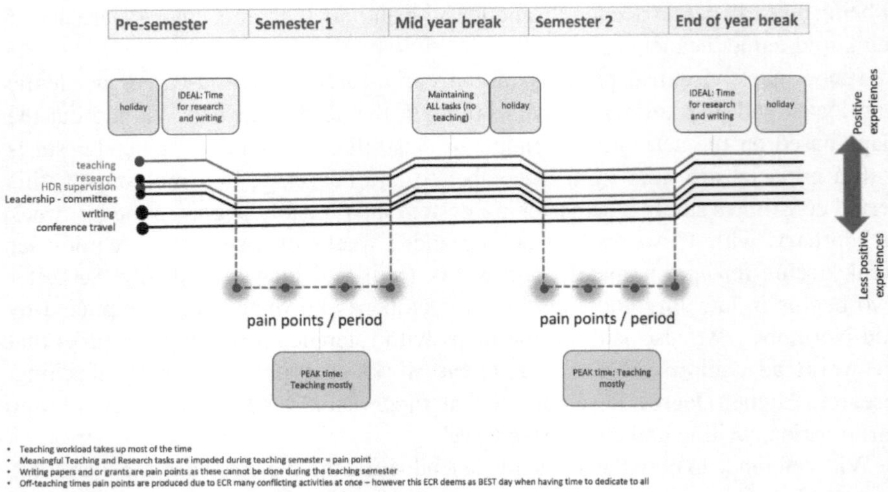

FIGURE 11.1 Early career academic journey map – Academic A

were impeded by the teaching load. Writing papers and grants are also pain points as they were challenging to complete during teaching semesters. Break periods or off-teaching times also become pain points as this is when Academic A is faced with conflicting activities at once. Overall, the journey map shows that Academic A is dissatisfied with their current academic workload.

When asked what tools Academic A uses to manage tasks, they referred to note-taking applications such as Apple Notes and Google Keep, the team management software Miro and the time-tracking application Jiffy. This academic also referred to their ability to adapt to emerging opportunities as a way to overcome the pressure of not having time, and to their capacity to execute without planning as a way of engaging in new opportunities despite the workload.

Figure 11.2 presents the early career journey map of one of the authors based on a retrospective account of entering academia a decade prior to Academic A. Similar to Academic A, the map indicates that, for Academic B, teaching tasks were a major portion of her workload. She managed to maintain her research activities throughout the academic year even though the workload allocation for research was minimal. Meaningful teaching and research tasks were impeded at various pain points, which resulted from conflicting tasks. For Academic B, writing papers and applying for grants were also considered pain points as these were difficult to achieve during the teaching semester. In contrast to Academic A, this academic used deadlines and to-do lists rather than digital applications to manage different tasks.

Reflecting on the academic journey maps of both Academic A and B, it is clear that this stage may be one of the most difficult in an academic career. It is particularly challenging as it is also the "making sense of everything" period. These early years are riddled with confusion over expectations, time pressures and competing interests and all create tension between high teaching loads and a desire to conduct more research.

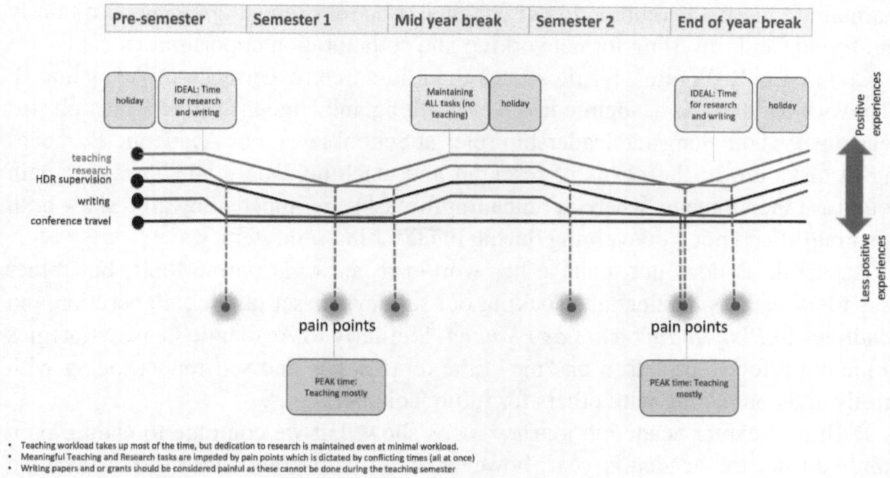

FIGURE 11.2 Early career academic journey map – Academic B

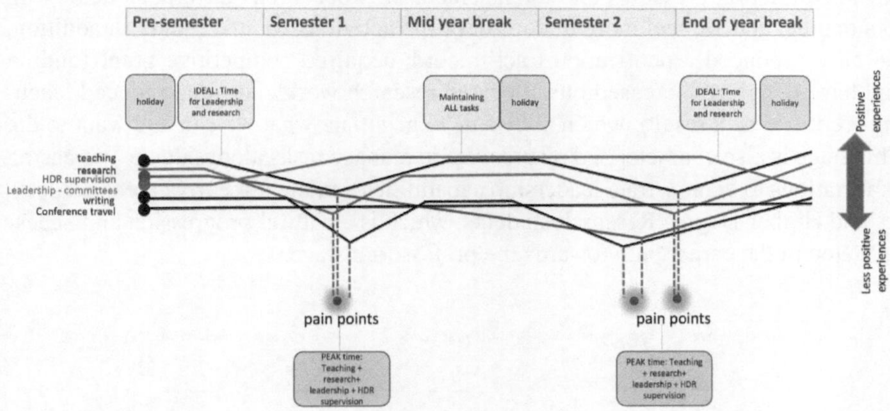

FIGURE 11.3 Mid-career academic journey map – Academic C

11.3.2 Mid-Career Academic Journey Maps

Figure 11.3 shows the mid-career experience of one of the authors, with a workload that includes leadership roles at School, Faculty and University levels. Her workload includes less teaching than the maps of the early career academics. For Academic C, meaningful tasks are leadership opportunities. The map shows that pain points are when her meaningful tasks conflict with others. This usually happens during the teaching semester. Academic C manages her workload, tasks and goals with various online tools (such as Trello and Slack). To *Plan and Execute as Planned*, she also employs strategies such as deadlines, Shut Up and Write and Pomodoro sessions, blocking time, collaborating with others and learning to delegate. Her strategies to maintain focus are a combination of personal time for exercising, being with family and friends and travelling for networking and collaboration opportunities.

Contrastingly, Figure 11.4 illustrates the mid-career experience of Academic B. The workload for this academic involves teaching and Higher-Degree Research student supervision alongside leadership roles at School level. For Academic B, meaningful tasks are in the nexus of research and teaching. The map shows that pain points are the moments where her meaningful tasks are impeded by all tasks – both meaningful and not – converging during the teaching semesters.

Academic B does not manage her workload tasks via online tools, but rather uses tools such as a calendar, blocking out set days or set tasks, collaboration and deadlines to *Plan and Execute as Planned.* Similarly to Academic C, her strategies to maintain focus are based on "me" time to actualise and self-reflect, being with family and connecting with others for future collaborations.

Both mid-career academic journey maps show that we continue to endure pain points during the academic year; however, these pain points are more acute than the early career academic ones. The pain points are not as widely spread during the teaching semester, and we agree that this is because both the load and pressure of undergraduate teaching is largely under control at the mid-career stage. Many of the conflicting pressures and management of expectations have been dealt with to a degree, and we feel more in control of the activities we undertake. In addition, we have produced a publication track record, acquired competitive grant funding and have therefore increased our allocated research workloads (and reduced teaching workload as a result), which allows us to negotiate what we can and want to do. This stage is also characterised by pressure to manage tasks alongside the increasing expectations to demonstrate leadership capabilities and mentor early career academics and Higher-Degree Research students, which is a natural progression and necessary step in the career path towards the professorial stage.

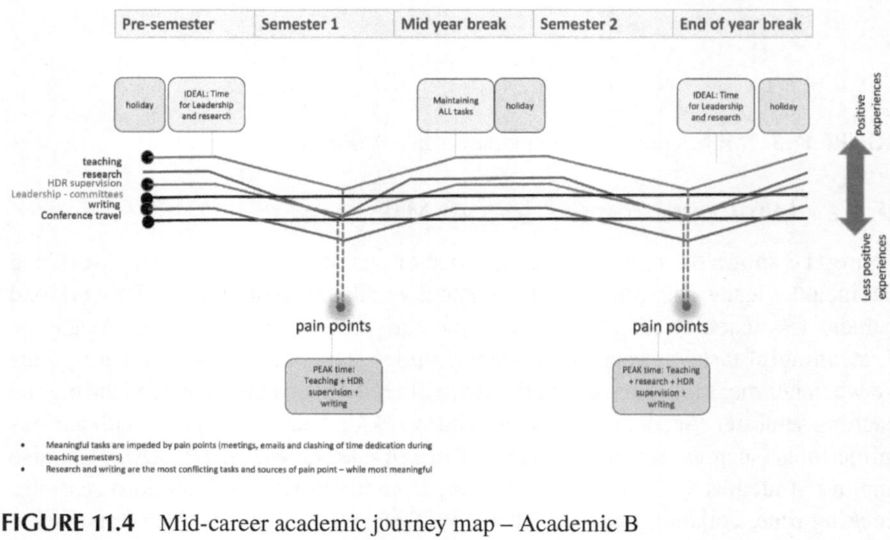

FIGURE 11.4 Mid-career academic journey map – Academic B

11.4 THE PROFESSORS' EXPERIENCES: INSIGHTS FROM THE INTERVIEWS

From the perspective of planning, we consider professors as the most experienced in "juggling it all." With more substantial leadership roles and management duties, the way they *Plan and Execute as Planned* includes both their personal goals and a responsibility of supporting and mentoring the people they lead. The responses to the interview question – "*In your role, how do you plan and execute as planned?*" – revealed the importance of having an approach to planning, of using organisational tools and of dedicating daily time to their health and wellbeing. Some of the strategies mentioned in the interviews with the professors were:

- Waking up early (4:30 AM or 5.30 AM) to dedicate time to academic writing
- Exercising first thing in the morning to focus and plan their goals
- Utilising their Annual Review to reflect on set goals
- Setting up external deadlines
- Blocking out time for specific goals
- Taking time off to recuperate and re-energise
- Making lists and annotating them all in one source of reference
- Setting long-term goals to keep focussed

To illustrate these strategies, we grouped them within three themes and offer the following excerpts from the interviews (Table 11.1).

All three professors also highlighted the need for flexibility and the maximum use of the little time they might have between tasks to dedicate themselves to their writing and their own projects. This correlates with the findings from the study of Dutch professors by Koens et al. (2018), where the participants stated that the focus on developing others' careers and university strategies in their workloads left little room for their own academic pursuits.

11.5 DISCUSSION: ENERGY AND PRODUCTIVITY IN RUNNING THE ACADEMIC MARATHON

The early career and mid-career academic journey maps and insights from the interviews with the professors reveals that being productive as an academic entails different approaches to *Planning and Executing as Planned.* We also observed that the type of workload and tasks involved different types of commitment and energy. Neuroscience literature tells us that the human brain is approximately 2% of the full body weight but consumes 20% of our energy (Koch, 2016; Mattson, 2012). Planning is part of the energy we use to be productive, and we need strategies to ensure we plan in the best way possible. Kogon et al. (2014) explain that people draw energy to be productive from three sources: their passion for the task, their sense of achievement and from their body.

TABLE 11.1
Professor Interview Excerpts

Approach to Planning	Tools for Getting Tasks Done	Maintaining Health and Wellbeing
Professor A: *I set short-term and long-term goals. So, if you get a long-term goal, you sketch out the process, of how many months or years it is going to take to achieve that goal. In the past, I've made like a five year plan…then you have the micro goals to get you through those five years. Something like, what you are going to publish, how you are going to pitch your teaching and research …*	Professor A: *I make like a list, multiple lists in Word, that sometimes I don't adhere to. And sometimes I put it in my whiteboard to knock down. For instance, as my research is now kind of falling into a secondary thing, although I'm trying to get it done, that is what I actually put in front of my screen, like these are the research metrics I need to do …*	Professor A: … *I go swimming in the morning, and that is when I get to think on stuff. When I'm swimming I think and I keep it in my head. And then I organise my day when I am having a swim, or if I cycle to work.*
Professor B: *PPR [performance review] is actually a surprisingly useful tool, because it does make you look back on everything once a year … it helps me keep up to date and reassess what I've [been] doing, what I've progressed in. And in terms of actually making a plan, that is not something that I think there are tools for.*	Professor B: *So daily tasks first. So I have a notebook, which is great, and I keep it with me, and it works … I have all my notes from meetings, and notes I need to remember, and notes from people I meet and then I have a list and I cross them out. So that is my main method.*	Professor B: … *I do get up at 5:30 AM and go on the treadmill and do all my reading. So I do marking, reading, editing on the treadmill. So I've put everything there and I have my red pen.*
Professor C: … *I know that I respond to deadlines really, really well … It's not about research first, sometimes … I will sit and look for calls for papers for special issues, because it's an external deadline [I] can respond to every few months … then I'll work towards that.*	Professor C: *I prioritise and set goals so I'm very much focussed on outcomes. So, okay, I want to achieve this, what is the outcome? So I'll block out time during my week to do things, but I know, more importantly, I know myself so I know how I respond to things.*	Professor C: … *I try really hard to take a Friday…a month … to balance my academic jobs and actually maintain my sanity … and so everyone's got this whole sanity and timing in a different way. But then you're a researcher, you know you're smarter, you can be much more productive doing great good hours and then recuperating.*

The first source of energy is passion, which is related to our intrinsic motivations. A definition of intrinsic motivation in relation to productivity is the motivation to perform an activity for its own sake in order to experience the pleasure and satisfaction inherent in the activity (Deci et al., 1989). The second source of energy is achievement; this is related to our extrinsic motivations. Extrinsic motivation in relation to productivity is focussed on attaining a desired outcome through the activity rather than on the activity itself (Gagné and Deci, 2005). The third source of energy is our body; the most productive people care for their health and wellbeing and include an exercise or meditation routine in their daily lives (Hunter and Scherer, 2009). Our academic journey maps and interviews show that Design academics are motivated by tasks that are meaningful to them, and therefore they *Plan and Execute as Planned* to accommodate such tasks no matter their workload. This observation is also supported by research which suggests that meaningful work motivates people when they believe they have responsibility for the outcome (Dysvik and Kuvaas, 2013). Academics' intrinsic motivations are driven by their passion for their work, and therefore their careers are driven by their desire to achieve their goals and the rewards at different stages (for example, academic promotion). Intrinsic motivation and passion are key components of running the academic marathon, and the way that academics navigate *Planning and Executing as Planned* is influenced by the way academics manage their own wellbeing.

At the early career stage, an academic's intrinsic motivation is often to establish a career and take on opportunities, and their ability to do this is limited by their high teaching load. Their energy is invested in getting the teaching load planned, prepared and delivered. In our academic journey maps, we found similar patterns between the experiences of an early career academic from a decade ago to one currently in this stage. Early career academics employ their extrinsic motivation of targeting their important goals by networking more than in the past, being very adaptable to the changing environment and by actively participating in interest groups (such as Shut Up and Write) in order to get work done. Their academic journey maps show that they need a constant investment of energy throughout the year to accomplish their workload and goals.

Our mid-career academic journey maps demonstrate that a more experienced academic might plan and get things done because their energy is invested in multiplying opportunities to work on meaningful tasks around their research brand and leadership roles. Different career paths are observed for those who are driven more by their teaching and mentoring experiences and those who are more interested in leadership positions. Their internal motivation drives their energy, which helps mid-career academics to deal with the constant struggle to fit it all in. For mid-career academics, external motivators are about moving on to the next level of the academic career (the professorial stage). Their wellbeing and emotional state is impacted by time-consuming tasks that converge during key pain points. Mid-career academics employ both online and offline tools to manage tasks and have a deep interest in setting career goals. Their lack of time makes them more selective of the networking tasks they engage in.

At the professorial stage, these academics are driven by their intrinsic motivation of being leaders and mentors for the people they lead. They feel responsible for their team's performance and their time is invested in developing courses of action for others. Their external motivations are vested in achieving university-wide performance metrics, and they are limited by the little time and opportunity they have to work on their own projects. While strategies for planning and executing are simpler and are shared by collaborators and their teams, their personal wellbeing is paramount for them to remain focussed and to accomplish their work. A commonality from the interviews of all three professors is that they know how they work, and, therefore, they are not trying new approaches, tools or strategies.

In the next section, we present our seven strategies for *Planning and Executing as Planned*. Some of these strategies include known approaches; we present them below in the way they work for the Design academic. These strategies that come from our own practices are now supported by the understanding gained from surveying and interviewing our Design academic colleagues and their different patterns of work.

11.6 STRATEGIES FOR PLANNING AND EXECUTING AS PLANNED

11.6.1 Strategy One: The Calendar Is My Sanctuary

Academics have a never-ending list of tasks. Running the daily academic marathon requires keeping track of many activities, and it would be unproductive to consider that we can keep up with it all in our heads. Ensuring we dedicate enough time to tackle these tasks requires a disciplined approach to the use of one calendar system. Not respecting the calendar leads to miscalculating the time needed to work on tasks, missing deadlines, feeling rushed and, ultimately, high levels of stress. The calendar also helps to annotate agenda items for the effective use of meeting times and to attach relevant information. An approach to a calendar system should include:

- Time for the task itself
- Time for commuting between tasks (in person or online). We cannot be on time if we do not allocate sufficient time to move from one venue to another
- Time to prepare. Some tasks (like writing or teaching) require preparation to work on the task
- Time for lunch or for a wellbeing practice. Skipping lunch to cram in more is not going to fuel our energy to get things done
- Time to assess daily "outputs." Set a goal for the day in the calendar and make an appointment with yourself at the end of the day to review whether or not you completed the goal

11.6.2 Strategy Two: Everything in Its Place

Patterns are important. Most academics work around the semester calendar, which indicates the teaching periods. Academics need to consider and plan their goals and decide how to accomplish them knowing their activity pattern. Some academics have found a good formula and compromise teaching loads and research by doing all their teaching in one semester and freeing up the other semester for research. In the case of one of the authors, this is unthinkable as her teaching–research nexus is important to her. Therefore, dedicating time where we can focus on our most meaningful goals and activities is important. Planning this ahead of time and using the calendar to block out these times will create a sustainable pattern of work that will help us execute our intended work and achieve our goals. This strategy includes:

- Understanding your own academic pattern
- Setting personal goals that align with your patterns, and according to the opportunity to work on them
- Using the calendar to block out time to work on those goals
- Tracking your progress and adjusting your plan accordingly

11.6.3 Strategy Three: One Hat at a Time, Including the "Me" Hat

Beyond being a teacher, a supervisor, a researcher, a mentor and a leader, the Design academic is also a designer, a social communicator, a connector and, in some instances, an entrepreneur. We wear many different hats. Knowing our patterns of work and our use of energy throughout the academic year can help us to identify where we need to dedicate our energy to focus on a particular mindset for particular periods. For example, during the teaching semester, our days would require a "demonstrator" or "problem-solving" mindset in order to teach and inspire students.

We know from the early career and mid-career academic journey maps that pain points are the result of conflicting tasks clashing. While most of us cannot specialise and dedicate our energy to an entire day for one type of work with only one type of mindset, we can be informed by our patterns and be prepared. Knowing the different "hats" that we will need to wear on the day would help us prepare the right techniques or tools to carry on with those roles in the most advantageous way. Being focussed on wearing one hat at a time can help us dedicate the right energy to accomplish a task. This is where it is important to remember the third source of energy management and the most forgotten one: our health and wellbeing. We tend to use all of our energy to "fit in" more tasks in the day. In their interviews, the most experienced academics – the professors – talked about the importance of exercise and other mindfulness activities to help them clear their minds, set goals and focus on their plans. In particular, awareness of the "me" hat is important to help us moderate the use of all other hats and to keep us well and productive. This strategy requires:

- Identifying the mindset required for the day
- Scheduling tasks in blocks that relate to a particular mindset
- At the start of the day or week, reviewing some tools or techniques for the particular mindset required from you on the day
- Always including the "me" hat or mindset every day, at the start and at the end of the day

11.6.4 Strategy Four: Reframe for Success and the 80/20 Ratio

When doing "too much," supervisors tell us that we are "spreading ourselves too thin." On the other hand, we also hear "do not focus on the bean counting" but strive to achieve higher-level goals that match the university performance criteria of success. What does it mean in terms of workload, goals and strategies to manage our days? Here is where strategies such as the 80/20 ratio work to help us identify the most valuable goals on which we should focus our energy and time. Koch's 80/20 ratio (2017) is based on the Pareto principle and it states that, in most cases, 80% of effects come from 20% of causes. This principle suggests that focussing on the important 20% of our regular activities would help us produce 80% of the desired results. By knowing the impact we want to achieve, we can select the most relevant activities.

Our early career academic journey maps revealed that academics in this stage do not have many choices for how they spend their time as their workload is largely teaching. Reframing in their case seems difficult, if not impossible, until they are able to obtain higher research workload allocations. Nevertheless, we can look at reframing in the context of our decision-making process of the different opportunities presented to us. In the survey, one of our early career colleagues expressed that opportunities are plentiful and early career academics need to be flexible and ready to adapt. It is in this context that the 80/20 ratio can be a helpful strategy in helping us choose how we invest our energy into projects that matter to our careers. Applying this strategy to our academic marathon requires:

- Setting goals
- Using a system to track our goals and identify steps towards those goals
- Putting into the calendar those goals according to our academic patterns
- Using the weekly calendar to track those goals according to the 80/20 principle

11.6.5 Strategy Five: Multitasking Ninja: A Gender Thing?

Women are multitaskers, men are not. Is that a stereotype? Recent science demonstrates that the human brain does not multitask but instead shifts from one task to another (Rosen, 2008). While science provides examples of how this happens, the reality is that women are the main care providers in families and their workloads reflect that. The juggle between family and work tasks shows their ability to segment the hours of the day into moments to accomplish it all (Wolfinger et al., 2009). How can academics multitask? We believe it is mostly about how we organise our minds to get on with the day ahead. Strategies one, two and three talk about how we need to know what we need to do in order to schedule it. It seems simple, but not taking the steps to cater for all aspects of the tasks often leads to not completing the work. Our strategy for multitasking is simple:

- Put everything into the calendar
- Know your pattern
- Consider the required mindset for the day

- Prepare accordingly and run the daily academic marathon
- Rest when and as much as possible

11.6.6 Strategy Six: The Urgent Is Not the Important

Attributed to Sir Winston Churchill, the notion of the important vs the urgent should be paramount for academics in this time of constant change in higher education. The difference between the two concepts is striking; the urgent activities that demand our attention are often about another person's goals. The important activities are the ones that lead to an outcome related to your personal goals. Clearly, our academic careers are filled with urgent tasks that make our workload. The academic journey maps show that the meaningful tasks are associated with intrinsic motivation – our drivers – and that the pain points take place during the time where there is no opportunity to work on those tasks. These meaningful tasks are our important activities and deserve our attention and the best of our energy. How does knowing the urgent and the important help us plan and execute our meaningful tasks? This strategy requires:

- Knowing and setting your goals and labelling them as priorities in your calendar
- Using the calendar to allow time for your priorities
- Using a system to track your goals
- Learning from mistakes and failures
- Celebrating the small wins or steps towards your career goals

11.6.7 Strategy Seven: Recognise Your Allies and Foster a "Circle of Niceness"

An academic career is a competitive one. In order to establish a research track record and to apply for competitive research funding we are required to submit our writing and grant proposals to peer review where criticism can often be harsh. As with any large bureaucratic institution, applying for roles and promotion is full of challenging hurdles and many people vying for the same positions. It is not unusual to face internal politics with peers and colleagues. Therefore, one of the most fundamental strategies that our colleague Associate Professor Inger Mewburn, also known as the Thesis Whisperer, has written about is the circle of niceness (2013). The circle is a safe space where we can be ourselves and find our support network. We elaborate on Mewburn's concept of the circle of niceness – which is about collegial support, being together and talking about ideas with honesty and openness – and add some simple approaches to implement it:

- Be vulnerable
- Be open
- Make time for the circle and listen
- Foster and nurture the circle by being nice to one another
- Be supportive of our wins and failures while running the academic marathon

11.7 CONCLUSIONS: TAKING THE NEXT STEPS

In this chapter, we have illustrated the three key career stages and productivity experiences of Australian academics in Design disciplines. We acknowledge that our analysis is limited in two ways: that our data is drawn from largely female academics with young families and by the small number of colleagues who contributed to our data collection. The experiences and academic journey maps might be limited in these ways; however, these have been developed with the aim of understanding the work patterns and experiences of Design academics, and therefore are valuable in illuminating their use of strategies to help their productivity at the early career academic, mid-career academic and professorial stages.

We also acknowledge that the strategies and experiences presented in this chapter are not recipes and that strategies would work differently for different individuals within different situations. We need to view the benefit of these strategies not only in the context of the stage of the academic career, but also with consideration of the academics' personal values and career drivers, as well as intrinsic and extrinsic motivations.

On this topic, we would like to state three points. First, that approximately half of full-time academic staff in Australia are over 45 years old (Bexley et al., 2013; Loomes and McCarthy, 2011). At our university, current statistics indicate that approximately 50% of our workforce comprises women and that mid-career academics are on average 50 years old. This indicates that mid-career academics in that age bracket would have approximately 15-20 years more of an academic career left ahead of them. What can be accomplished in that time? How can we *Plan and Execute as Planned* accordingly? As literature indicates that it takes about five years to accomplish a big project (Bergman, 2011), this means that mid-career academics in this age group would have only four to five big projects to accomplish in their remaining academic careers. Putting strategies in place to set these projects as future goals would work efficiently as extrinsic motivations, and would help to keep these mid-career academics focussed on tasks that effectively contribute to accomplishing them.

Second, that the way academics set goals is driven by intrinsic motivations and personal values. At its core, intrinsic motivation is what makes up the distinctive ways in which we conduct ourselves in planning and executing. For some, their personal values might mean that they desire to lead careers where key drivers are recognition; for others it might be collaboration. It is important to acknowledge these values in order to be more effective in setting, planning and executing tasks towards achieving our goals.

Third, our reflections about *Planning and Executing as Planned* come from a singular perspective and do not encompass the many implications that collaborative work might have on our planning. Collaboration is a key part of academic work, but not all collaborations lead to productive experiences. *Planning and Executing as Planned* are always influenced by our own value set, which involves delicate issues such as negotiating priorities and agreeing on the important vs the urgent. How can we adapt our planning and executing strategies when working with others?

It is important to acknowledge that different tools work for different people in different ways, and that strategies such as the circle of niceness are needed in order to support the collaborative effort. In our experience, we find that our best collaborators are part of our circles of niceness to some degree where meaningful collaboration is built upon trust, respect and being nice to one another.

Currently, academics engage in a multi-faceted workload that reflects the constant changes in higher education. In order to accomplish the important tasks that bring value to our careers, it is important that we take control of our days and *Plan and Execute as Planned*. In this chapter, we have demonstrated how you could map your academic career and understand your own patterns, and use this as a self-reflection tool to devise ways, tasks and goals that enable a more efficient advancement of your academic career.

This chapter has helped us to understand and come to grips with the fact that the academic career is a marathon and it cannot be sprinted. The stages of the academic career need to be thoroughly experienced and this can only be done when we realise that, for most of us, it all takes time. Often the expectations we place upon ourselves are unrealistic, and the sooner we are able to make more reasonable choices, commitments and set appropriate goals the better our overall wellbeing and performance. We hope that this chapter will resonate with our academic colleagues across the different stages of their careers. We also hope it will help to highlight that, whatever stage you might be in, it is important to *Plan and Execute as Planned* in order to make the most of it. We do what we do because we love it and, as we see it, working in an academic environment is full of challenges, opportunities and inspiration.

REFERENCES

Aiston, S.J., & Jung, J. (2015). Women academics and research productivity: An international comparison. *Gender and Education*, *27*(3), 205–220.

Bergman, P. (2011). *18 minutes: Find your focus, master distraction and get the right things done*. Hachette, UK.

Bexley, E., Arkoudis, S., & James, R. (2013). The motivations, values and future plans of Australian academics. *Higher Education*, *65*(3), 385–400.

Chase, J.D., Topp, R., Smith, C.E., Cohen, M.Z., Fahrenwald, N., Zerwic, J.J., Benefield, L.E., Anderson C.M., & Conn, V.S. (2013). Time management strategies for research productivity. *Western Journal of Nursing Research*, *35*(2), 155–176.

Clark, A., & Sousa, B. (2018). *How to be a happy academic: A guide to being effective in research, writing and teaching*. Sage.

Deci, E., Connell, J.P., & Ryan, R.M. (1989). Self-determination in a work organization. *Journal of Applied Psychology*, *74*(4), 580–590.

Deem, R., Hillyard, S., & Reed, M. (2007). *Knowledge, higher education, and the new managerialism: The changing management of UK universities*. Oxford University Press.

Dysvik, A., & Kuvaas, B. (2013). Intrinsic and extrinsic motivation as predictors of work effort: The moderating role of achievement goals. *British Journal of Social Psychology*, *52*(3), 412–430.

Evans, R., Kelly, N., & Kerr, J. (2021). Being a Design Academic: Design Process, Practice, Thinking. In Blackler, A. & Miller, E. (Eds.), *How to be a design academic*. Taylor & Francis, 13–28.

Gagné, M., & Deci, E.L. (2005). Self-determination theory and work motivation. *Journal of Organizational Behavior, 26*, 331–362.

Gillespie, N.A., Walsh, M., Winefield, A.H., Dua, J., & Stough, C. (2001). Occupational stress in universities: Staff perceptions of the causes, consequences and moderators of stress. *Work & Stress, 15*(1), 53–72.

Gorczynski, P., Hill, D., & Rathod, S. (2017). Examining the construct validity of the transtheoretical model to structure workplace physical activity interventions to improve mental health in academic staff. *EMS Community Medicine Journal, 1*(1), 002.

Hall, J.G. (2017). Reflections on an academic career. *Molecular Genetics & Genomic Medicine, 5*(3), 187–195. doi: 10.1002/mgg3.298

Hall, R., & Bowles, K. (2016). Re-engineering higher education: The subsumption of academic labour and the exploitation of anxiety. *Workplace: A Journal for Academic Labor, 28*, 30–47. doi: 10.14288.workplace.v0i28.186211

Hunter, J., & Scherer, J.S. (2009). Knowledge worker productivity and the practice of self-management. In C.L. Pearce, J.A. Maciariello & H. Yamawaki (Eds.), *The drucker difference: What the world's greatest management thinker means to today's business leaders* (pp. 175–194). McGraw-Hill.

Koch, C. (2016, January 1). Does brain size matter? *Scientific American, 27*, 22–25. www.scientificamerican.com/article/does-brain-size-matter1/

Koch, R. (2017). *The 80/20 principle: The secret of achieving more with less.* Hachette.

Koens, L., Hofman, R., & de Jonge, J. (2018, July 9). *What motivates researchers? Research excellence is still a priority.* Rathenau Instituut. www.rathenau.nl/en/vitale-kennisecosystemen/what-motivates-researchers

Kogon, K., Merrill, A., & Rinne, L. (2014). *The 5 choices: The path to extraordinary productivity.* Simon and Schuster.

Lewis, L.S. (2018). *Scaling the ivory tower: Merit and its limits in academic careers.* Routledge.

Lieff, S.J. (2009). Perspective: The missing link in academic career planning and development. *Academic Medicine, 84*(10), 1383–1388.

Loomes, S., & McCarthy, G. (2011). The aging workforce: How can Australian universities address future workforce challenges? [Paper presentation]. *Australian and New Zealand Academy of Management (ANZAM) Conference 2009*, Wellington, New Zealand. https://scholars.uow.edu.au/display/publication44523

Mattson, M.P. (2012). Energy intake and exercise as determinants of brain health and vulnerability to injury and disease. *Cell Metabolism, 16*(6), 706–722.

McCluskey, F.B., & Winter, M.L. (2012). *The idea of the digital university: Ancient traditions, disruptive technologies and the battle for the soul of higher education.* Westphalia Press, Washington.

Mewburn, I. (2013, February 13). Academic assholes and the circle of niceness. *The Thesis Whisperer.* https://thesiswhisperer.com/2013/02/13/academic-assholes/

Miller, E. (2021). Developing the resilience to 'dare greatly' in academia. In Blackler, A. & Miller, E. (Eds.), *How to be a design academic.* Taylor & Francis, 177–188.

Morrish, L. (2019, May 23). *Pressure vessels: The epidemic of poor mental health among higher education staff.* Higher Education Policy Institute Occasional Paper 20. www.hepi.ac.uk/wp-content/uploads/2019/05/HEPI-Pressure-Vessels-Occasional-Paper-20.pdf

Mukerjee, S. (2014). Agility: A crucial capability for universities in times of disruptive change and innovation. *Australian Universities' Review, 56*(1), 56–60.

O'Connor, P., & O'Hagan, C. (2016). Excellence in university academic staff evaluation: A problematic reality? *Studies in Higher Education, 41*(11), 1943–1957.

Pencavel, J. (2014). The productivity of working hours. *The Economic Journal*, *125*(589), 2052–2076.

Rafnsdóttir, G.L., & Heijstra, T.M. (2013). Balancing work–family life in academia: The power of time. *Gender, Work & Organization*, *20*(3), 283–296.

Robinson, E. (2009). *Why crunch model does not work? Six lessons*. IGDA February 2009:139. www.igda.org/page/crunchsixlessons

Robinson, S. (2012). *Bring back the 40hour work week*. www.salon.com/2012/03/14/bring_back_the_40_hour_work_week/

Rosen, C. (2008). The myth of multitasking. *The New Atlantis*, *20*, 105–110.

Scharoun, L., & Muratovski, G. (2021). Leadership of a design school. In Blackler, A. & Miller, E. (Eds.), *How to be a design academic*. Taylor & Francis, 307–326.

The World Universities Rankings. (2018). *Work-life balance survey 2018: Long hours take their toll on academics*. www.timeshighereducation.com/features/work-life-balance-survey-2018-long-hours-take-their-toll-academics

Winefield, A.H., & Jarrett, R. (2001). Occupational stress in university staff. *International Journal of Stress Management*, *8*(4), 285–298.

Wolfinger, N.H., Mason, M.A., & Goulden, M. (2009). Stay in the game: Gender, family formation and alternative trajectories in the academic life course. *Social Forces*, *87*(3), 1591–1621.

Section 5

Leading Others

This section focuses on what it is like to lead others as a design academic, outlining an array of different activities, initiatives and approaches. Chapter 12 starts with a deep exploration of what it takes to become a research leader, outlining some strategies to lead globally within the international research community, as well as locally within your own institution. Continuing the focus on research leadership, Chapter 13 documents different ways to develop meaningful research partnerships with industry, outlining a design-led framework to enable and guide research collaborations. Drawing on three case studies – technology innovation in youth mental health, industrial design research and education, and digital innovation projects in teacher education – this chapter examines each partnership through the lens of co-design. Chapters 14 and 15 focus on educational leadership. Chapter 14 documents the process of large-scale curriculum development and change, sharing the challenges of negotiating institutional approval processes and engaging stakeholders to design a future-focused curriculum. Chapter 15 discusses the learnings from setting up a student peer mentoring group and how such initiatives help to positively transform the culture. Combined, the practical tips in Section 6 are designed to help design academics lead others, whether they be students, colleagues, peers or industry partners.

12 Research Leadership

Alethea Blackler

CONTENTS

12.1 INTRODUCTION

I have built my research profile from the ground up. I have established thorough methods and conducted rigorous empirical research backed up with quantitative analysis in an area which is not traditionally receptive to that. I forged my own way through finding interdisciplinary and industry collaborators, leading my own publications and leading funding applications, starting even before graduating with my PhD. Outcomes of my research have been hard-won but highly impactful, and I now have a strong international reputation in design research and especially intuitive interaction research. Now I am continuing to work as a researcher, but I also act as a leader for other researchers both locally and globally. So, according to Burge (2015), I am in "cohort 2" of 3 cohorts of the academic lifespan; achieving success of my own and able to promote the success of others as well. This is an ideal position to be in as a research leader.

This chapter will discuss some of the tactics of an experienced research leader in two sections, Section 12.2, leading globally within the international research community, and Section 12.3, leading locally within your own institution. Whether you want to increase your reputation as a researcher and your leadership credentials for the love of giving back, for a promotion or tenure application, to apply for funding, or you want to move on to another institution or role, this chapter will help you to formulate some strategies to become a research leader.

12.2 HOW TO BUILD YOUR OWN RESEARCH AREA INTERNATIONALLY

Doing the seminal work is a great advantage as then the work is indisputably yours and everything else done in that area is based on your work (if it is done properly and rigorously and if you have done your job of making your research easily available to all who may need to read and cite it – more on that below in Section 12.2.4). It does not have to be the first work in a whole field – it can be the first offshoot of something or the first specific discovery of something based on a well-established field.

Almost every PhD should have the potential to make a unique contribution like this. Burge (2015) mentions that researchers should "locate some aspect of [their] field and know everything there is to know about it" (p. 73). He explains that this means not knowing everything about the whole discipline, which is impossible and overwhelming to even attempt, but being the expert known for a particular speciality. Gatrell (2019) is an example of another academic who, like me, has stuck with her PhD topic for more than 20 years and made her mark in that area, even though she was encouraged by mentors early in her career to move away from it as her research on parenting and work was seen as not being central to the management literature. She had to fight to continue working in her area of expertise and passion and now the issue has become much more central to our working lives, especially during the COVID-19 crisis where so many children were learning from home alongside their parents working from home. If you love your research area, focus on building it and fight for it if needed.

BUT an important caveat is that you need to be willing to keep working in this area for another decade or more after you finish your PhD if you want to build and lead the field. Some students graduate completely fed up with their PhD topic and want nothing more to do with it. Feeling like this towards your research is understandable, but it is not conducive to building it into a field of expertise and leading it globally. Burge comments that self-motivated researchers pursue subjects that spring from their own interests, so forcing yourself to work on something you are no longer interested in will not serve you well.

Gatrell emphasises that changing direction should be your decision, not something forced on you. If you feel this way, find another research area where you can transfer and apply your skills and make an impact in a way that resonates more with your current interests. Miller (2021b) offers a good example of how this can be done very successfully, and Evans, Kelly and Kerr (2021) also offer different ways of thinking about yourself as a design researcher that may be helpful in this situation. However, leaving behind your PhD topic should not mean giving up on research altogether – it is a vital part of any academic career.

Make sure the work is rigorous and defensible, especially if it is in a completely new area for your discipline. For example, intuitive interaction had not been studied before in any serious way; 20 years ago, only a couple of very short papers (Raskin, 1994; Spool, 2005) and 1 Master's thesis (Okoye, 1998) were available on the topic. No empirical work had been done. Therefore, I based my definitions of intuition and intuitive interaction on a thorough reading of the psychology literature and prepared a very detailed research plan, which allowed me to definitively confirm or deny my ideas so I could defend my position.

In my case, this meant designing careful experiments, recruiting as many participants as possible, carefully balanced for gender and experience, and learning and running statistics to show my results very clearly. I developed new ways of coding video data in order to identify when people were using things intuitively, very clearly based on the literature. Methodological contributions can be at least as important, if not more so, than the actual findings, especially in a young field and if you are looking to be the leader in that field. I published my methods from early on (Blackler et al., 2011; Blackler, Popovic and Mahar, 2004), which has helped to build a strong and rigorous field (Blackler, Popovic and Desai, 2019).

12.2.1 Theory Making

Present your ideas in ways that allow you to build theory so that other people have a basis to work from; for example, by making a framework or model that can be used by practitioners or other researchers to apply your research to practice or to build further on it. Test your theory through more research. I developed a conceptual model for designers to use during the design process, which incorporated a continuum of intuitive interaction that was part of my theory making. I later published on how my continuum related to another one in intuitive interaction (Blackler, Desai, et al., 2019; Blackler and Hurtienne, 2007), alongside some of those involved in developing the other continuum. I also tested my conceptual model with designers to assess its viability, and published that also (Blackler, Popovic, and Mahar, 2014). In the absence of an existing framework that your research area fits within, creating theory or frameworks that others can build on enables the field to grow. Where there is an existing viable theory or framework, be explicit in your publications about where your contribution fits within this and how it builds on it, so that people can build further on whatever you have added.

12.2.2 Expanding the Field

After my PhD, I expanded my initial work in defining intuitive interaction into various areas – for older people, for people living with dementia, for children, and in a variety of tangible and intangible interfaces (along with some mixed reality ones), video games and websites. I also worked on intuitive navigation in complex spaces, and on understanding the development of intuitive expertise. All of this was achieved by joining with others from various disciplines, through funded research projects and through PhD supervision. I advertised for PhD students interested in researching intuitive interaction and attracted high-quality students to my research program through our university's scholarship program. I also began a collaboration with a statistician from the School of Mathematical Sciences in order to assist in further building quantitative research and modelling.

The first area I expanded into was designing for older people. This was a very strategic move as I learned during my PhD that enabling a fitter and more involved older population was a national research priority, and based on the demographics of much of the world would remain an issue throughout my career. I saw during my PhD a potential trend in my data suggesting that older people had particular issues with interface use, and so, after PhD graduation, I applied for and received competitive funding (ARC Discovery Project) to investigate why older people were using things less intuitively and how we might be able to mitigate it. So, although my research started out not being a national research priority, I was able to expand it in a direction that exploited an interesting issue and built me a track record in a high-profile area with a better chance of funding.

Make physical facilities if you can – this makes it easier to attract funded projects and PhD students, as well as visiting scholars, and makes it easier to do the actual work. We were lucky enough to be able to get internal funding to set up a purpose-built lab (with associated equipment and software) at our institution, which has served all of the work in intuitive interaction and much other work in HCI and human-centred design since it was built in 2005. This facility contains equipment and software for recording and analysing participants using a whole variety of devices, interfaces and systems, in the field or in the lab, alone or in pairs or groups.

You can lobby or apply for funding for this sort of facility based on the income it could generate through commercial research, increased research students or large funded projects. However, it is not impossible to build your area without expensive facilities as work becomes more mobile and cheap or free apps and analysis software become available (e.g. Otter.ai, which works as a phone app; fireflies.ai, which plugs into Zoom; and Atlas Cloud, which is free in the beta version). Another alternative is to team up with others who do have facilities and may be looking to build their research profiles or research group members.

12.2.3 Publication

Publish early and often, including your methods. Claim your space. And self-cite when appropriate. Share your publications on a repository so that they are easy for people to access, read and cite. If I had put off publishing my early work by even a few months, others who were working in a similar vein would have got theirs published first.

In many institutions and cultures, journal papers are the most highly prized publications. In design generally there is not a large number of quality journals available to publish in, although this is improving, and some specialities such as HCI and architectural science are better served than other areas in design by well-established journals and/or high-impact conferences. However, for much design research, you can look at publishing in interdisciplinary journals as an alternate route if you cannot find a design journal that suits your work. Also, Gatrell (2019) recommends investigating where the top scholars in your area are publishing in order to gauge which journals might be most valued in your specific field.

A journal paper can be a daunting challenge, especially if you are going it alone or have no experienced co-authors or mentors, but there is help available. Many of the bigger publishers offer a lot of free help and advice on writing for publication (in addition to the standard style guides and rules). For example, Taylor and Francis offer a clear step-by-step guide (Taylor and Francis, 2020), and Elsevier have a comprehensive support site (Elsevier, 2020). Springer has a whole suite of online tutorials to help you get to grips with academic publishing from start to finish for both authors and reviewers (Springer, 2020). There are also books on academic writing (Sword [2012] comes highly recommended) as well as online support and retreats such as #RemoteRetreat on Twitter and others offered by Mazak (2019). So, take the challenge, send your work in, listen to reviewers, and when you get rejected (and you will – we all do), re-work and re-submit. Nothing is ever accepted straight away or without changes, so see the publication process as a long game in which the most important characteristics are persistence and resilience. Miller (2021a) has more advice on building your resilience in academia.

Even after you finish your own PhD, use auto searches and alerts to keep up to date with where the field is going and who is publishing what. Also, review papers and examine theses so you know what is happening in the field. (And if anyone is misunderstanding, plagiarising or grossly misrepresenting your work you can take steps to fix it, such as pointing out to a journal editor or thesis supervisor if things are incorrectly cited, not cited or directly quoted without attribution.)

Publish with others who are interested in your field. I connected initially with others in intuitive interaction research via email after they read my papers, and I have mainly built my intuitive interaction network in this way – they contact me after reading my work, or attending my conference presentations, or vice versa. Mazak (2019) also highlights the critical importance of amplifying your voice through publication and making the messages that come through your writing the centre of your career. When I edited a special journal issue there was an open call for papers sent through the journal and through my networks, so I also made some more contacts with authors who wrote for that issue. One of my most highly cited publications is a comparison of intuitive interaction research in the two different groups who were then active in the world in this field, which I wrote with the leader of the German group (Blackler and Hurtienne, 2007). This paper has been a useful reference for other researchers to use in order to understand the foundations of the field and the various approaches to it in the early days. It is probably now superseded by one of my recent book chapters, which contains an updated and expanded discussion of theories and ideas from around the world (Blackler, Desai, et al., 2019).

This shows the importance of building networks through publication; there are now many other platforms we can use to connect with people who share our research interests, but connecting through interest in published work means connecting with people who are serious about conducting and sharing research in your field rather than just reading, teaching or applying it. These are the people who are also building their track record in your area, so will make the best collaborators.

Edit a journal issue or book in your niche field, or co-author something with others in your area. This type of opportunity may be offered to you based on your previous work and your efforts in creating a network, but if it is not offered, apply for it anyway if you see an opportunity you would like to build. Don't wait to be asked, says Gatrell; the worst that can happen if you put yourself forward is that you are turned down and don't get to do the project (this time around), which is what will happen anyway if you don't put yourself forward. If you don't get accepted, ask again (Gatrell, 2019). Through this kind of editing project, you can mentor others in your field, increase the influence of your work, get to know what others are doing and be very visible as the leader in the field. If others offer co-editing of a relevant book or special issue, accept it. If others are already editing publications of this sort, make sure you submit something to be part of the body of work.

In 2015 I published a special edition of Interacting with Computers as guest editor (Blackler and Popovic, 2015) (no, I did not wait to be asked – I wrote a proposal and sent it to a suitable journal). In 2019, I released the first edited book on intuitive interaction (this time the suggestion came from another researcher who had read and cited my work, and who even made the initial contact with a suitable editor before encouraging me to go for it). These editing projects are time-consuming but they have allowed me to thoroughly review and understand almost all the work in my area, to be seen as the leader in the field, to curate the papers and chapters into a digestible format for the world to easily grasp, and to make contacts around the world who are interested in intuitive interaction. I continue to work on joint publications with some of these researchers.

Finally, watch your metrics – keep an eye on who has cited you, where and for what purpose, and let that inform your publication strategy. This can be as simple as setting up a free Google Scholar profile and checking it every couple of months.

12.2.4 Collaborate

I hope it is already clear that none of the expansion and building of intuitive interaction research would have been possible without collaboration, both across disciplines and around the world. As the first researcher to seriously address Intuitive Interaction, I consider it part of my leadership role to maintain international contacts with other researchers in the field. I now have an international network of researchers in Intuitive Interaction in Australia, Europe, the UK, Canada, the US, Japan, Taiwan and Africa, and have collaborated on publications and led publication projects, as mentioned above. I have hosted and worked with visiting researchers from the UK, Europe, Canada and Taiwan just over the past few years. I have also visited some of them and undertaken intensive work with them. I addition, I have built Industry collaborations for research in intuitive interaction; for example, when I included industry contributors from IBM US and Africa, and doProcess, Canada, in my recent book (Blackler, 2019). Zelenko, Gomez and Kelly (2021) offer useful advice on how to build and maintain collaborations with industry partners for research.

I have made an active effort to build interdisciplinary links. For example, even as a PhD student I had a cognitive psychologist as my associate supervisor, who I recruited myself onto my supervisory team. Since then I have collaborated across disciplines and across the world as most complex problems require more than one set of knowledge and approaches. I have collaborated with colleagues at my institution, elsewhere in Australia and overseas from a variety of disciplines, among them Robotics, Psychology, Games Design, Interaction Design, Business, Usability, Informatics, Architecture, Biological Sciences, Materials Science, Maths, Aged Care, Nursing, Medicine and Occupational Therapy. For example, I initiated a supervisory collaboration with IT (Gaming), and a collaboration with a statistician from the School of Maths in order to assist me to further build quantitative research and modelling. I have also worked with our robotics research centre on a high-profile project to develop an intuitive control interface for an underwater autonomous vehicle designed to monitor the Great Barrier Reef.

Make the most of opportunities to boost your research output by working with others. Create a network, and lead it. Contact people who are also interested in your area or niche and are publishing in the area. Meet them, if possible, at a conference or similar event. Organise or attend a symposium in your niche area, or a stream or session at a larger conference. Organising a successful event such as a conference stream, symposium or SIG (special interest group) conference is a lot of work. However, it enables you to build strong networks, learn a lot about what is going on in your field in a short period of time and puts you at the forefront of people's minds when they think about your research area.

Finally, be generous. If people contact you asking for advice or want to use your research instruments (such as questionnaires, models, measures, coding schemes, etc.), to understand more about your methods or to build on your work, respond promptly and be helpful – it will help to further build the field, and prevent any miscitations or poor explanations of your work. I regularly have people asking if they can use my Technology Familiarity (TF) questionnaire, which I developed in 2000 and have since adapted and refined as needed over the years. It was published in my thesis and as supplementary material in several papers, so sharing it is just a matter of sending a link to the publication on my institution's repository and answering any questions they may have about adapting and using it. I just ask that they cite the appropriate work.

Also, you may be asked to act as an examiner or reviewer for PhDs or other research degrees, to attend vivas or be an external or internal supervisor. This is time-consuming but allows the research field to grow and potentially advance into other fields. For example, I was recently invited to comment on a PhD milestone seminar by a colleague at another university. This was online, because of COVID-19, so it was easy to accept and attend the seminar remotely and learn about how my work had been applied to the field of AR/VR (where not much has been done so far). This has linked me up with researchers who are starting to work in the area and allowed me to assist them along their research pathway, as well as letting me in on what they are doing and helping to build the field in a new direction. It took an hour or two of my time, so it was well worth it from my perspective.

Sharing in this way allows me to stay at the forefront of intuitive interaction research, collect more citations, and it means that the field can grow more quickly as there is some consistency in the measures used.

12.2.5 Summary

This section has described my experience in building my research field from my own vague and initial ideas during my PhD proposal preparation into an international group of researchers. This has taken the form of two main approaches – conducting research and publishing research. Conducting a growing research program requires facilities, collaborators, funding and postgraduate students. On the other hand, publishing your research, if you can leverage it right, leads to opportunities such as funding, collaborators, co-authors, contacts and citations. Therefore, the conducting and publishing of your research are both essential factors in building your research field or niche. Your field may be quite different from mine, but some of the suggestions here should help you to build it into something bigger than you could manage simply working away on your own, and enhance your own profile, reputation and research track record in the process.

12.3 HOW TO BECOME A RESEARCH LEADER WITHIN YOUR INSTITUTION

This section is intended to help you to develop opportunities to become a research leader in your Department, School or Faculty. I reflect on the local initiatives I have led, of varying duration, effort and impact, and why they worked. You may have applied (or intend to apply) for a leadership role, or you may have chosen to enhance your research leadership in more informal ways. I have held formal leadership roles in teaching, research and now in staff supervision (which really covers both), but all of these ideas can work for formal or informal leaders. I have used these approaches to enhance my leadership experience and ability, as well as my leadership profile within the School.

You may find some of these things are already happening in your institution, and so you could offer to join in or take over the leadership of them, or you may need to set up new initiatives. There are advantages to both of these approaches. Things that are already happening require less effort from you to get going and less work to get people on board, and you may already have a ready-made team to work with. However, if you set up your own initiatives, you have ownership of them, can shape them as you see fit and claim them as your own. Even so, I would recommend recruiting deputies who can take over sessions when you cannot be there.

12.3.1 Supporting Staff and Research Students in Writing Activity

"Publish or perish" is a not a cliché for nothing, and it is as true now as it has ever been. The pressure for staff and research students to publish research outcomes can be intense. Often writing is not the main skillset of academics in design, even those with quite extensive research experience, who may rely heavily on non-traditional research outputs. Swann and Rezayan (2021) found that writing can be a major challenge for new academics, and Williams and Satherley (2021) highlight the issues that design academics face with the pressure to publish. English is also often not their first language – in fact, I am in the minority in our School as a staff member who can speak only one language. Therefore, design staff need support in enabling them to write for publication in order for the school to be successful, and this can create a culture of research and writing in the school just as many schools have a culture of teaching, because it is a shared endeavour of all academic staff. Supporting writing offers an ideal opportunity for research leadership. I have supported staff writing endeavours in several ways over the years.

12.3.1.1 Shut Up and Write

Shut Up and Write (SUAW) emerged from the café scene in San Francisco. It is a group that brings writers together at a specific time and place to write side by side, making writing practice more social (Mewburn, Osborne, and Caldwell, 2013). I have run weekly SUAW sessions for several years for staff and research students in the School. SUAW typically involves three 20–25 minute "Pomodoro" (timed) sessions of writing, with 5 minutes of chat and sharing in between. The informal chat allows people to share ideas about publication venues, hints and tips on revisions and resubmissions, etc., and to encourage each other.

Often, I give advice as the senior staff member and leader of the group, but, on other occasions, other group members offer useful suggestions based on their experience. I also often take refreshments to entice people along and to encourage them to stay for the whole session even though it typically goes over lunchtime. Dark chocolate goes down well and, as an added, bonus it contains caffeine, a low-to-moderate dose of which can help cognitive performance in terms of alertness, vigilance, attention and reaction time (McLellan, Caldwell and Lieberman, 2016). Sometimes other group members also bring something to share (for example, when I first drafted this at a SUAW session we were enjoying pistachios from a group member's family farm). Often SUAW sessions happen at cafés (Mewburn et al., 2013), so the refreshments are outsourced but the noise levels may be prohibitive for some.

Due to the COVID-19 crisis in 2020, I had to transition SUAW online. I was not sure if participants would feel this was needed but offered it after being asked to do so. We used Zoom and ran three Pomodoro sessions as we had previously. During each 20–25 minute writing session, I set a timer and everyone turned off video and muted their microphones. I un-muted and interrupted the session once the alarm sounded. I also sometimes shared the screen with the timer on it so all could see the countdown. At times I also put people who were working on writing projects together into breakout rooms for the Pomodoro times so they could work and talk together without interrupting others, but still have the contact of the chat sessions in between. Therefore, in some ways online SUAW was more flexible than the on-campus version. Numbers were stable or slightly higher than previously, with some new members joining who had never attended in-person sessions. In addition, during the COVID-19 isolation period, two new SUAW write sessions per week were started due to demand from people who could not always manage the time I was offering.

SUAW online worked successfully because, online or offline, the accountability is still there. Writing is like running a marathon and the regular time allows you to run it slowly. Regularity, maintaining the practice and habit of writing, and keeping it going (even if more slowly than before) were all important to the group members. Maintaining the practice meant creating boundaries to protect writing time, which helped to deal with guilt about being away from kids who were learning at home, students sending constant emails, etc. People felt that the other group members were on their side in working towards writing goals. One member commented that during home-schooling, SUAW was the only time she had for writing as it was seen by the family as "protected" time with an official diary entry and other people attending. For all of these reasons, both online and in person, SUAW works better than just having a slot in a calendar and writing alone.

Once we go back to "normal," a Zoom option for SUAW is something I could consider for people who are working from home or remotely during the regular timeslot (especially when people are split between campuses). Some also expressed that they preferred the online version of SUAW as they could stay in their own space with all their own resources, large screens, etc., rather than having to bring things to another location, and they could write undisturbed by the tapping of others' keys.

The great advantage of SUAW for me is that it allows me to make a contribution to supporting my colleagues and building research and writing culture but does not take any extra time out of my day as I am writing too. And it forces me to write for around 1.5 hours each week, even if I get no other writing done that week. For all these reasons, I highly recommend initiating a SUAW writing group within your discipline, school or faculty.

12.3.1.2 Writing Retreats – On and Off Campus

I have also been involved in setting up on- and off-campus writing retreats and workshops. These allow a longer block of time for writing tasks (typically one whole day for a workshop or, for a retreat, two or more days), and also sometimes allow for a facilitator who is a professional editor or writing educator to run group activities and provide one-on-one support and feedback. Writing workshops and retreats typically provide catering, as well as a structured mix of writing time and group activity time. They can be run quite cheaply on campus, or with a larger budget at an overnight location.

Retreats can also be run online – for example, #RemoteRetreat on Twitter, which was built on Murray's (2019) social writing framework. Mazak (2019) also offers writing workshops remotely to people all around the world. Retreats offer the opportunity for people to focus a larger block of time on their writing and make significant progress, as well as building a collegial atmosphere of support for writing. Attending and/or organising these events as a research leader means that you will get the benefit of them alongside others who attend, and you will be leading by example as someone who makes time for writing and always has something new to learn about it.

Murray (2019) described how she set up a social writing framework for writing retreats and used it to both run retreats and create her own writing community away from the day-to-day sexism she encountered in her workplace, which helped her to build a career. Her framework is one approach to setting up and running retreats, focussing on the physical, social and cognitive sides of writing. Running retreats on an ongoing and regular basis helped Murray to develop her profile as a research and writing leader, to build her self-worth as a writer and, of course, to carve out time to write.

12.3.1.3 Women's Writing Support Group

Women in design and other STEM and related disciplines often face particular challenges in academia, and while the jury is still out on the exact reasons for this (Schleunes, 2020), helping them to get into a writing habit and stick to it despite family commitments, service roles (often women are over-represented in these) and other demands on their time is important for the progression of their careers. Murray (2019) discusses the fact that a lot more women than men attend writing retreats, and emphasises both the need for women to be a part of this sort of collegial community and the benefits they may reap from it – such as group support, discussion and debate, academic networking and team-building (including across disciplines). Murray (2019) is passionate in her advocacy for women's writing support as a way to create a collegial space, and Gatrell (2019) also mentions how important a women's writing group was for her to build a support network of women around her writing practice.

In 2019, with Evonne Miller (the co-editor of this book), we set up a women's writing support group in the School which was intended to help us make each other accountable to our own writing goals. We run this through setting goals and updates on Trello and meeting every few weeks to account for where we have got with them, as well as to run a writing Pomodoro if there is time. The online Trello board that we all use to track our progress can be accessed by any of us from anywhere, so it was handy tool to have already set up when everything went online during 2020, and allowed us to continue to host, and even grow, the writing support group. Continuation of the group was particularly important for the women in the group who suddenly found themselves trying to home-school and/or overloaded with online teaching preparation and delivery. A reminder of more "normal" times and touching base with their often less pressing but still vital writing goals helped them to remember and stay somewhat on track with some of those goals in difficult circumstances.

12.3.1.4 The Book Project

I initiated and co-edited this book project involving academics across our School, partly as an avenue to get more people writing and publishing in a supported and structured way, and to build new co-authorship teams outside of the normal discipline silos. It was part of a broader research initiative to encourage research writing collaboration and publishing in our school. This has been instrumental in encouraging less experienced writers to submit chapters, and an interesting exercise in understanding the writing approaches and styles of our colleagues, working out what support they needed and providing it. We have run drop-in sessions for book authors to come with questions or for advice every few weeks, and have timed these to happen right before SUAW in order to encourage authors to attend that session also, and some have now become SUAW regulars.

Drop-in sessions helped to provide people with stage-by-stage deadlines as they worked on their chapters, and allowed them to have a time when they could talk to us directly if needed, which is often far simpler than a long email thread. It also allowed chapter authors to become more aware of what authors of other chapters were writing about, which we hope has helped to give the book cohesion. These have been voluntary, and so have been more useful for some than others, as they do not work if they are not utilised, but it also helped us to set dedicated time aside to shape and edit the book as all the chapters progressed. We also offered one-on-one consultations and email advice, plus detailed editing through comments and changes in chapters, and cross reading of chapters by authors whose subject matter might resonate, to enrich the content and provide support for each other.

12.3.2 Mentoring

Mentoring staff in research matters, formally or informally, those you may officially supervise and those you don't, is a good way to build your experience and reputation as a research leader. I have acted as a mentor for staff I supervise (e.g. as part of the performance planning and review process), on request from staff, and a couple of times as part of formal training programs where participants were required to find a mentor. I have also been instrumental in mentoring at university-wide women's writing retreats, run through the women in STEM and women in leadership programs. I have been lucky enough to get excellent professional development in coaching at my university, which has contributed to my skills as a mentor.

When mentoring one-on-one, it could be valuable to structure conversations so that you make sure you are dealing with issues of real importance rather than simply addressing administrative or process issues. You can access examples of prompting questions that really probe your mentees' values and goals in a wonderful book by Michael Bungay Stanier (Stanier, 2016) or online (e.g. The Balance Careers, 2020).

And finally, get a mentor for yourself. They could be a senior researcher or research leader (who does not have to be in your niche area if that is not possible), and needs to be sympathetic to your aims and able to offer you mentoring in how to achieve them. I received mentoring (formal and informal) from senior staff in my School as an early career academic and this was instrumental in helping to formulate and work to meet my research goals. Finding a mentor may not be easy, so you could ask someone to try out a short trial period to see if you suit each other. And, if you feel you have both got as much as you can out of mentoring for now, or that one person cannot support all the areas of your work, find someone else who can fill those gaps – after all, it is not a marriage.

In our School, we have set up a suite of mentoring offerings, including the women's writing support group and SUAW (mentioned in Section 12.3.1), a promotion club for those looking to get into the professoriate (see more below in 12.3.3), a PhD club for those completing PhDs alongside working as academics (inspired by Williams and Satherley [2021]), and a buddy system for new staff so they have an experienced person to meet with regularly and ask all those random questions. Some of these are new initiatives and some have been running for a while, but we recognised how hard it is for staff to seek out help at times, and identified groups who might need particular support so that we could target each scheme for those who need it.

12.3.3 Promotion Club

Another form of mentoring is setting up a special interest group of people who have a particular need or goal. For example, I set up and run "*Promotion Club*" to assist staff in my faculty applying for personal promotion to Associate Professor or Full Professor (the top two of the five levels of academic staff in Australia). Promotion Club helps staff in deciding when to apply, preparing promotion applications and practising for promotion interviews. I have run support, information and writing sessions, invited relevant guests to give advice (e.g. a research metrics expert, the Faculty Director of impact and engagement and a Higher Education Academy mentor), had mentoring conversations and given feedback on application documents. I have also recruited past successful members to act as mentors to current applicants. We had 80% success rate of those Promotion Club members who applied in 2019 and 2020. I have now extended invitations to others around the university outside of our School, and also "franchised" the model to the Faculty of Law, who ran Promotion Club for the first time in 2020. We are also looking to set up a Promotion Club for those looking to move from Lecturer to Senior Lecturer, which will likely run along similar lines.

Promotion Club powerfully disrupts the traditional narrative around promotion, which is that it is a private endeavour to be tackled alone. However, once people get past that barrier and share their intentions, they can find the support of others in the group just as valuable as support from myself and other mentors I have brought in. Promotion Club works because people share lots of little tips and tricks on sticking to the word limit, making the narrative flow, curating online records of publications and funding, etc. No one has to find out everything for themselves and people can feel they are not alone and that what they are attempting has been achieved by others just like them.

12.3.4 Summary

This section has talked about the various ways in which you can grow as a research leader within your institution. One side of this is providing writing support, such as writing groups, leading by example at writing events and creating publication projects. The other side is mentoring, which includes whole School/department programs, groups with particular needs or goals (e.g. Promotion Club), formal and informal mentoring of staff that you supervise or work with. These activities can be time-consuming, but they are also hugely rewarding and can be leveraged to build your own career as a leader, as well as the good will of your colleagues. I hope this helps you to consider adopting some of these ideas and adapting them to your situation, or to start your own initiatives, in order to start or progress your journey as a research leader.

12.4 CONCLUSION

This chapter has offered advice on becoming a research leader – both locally and globally, even in challenging times. I have offered tactics for leading your colleagues in your School or Faculty who may have quite diverse research interests, as well as those around the world who share your particular research obsessions. The suggestions here are intended to guide you in developing your own research leadership profile within your context, so adapt them as needed. However, there is no point in becoming a wonderful research leader if it comes at the expense of being an actual researcher. Always keep your eye on your research goals and how you will achieve them, and don't allow your leadership and mentoring work to replace the research you are passionate about.

REFERENCES

Blackler, A. (Ed.). (2019). *Intuitive Interaction: Research and Application.* Boca Raton, FL: CRC Press.

Blackler, A., Desai, S., McEwan, M., Popovic, V., & Diefenbach, S. (2019). Perspectives on the Nature of Intuitive Interaction. In A. Blackler (Ed.), *Intuitive Interaction: Research and Application* (pp. 19–39). Boca Raton, FL: CRC Press.

Blackler, A., & Hurtienne, J. (2007). Towards a Unified View of Intuitive Interaction: Definitions, Models and Tools across the World. *MMI-Interaktiv, 13*(Aug 2007), 37–55. Retrieved from www.mmi-interaktiv.de/fileadmin/mmi-interaktiv.de/Ausgabe_13/mmi-ue.pdf

Blackler, A., & Popovic, V. (2015). Special Issue: Intuitive Interaction. *Interacting with Computers, 27*(3).

Blackler, A., Popovic, V., & Desai, S. (2019). Research Methods for Intuitive Interaction. In A. Blackler (Ed.), *Intuitive Interaction: Research and Application* (pp. 65–88). Boca Raton, FL: CRC Press.

Blackler, A., Popovic, V., Lawry, S., Reddy, R. G., Doug Mahar, Kraal, B., & Chamorro-Koc, M. (2011). *Researching Intuitive Interaction.* Paper presented at the IASDR2011, the 4th World Conference on Design Research, Delft.

Blackler, A., Popovic, V., & Mahar, D. (2004). *Studies of Intuitive Use Employing Observation and Concurrent Protocol.* Paper presented at the Design 2004 8th International Design Conference, Dubrovnik, Croatia.

Blackler, A., Popovic, V., & Mahar, D. (2014). Applying and Testing Design for Intuitive Interaction. *International Journal of Design Sciences and Technology, 20*(1), 7–26.

Burge, G. M. (2015). *Mapping Your Academic Career: Charting the Course of a Professor's Life.* InterVarsity Press.

Elsevier. (2020). Author Resources. Retrieved from www.elsevier.com/authors/author-resources

Evans, R., Kelly, N., & Kerr, J. (2021). Finding Your Identity as a Design Academic: Design Process, Design Practice and Design Thinking. In A. Blackler & E. Miller (Eds.), *How to Be a Design Academic.* Boca Raton, FL: CRC Press, 13–28.

Gatrell, C. (2019). "You Must Wait to Be Asked": Career Advancement and the Maternal Body. In R. Murray & D. Mifsud (Eds.), *The Positioning and Making of Female Professors* (pp. 49–73). Palgrave Macmillan.

Mazak, C.M. (2019). Babies Taught Me How to "Do" Academia: Crafting a Career in an Institution that Was Not Built for Mothers. In R. Murray & D. Mifsud (Eds.), *The Positioning and Making of Female Professors* (pp. 75–87). Palgrave Macmillan.

McLellan, T.M., Caldwell, J.A., & Lieberman, H. R. (2016). A Review of Caffeine's Effects on Cognitive, Physical and Occupational Performance. *Neuroscience & Biobehavioral Reviews*, *71*(December 2016), 294–312. doi: 10.1016/j.neubiorev.2016.09.001

Mewburn, I., Osborne, L., & Caldwell, G.A. (2013). Shut Up & Write! Facilitating Informal Learning in Doctoral Education (with 'No Critiquing, Exercises, Lectures, Ego, Competition or Feeling Guilty'). In *Writing Groups for Doctoral Education and Beyond: Innovations in practice and theory , Claire Aitchison and Cally Guerin (Eds), Routledge* (pp. 399–425).

Miller, E. (2021a). Developing the Resilience to 'Dare Greatly' in Academia. In A. Blackler & E. Miller (Eds.), *How to Be a Design Academic*. Boca Raton, FL: CRC Press, 177–188.

Miller, E. (2021b). Finding Your Place as a "Non-designer" in a Design School. In A. Blackler & E. Miller (Eds.), *How to Be a Design Academic*. Boca Raton, FL: CRC Press, 45–55.

Murray, R. (2019). Writing Myself into an Academic Career. In R. Murray & D. Mifsud (Eds.), *The Positioning and Making of Female Professors: Pushing Career Advancement Open*. Palgrave Macmillan (pp. 89–111).

Okoye, H.C. (1998). *Metaphor Mental Model Approach to Intuitive Graphical User Interface Design*. Doctoral dissertation, Cleveland State University, Cleveland.

Raskin, J. (1994). Intuitive Equals Familiar. *Communications of the ACM*, *37*(9), 17–18.

Schleunes, A. (2020, February 17). Scholars Debate Causes of Women's Underrepresentation in STEM. *The Scientist*. Retrieved from www.the-scientist.com/news-opinion/scholars-debate-causes-of-womens-underrepresentation-in-stem-67117

Spool, J.M. (2005). What Makes a Design Seem 'Intuitive'. Retrieved from http://uie.com/articles/design_intuitive/

Springer. (2020). Author and Reviewer Tutorials. Retrieved from www.springer.com/gp/authors-editors/authorandreviewertutorials

Stanier, M. (2016). *The Coaching Habit: Say Less, Ask More & Change the Way Your Lead Forever*. Box of Crayons Press.

Swann, L., & Rezayan, L. (2021). Exploring the Mindset of a New Academic. In A. Blackler & E. Miller (Eds.), *How to Be a Design Academic*. Boca Raton, FL: CRC Press, 83–111.

Sword, H. (2012). *Stylish Academic Writing*. Cambridge, MA: Harvard University Press.

Taylor and Francis. (2020). Author Services. Retrieved from https://authorservices.taylorandfrancis.com/

The Balance Careers. (2020). 70 Coaching Questions for Managers Using the GROW Model. Retrieved from www.thebalancecareers.com/coaching-questions-for-managers-2275913

Williams, T., & Satherley, S. (2021). Getting that PhD: How Hard Can it Be? In A. Blackler & E. Miller (Eds.), *How to Be a Design Academic*. Boca Raton, FL: CRC Press, 59–82.

Zelenko, O., Gomez, R., & Kelly, N. (2021). Research Co-design: A Design-Led Framework to Enable Meaningful Collaboration in Research. In A. Blackler & E. Miller (Eds.), *How to be a Design Academic*. Boca Raton, FL: CRC Press, 227–244.

13 Research Co-Design
Meaningful Collaboration in Research

Oksana Zelenko, Rafael Gomez and Nick Kelly

CONTENTS

13.1 INTRODUCTION

Research impact is defined as the contribution that research makes to the economy, society, environment or culture, beyond the contribution to academic research (Australian Research Council, 2019). In 2018 Australia held its inaugural impact assessment on the back of the United Kingdom's Research Excellence Framework (Khazragui and Hudson, 2015), which aimed to capture the value that funded research has made, and create greater accountability of researchers to the public, as well as funders and collaborators. The shift to capture value in terms of tangible impact on industry, community and individuals has prompted institutions to pause and reflect on how they approach research partnerships and to engage in more meaningful – and more impactful – collaborations.

Beyond the national level, larger-scale, world-wide efforts to measure the real-world impact of research have resulted in world-wide assessments such as the *Times Higher Education* (THE) University Impact Rankings study (Times Higher Education, 2019). This study presents the only global performance indicator that assessed 476 universities in 76 nations against the United Nations' 17 Sustainable Development Goals, which include individual Universities' contributions to good health and wellbeing, gender equality, and peace and justice, amongst other domains of public good. The results signalled an important shift to what research *value* might look like: University of Auckland (UoA) topped the list. By contrast, in the annual scholarly ranking by THE (which focusses on outputs rather than impact), UoA was in the 201–250 bracket (Times Higher Education, 2020). This outcome is significant because impact assessment has prompted universities to once again foreground research that makes a difference in the society – whether health, education, culture or economy.

The growing emphasis on impact is significant for design academics and for the research partnerships that they develop. Institutions have been introducing capacity-building programs that train researchers in formal aspects of identifying, tracking, measuring and communicating research impact, including development of research impact indicators and tools to understand and track social change (Centre for Social Impact 2018). However, while there has been a rapid increase in formal tools, resources and training packages to measure impact of completed research, there has been much less emphasis on effective engagement and research design strategies for developing research partnerships that can lead to impact.

While there is consensus around the definition of impact, there is significant scope for interpretation within this, which needs to be negotiated when working with partners. For instance, in the corporate sector, impact can often refer to an increase in financial profit, reduced operational costs and meeting key performance indicators (Canhoto, Quinton, Jackson and Dibb, 2016; Ulaga, 2003), while in the not-for-profit sector impact tends to focus on social outcomes such as improvements in broad social and cultural wellbeing, support and advocacy (Productivity Commission, 2010). This chapter attempts to theorise research partnerships and the way that this negotiation process can be navigated, by presenting a design-based framework for deepening engagement with partners. It presents three case studies – one each in industry, government and not-for-profit sectors – that illustrate how elements of the framework can be applied in practice to create a shared understanding of impact, and collaborations with mutual reciprocity.

For successful collaborations or partnerships, it is important that all relevant stakeholders have a common frame of reference to begin defining a shared goal. Within this context, we suggest that university–industry partnerships would benefit from framing as *relationships*, in which the development of trust might form the basis for achieving successful outcomes and impact. A relationship implies something deeper than merely collaboration; it suggests a meaningful connection between parties that might not be created immediately but rather fostered over a length of time. While development of a long-term collaboration may ultimately be a goal shared by all parties involved, there could arise at any point a potential mismatch between partner goals and academic goals, creating areas of tension. One example would be where commercially driven partner goals may lead to ethical challenges in the context of academic research, resulting in careful considerations of implications for ongoing collaboration. Incorporating tools like values-mapping may assist with identifying such points early on in the partnership development cycle.

Models for developing trust in collaborative partnerships are an established field in business and organisational psychology with applications across a range of contexts from health to business to education (White-Cooper et al., 2009). A widely applied approach in research is community-based participatory research, in which researchers and community-based partners collaborate as equals on research related to novel design solutions, an approach most commonly found in health research (Hacker, 2013). Other approaches apply principles of attachment theory, which uses relational strategies to build partnerships and is based on the premise that, where a secure attachment is present, partners "can more easily take risks, assume appropriate responsibility for problematic situations" (Rappaport et al., 2008: p. 695).

Design academics frequently develop research projects (research partnerships) with non-academic organisations (e.g., government, industry and non-government organisations) by bringing their design discipline or process expertise to create research outcomes, such as in design-led innovation (Wrigley, 2016). Although there is a variety of user-centred design approaches, one with potential here is co-design as it begins by situating people as the key drivers. Co-design is a form of participatory design approach and attempts to actively involve all stakeholders in the design and development process to ensure the outcomes meet all necessary needs. From this perspective, in the forthcoming sections of this chapter, we can begin to utilise the co-design method as a framework to consider the emergence of a successful university–industry relationship.

13.2 FROM COLLABORATION TO RESEARCH CO-DESIGN: INTRODUCING A FRAMEWORK FOR DEEPENING RESEARCH ENGAGEMENT

Co-design, as a form of participatory design research, is a methodology with democratic foundations and roots going back to work in Scandinavia in the 1970s (Chisholm, n.d. Sanoff 1990). It is based on the premise that people affected by design decisions should take part in the making of those decisions. Co-design therefore is a method that gives prospective users of a given service or technology equality in the creative decision-making process. It is a methodology that has found wide-spread application across multiple fields and sectors, from education, to business innovation, to health and wellbeing (Greenhalgh et al., 2016). In their review of co-creative methods, Greenhalgh et al. (2016) state that co-creation, as an established methodology, arose across numerous fields of research including business studies, design science, human-computer interaction and community development.

Co-design principles can be used when developing new products and services, where they are applied across the phases of a research project, with each phase subject to dialogue and renegotiation between researchers and industry partners. Co-design brings with it a sort of inversion of expertise – academic researchers with knowledge domain specialisation take a back seat and become learners, while collaborating partners become educators. Co-creation, co-design and participatory approaches to collaboration are widely believed to increase research impact.

In the context of research partnerships, co-design applies to the design of the research itself (Slattery et al., 2020) and can be considered as *a meta-level process*, positioned either as an open-ended process, or a process for *designing designing* (Zelenko and Bridgestock 2013). It is simultaneously directional and non-prescriptive, applicable to multiple contexts, yet devoid of its own contextual references. Such an approach of applying co-design at meta-level to the context of building partnerships and deepening engagement – that is, as research *on* research – is an emerging area of study but is as yet underdeveloped. While co-creative and co-design frameworks have been acknowledged as enabling meaningful involvement of end-users in research, widely applied across the board, their effectiveness in research impact requires further documentation and interrogation (Greenhalgh et al., 2016).

To explore how co-design might support development of meaningful partnerships, we introduced an existing co-design framework, the Participatory Design-led Engagement (PD-E) framework, iteratively developed by the first author over the course of 14 years and applied nationally and internationally in over 18 projects in contexts spanning youth mental health, education, ageing, corporate education, youth and community leadership (Zelenko 2012, Zelenko et al., 2018).

13.2.1 Unpacking Co-Design Framework in Detail

Within the PD-E framework (Figure 13.1), each research phase of a project is reconceptualised as a site of co-creation – the phases need not be in order, but often are.

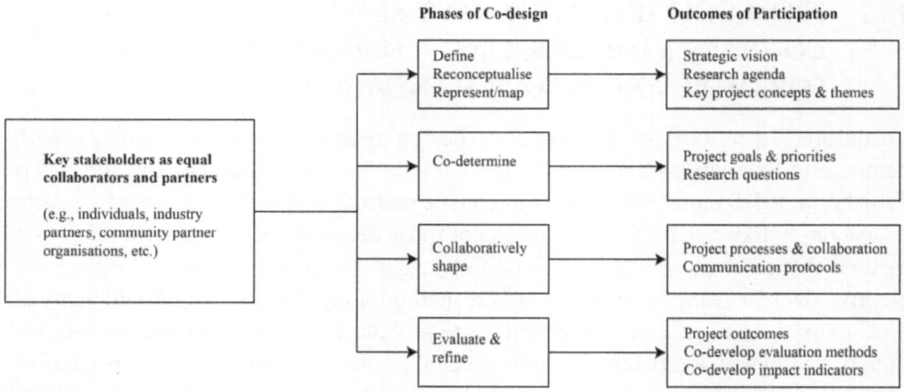

FIGURE 13.1 Participatory Design-led Engagement (PD-E) framework: overview of co-design phases and outcomes of participation

Firstly, in Phase 1 there is a focus on project conceptualisation, scope definition, defining the research problem that will become a process of co-conceptualisation, and collaborative project definition. Throughout this initial phase, the industry or community partners' needs take precedence in determining the problem or the social need. Phase 2 builds on outcomes of Phase 1 by deepening engagement to include joint determination of project goals and priorities, as well as co-design of research questions. Phase 3 involves collaborative shaping of the research process, including communication and interaction protocols. Lastly, Phase 4 may involve collaborative evaluation strategies, joint decision-making on evaluation tools and methods, or co-design of impact indicators. Importantly, a core underpinning principle of this model is that industry or community partners, while not necessarily knowledge domain experts, *are experts in their operational realities*, and often experts in sector needs, challenges and opportunities. By opening each research phase as a site of dialogue, co-design deepens reciprocity and knowledge exchange, and proliferates opportunities to create shared value and mutual benefit. In the context of research collaborations, the principles of co-design are framed to deepen the partnership and, over time, increase the potential for relationship-building.

Researchers often commence research with an explicit commitment to create impact and develop research that responds to actual industry or community needs. However, despite the wide-spread use of approaches like design thinking to create empathy or co-design methods to engage stakeholders as partners, the message from our research partners at a recently held symposium on research impact was clear: researchers ought to become better at listening to partner needs (QUT Creative Industries Faculty Research Impact Symposium 2019).

13.2.2 Engaging Stakeholders

Approaches to building trust and strategies to enable mutually beneficial partnerships prompt researchers to ask: who benefits from the research? Processes like stakeholder mapping, needs analysis, experience and values-mapping are often used to identify not just one primary group but several interconnected tiers of stakeholders intended to draw value from the partnerships (Thorngren and Wincent 2010). Depending on the complexity of the collaboration, any number of stakeholders may be involved at any stage of the co-design process. Furthermore, depending on the needs of the project, some stakeholders as research collaborators and contributors may be involved in specific phases only, while others are involved across the full spectrum of the co-design phases. In this chapter we describe them as primary, secondary and even tertiary layers of beneficiaries, with groups' alignment to each layer determined by the primacy of their needs and the position of priority in the area/s of intended impacts. Applying such multi-tiered stakeholder mapping (see Table 13.1, for example) creates new opportunities to formulate a shared frame of reference among partners. Outcomes of this process should lead to alignments with any values-mapping researchers may also conduct to explore stakeholder needs. For partnerships spanning a significant timeframe of several years (or even decades), researchers may also shift from the initial intent to understand a network of stakeholders critical to a

TABLE 13.1
Mapping Three Tiers of Research Beneficiaries

Primary	Secondary	Tertiary
Value to Service Clients (Young People)	**Value to National Service Provider**	**Value to University Researchers**
Burning issue: Where can I get help on issues important to me, at a time I need, "on my terms"? • Tailored digital help-seeking toolkit and resources • New and alternative pathways to seek help online • Access to evidence-based advice and additional pathway to connect to a counsellor	*Burning issue: How can technology be used to better support young Australians seeking help online?* • Leading technology-enabled service innovation • Strategic commitment to innovative digital solutions for young people	• Social impact • In-depth understanding of sector needs • Growth in research capabilities in health technology innovation • Lead innovative research in design for digital Mental health

given partnership towards an understanding of stakeholders' entire communicative ecology and needs, including an in-depth understanding of how their strategic priorities may shift over time in response to changes in the sector.

In the following section, the authors present three case studies exploring how co-design principles were applied across three distinct contexts of design research to build trust in research partnerships and generate research impact.

13.3 CO-DESIGNING MEANINGFUL PARTNERSHIPS: THREE CASE STUDIES

The case studies presented are positioned in design research and span the contexts of technology innovation in youth mental health, industrial design research and education in the corporate sector with global outreach, and a digital innovation project in teacher education. The three contexts were selected in order to examine the partnerships through the lens of co-design, while also reflecting on the applicability of the co-design principles to creating meaningful research partnerships.

13.3.1 Case Study 1: Co-Designing Research With Services and Young People in Digital Mental Health

Our first case study examines how co-design principles detailed in the model above (Figure 13.1) were applied by researchers to build a meaningful partnership with a national youth counselling service provider and young people (aged 16–24) to co-design an evidence-based digital help-seeking app (Zelenko et al., 2018). Led by the first author, the project was funded by a nationally competitive research grant (Australia Research Council Linkage Project scheme, 2016–2019), conducted in partnership with KidsHelpline, a major national Australian youth counselling service, and culminated in a national launch of the award-winning mHealth app *niggle*. The project was a key milestone in a research partnership already spanning 19 years, focussing on design-based innovation in developing novel digital tools and resources for young people seeking help online. In this project, elements of the PD-E framework were applied at two levels: first to engage the service provider in co-determining the research focus, agenda and direction; and secondly to co-design the new help-seeking app with young people. Key to engaging both stakeholder groups was connecting with them as early as possible to collectively shape the project focus, process and outcomes.

Outreach to the partner organisation commenced early and was fairly informal in nature, based primarily on open-ended discussions of shared values and priorities to identify meaningful and authentic alignment. Within these initial discussions, following the principles of the co-design approach, research and specialist knowledge domain expertise took a back seat. Here, the partner organisation was positioned outright as the expert in their own strategic priorities with respect to the use of technology to improve outcomes for young people. While this may seem like common sense, especially in the context of design research and social impact, the default in research far too often falls back on engaging partners around an already pre-determined research agenda (or at least partially pre-determined). The reasons may be operational: tight funding application deadlines or other academic commitments like teaching, which significantly reduce the time available to establish a relationship with research collaborators that treats it as beyond episodic.

Our approach to the initial partner outreach for this project, regardless of the pre-existing longevity of this research partnership, was to start early and to start with a blank slate. Example questions included: what is important to your organisation now? Broadly, what are your strategic goals and priorities? In the short-term, mid-term and long-term? What is your future vision and ambition as an organisation for the next five years? Insights from the answers provided then informed how the research team aligned their research strengths, interests and capabilities *in support of* the partners' agenda and needs. It is important to note that this approach did not aim to devalue research expertise, but rather to reposition it in support of designing research that is partner-led.

Two critical outcomes emerged from the initial discussion. First, although the service provider was the primary collaborator in research, they were not the primary beneficiary of the research outcomes – Australian young people were. A clear distinction was made between primary and secondary beneficiaries. De-emphasising knowledge domain and research expertise enabled the team to adopt a learner perspective that led to a critically engaged, considered and systematic approach to planning for research impact based on a shared understanding among collaborators for whom the research ought be designed to benefit/impact. Documenting the different tiers of beneficiaries and explicitly articulating the *anticipated value of proposed research* to each group, as shown in Table 13.1, enabled researchers to identify the primary beneficiary. The outcomes of research were clearly first about young people's wellbeing, and secondly about the technology-led service innovation to enable it. Therefore, outcomes for the partner organisation were measured by successful outcomes for young people.

In addition to service providers, the project team also engaged young people – primary beneficiaries of the research – in further co-determining the research focus. As young people's needs, lived experiences and priorities in seeking mental health help and support took precedence, the PD-E framework was implemented in full by engaging them directly in contributing to shaping the focus, content, conceptual framework and iterative refinement of the final digital solution. Table 13.2 summarises young people's engagement across each co-design phase and the outcome of their participation.

In total, ten co-design workshops were conducted with young people to develop the app across all co-design stages, with additional ongoing consultations throughout the final iterative evaluation stages. As young people were the primary beneficiaries of the research outcome – the help-seeking app – their direct participation was critical not just at the commencement of the project (as was the case with the partner organisation), but at *all stages of the research process.* The quality of the research partnership between the university researchers and the partner organisation was contingent upon meaningful engagement of young people of the app design, i.e., in prioritising their voices within the shared decision-making which was intentionally woven into the research process. This process illustrates that co-design could be applied at the outset of research to co-determine outcomes at each key research phase, including research direction, focus and research questions, and further help re-position researchers' own agenda as secondary or even tertiary. Deploying the co-design approach has led to multiple modes of engagement with the same service provider, spanning undergraduate and postgraduate teaching, corporate education training in co-design and an ongoing program of future research projects of varying scale.

TABLE 13.2
PD-E Framework Applied to Engage Young People in Co-Designing a Mental Health App

	Phase of Co-Design	Outcome of Participation
Using participatory tools in co-design workshops, young people	1. Conceptualised/ represented/visualised	Definitions of wellbeing, help-seeking and self-care pathways, journey of recovery, sources of support, grounded in young people's lived experience
as co-researchers and co-designers	2. Co-determined	App purpose, content and priorities
	3. Collaboratively shaped	Visual, interface and interaction design of the app participatory research process and subsequent co-design workshops
	4. Iteratively evaluated and refined	App lo-fi (paper) and hi-fi (digital) prototypes

13.3.2 Case Study 2: Building Trust and Co-Designing a Vision With a Global Corporate

This case study centres on the relevance of relationship-building through trust via three key factors including *academic buy-in*, *face-to-face relationships* and *help your partner first*. Reflecting on this engagement, it is evident that trust is built as a result of a co-design process reflected in the PD-E framework. This particular case, led by the second author, involves an international partnership between Queensland University of Technology (QUT) and BMW Group in Germany. The collaboration began 5 years ago with a student internship program in which high-achieving students were provided with a 6–12 month work placement at BMW Group in Munich. Since then, more than 20 students have been involved in the internship program, with 1 student securing a lifetime contract with BMW Group as a result. Further, BMW Group has partnered with the final-year industrial design honours program through project collaborations. A Memorandum of Understanding was signed between both parties in 2018 and a Statement of Intent for further collaboration was signed in 2019. In early 2020 the BMW Group + QUT Design Academy was launched. The academy is a lab space which focusses on undergraduate design education, postgraduate research and collaborative R&D projects.

Although introductions between the QUT's Science and Engineering Faculty (SEF) and BMW Group management began around nine years ago, it was not until QUT's School of Design was involved that a deeper and more trusting relationship was built between the organisations. Initially there were three key elements that helped foster trust between the parties, which ultimately permitted the implementation of a strong partnership.

Academic buy-in: buy-in of a full-time academic within the School of Design to help drive the collaboration was one key element. "People determine the success or failure of industry–university partnerships" (Edmondson, et.al., 2012, p.10) and this was certainly the case in this instance. Although there had been some progress between SEF and BMW Group, this was at the administrative level, and did not involve an academic leader. Within the School of Design, however, the academic leader had more time and resources to devote to the partnership because of the long-term potential the partnership could produce.

Face-to-face relationship: another element was regular face-to-face contact between the QUT lead academic and key managers at BMW Group. This was a crucial part of the relationship-building process that is difficult to replace by other means even in this modern era of electronic communication (Werker and Ooms, 2019). Building that physical connection cultivates a distinct relationship between people that assists in all areas of the partnership including building rapport, trust and resolving issues. As BMW Group is headquartered in Germany, this required overseas trips, meaning time and financial investment – an ancillary way to show commitment and help build trust.

Help your partner first: helping the organisation achieve industry impacts, rather than academic goals, was another piece of the puzzle. That is not to say that academic goals were completely ignored or forgotten, but rather that it was necessary to achieve some wins for the industry partner so as to gain traction and build trust (Ulaga, 2003; Edmondson, et.al., 2012). Having experience in industry, the academic lead understood that a global leading corporate like BMW Group had no shortage of potential partners, but it was the partners that would help it achieve their goals that would stand out.

Having a long-term strategy and sharing that vision with the industry partner early, to collectively shape it, was important (Edmondson et al., 2012; Rybnicek and Königsgruber, 2019). Although trust is not established at the early stages, and discussions might be speculative, it's important to begin to cultivate ideas together that might flourish later down the track. As trust is building and the shared vision is being formed, the co-design model begins to come into focus. During this part of the relationship, the industry partner became a co-designer and co-contributor in the long-term vision, but also a partner of the short-term goals of the collaboration.

Some key features of this process were that long-term vision and shorter-term goals were always evolving and adapting. If circumstances changed, and this impacted the goals, then both parties would adjust accordingly and co-define a new strategy to achieve the long-term vision. This honest, flexible and adaptive approach is important (Barnes, Pashby and Gibbons, 2002) and continues to happen to this day.

Although the co-design model was not expressly considered from the outset, in retrospect, the PD-E framework presented above fits this case well. As Table 13.3 outlines, the PD-E framework fits well with the various stages of the partnership as it evolved. At the initial phase the long-term goals were conceptualised together, with a clear understanding that these were able to shift and evolve over time – flexibility was important here. Likewise, in the early stages, it was important to have an understanding of what each partner wanted to achieve as part of the collaboration, and respect these goals. As the partnership evolved, it was important to collaboratively shape the short- and medium-term milestones and iteratively refine these regularly.

TABLE 13.3
PD-E Framework Applied to Relationship-Building through Trust in University–Industry Partnership

	Phase of Co-Design	Outcome of Participation
Corporate industry partner; co-design an evolving strategy and vision as a result of trust	1. Conceptualised	Long-term 3–5 year vision reviewed every 12 months
	2. Co-determined	Understand key goals for each partner (what each wanted to achieve). Stating and respecting individual interests
	3. Collaboratively shaped	6–12 month milestones
	4. Iteratively evaluated and refined	Ongoing evaluation and adjustments of shared goals and vision

13.3.3 Case Study 3: Co-Determining Research Impact Emerging Over Time in the Education Sector

Our third case study is in the domain of education and outlines the way that co-design principles emerged over time through a seven-year ongoing partnership. The partner organisation was the teacher registration body for an Australian state, with a membership of >100,000 teachers. In Australia, teacher registration bodies are non-government organisations that operate under state government legislation. They are primarily responsible for teacher registration, accreditation of teacher education programs and helping to support teachers – in many ways, but with a focus on compliance to professional standards and legislation. Led by the third author, the project was initially developed as a "proof of concept" with internal university funds in 2013 (Kelly, Clará, and Kickbusch, 2015); this led to a funded research project with the partner (ongoing since 2014) to create an online community platform for teachers called *TeachConnect* (Kelly, Clará, Kehrwald, and Danaher, 2016; Kelly, Kickbusch, and Dawes, 2019; Kelly et al., 2018).

13.3.3.1 Phase 1 Conceptualisation

From the outset, there was a broad agreement between researchers and partners that the vision was to co-design a platform with teachers to enable three goals: (1) teacher peer support; (2) a shared pride in the profession; and (3) a focus on support for teachers to meet professional standards. In this respect, the *primary beneficiaries* were beginning and early career teachers within the state (represented in the first two of these goals); the *secondary beneficiary* was the partner organisation in fulfilling their organisational goals (who benefit indirectly from these first two goals and for whom the third goal is a priority); and the *tertiary beneficiaries* were the universities involved in the project.

This vision of the project has remained unchanged throughout the arc of the partnership. However, it was only through the passage of time and multiple phases of development, implementation and testing that it became apparent that *neither researchers nor partners were aware of the subtleties involved in co-designing to achieve this vision.* This can be seen in the way that the remaining three phases of co-design – co-determining, shaping and evaluating – shifted over time during the project, with old agreements requiring renegotiation in light of new knowledge and of changing circumstance.

13.3.3.2 Phase 2 Co-Determination

This project shifted as the understanding of key stakeholders changed. The project was initially conceived as having a focus on pre-service (those teachers still studying) and early career (those in their first five years of teaching) teachers, and the project used surveys and participatory workshops to understand the needs of these users. It also involved teacher educators (academics involved in teaching pre-service teachers in tertiary settings) at every stage in the design process as experts in the challenges of becoming a teacher.

It became clear through these workshops and ongoing evaluation that this notion of a platform for "teachers" was a foundational misunderstanding (on the part of the researchers) of the way that these stakeholders self-identify. Another way to put this is that *it was only through co-design with participants that it became clear that stratification of co-design was going to be required.* This is an example of the way that it is not always clear *who* needs to be involved in co-design at the outset.

In this project, primary teachers made it clear that they would feel far more motivation if they were involved as "primary teachers," not as part of a greater community of teachers; and, even further, secondary teachers would only want to be involved if they were associated with their subject area, e.g., as a "science and mathematics teacher," not as a secondary teacher. As these subtleties were realised, a pivot was required: co-design with teachers needed to be stratified into these different self-identity groups. It also become clear that current, experienced teachers (as the authorities within these spaces who provide legitimacy for any platforms) also needed to be involved, even though these teachers were not the primary beneficiaries. This necessitated a re-design of the project to focus on *supporting communities of teachers* rather than *a single platform to support the greater community of teachers*; this re-design work is ongoing.

Allowing these subtleties of the research to emerge over time has deepened the relationship between researchers and partners, with additional trust (from both parties) being developed: challenges had been presented and both partners and researchers remained committed to the overall vision, even though this necessitated co-determination of new goals and priorities.

13.3.3.3 Phase 3 Shaping

The shape of the collaboration and co-design also shifted from initial conceptions in this project, due to a change in the research context. TeachConnect initially had small-scale funding from partners to co-design and facilitate the online platform by adapting open-source software and working remotely with university partners in co-designing and implementing it. However, TeachConnect came to be included as a core component of a $3,200,000 (AUD) STEM teacher education project (StepUp; Dawes et al., 2017). Funding for TeachConnect was thus scaled up by an order of magnitude and new external accountabilities (beyond partners and researchers) were introduced. This shifted the relationship in that it opened up new avenues for co-design, where the research team was now travelling around the state, meeting face to face with hundreds of teachers, running workshops weekly and working with a commercial development team to create a platform with a much broader scope (described in Kelly et al., 2019). The shape of the project then needed to change back to a more sustainable format once the StepUp funding ended.

Once again, this shows that, despite efforts at the outset to co-determine the shape of the collaboration, situations do change over time. However, through ongoing commitment to the vision the partnership has adapted to changing circumstances, renegotiating the shape of the project as needed and deepening trust in the process. One question that may emerge here is: when is a project considered completed (and a new project commenced)? In this long-running project the overarching methodology was of *Design-Based Research* (Collins, Joseph and Bielaczyk, 2004), in which a single project has multiple phases of re-design (with stakeholder engagement) and each phase builds upon the last. In other circumstances a researcher might conceive of such a long-running project as multiple smaller projects – ultimately, the way that the research is conceptualised is yet one more area that requires negotiation between researchers and partnerships.

13.3.3.4 Phase 4 Evaluation

Finally, the approach to evaluation in the project also emerged over time due to changing government agendas and available methods. A shift in research policy from the government caused the research partner to shift from an evaluative focus on how the project has led to positive outcomes for *teachers* to a need to demonstrate how these positive teacher outcomes have led to positive outcomes for *students*. Currently the researchers are still in discussions about how the evaluation of the platform can expand its scope to include this broader impact – effectively expanding the primary tier of beneficiaries. Further, methodological developments for the evaluation of online platforms for teachers have developed over the seven-year course of the project, leading to more nuanced forms of evaluation (Kelly and Antonio, 2016; Lantz-Andersson, Lundin, and Selwyn, 2018). Again, these changes only emerged through time over the course of the project and required ongoing discussion with partners.

13.4 CRITICAL INSIGHTS

Of course, these are three separate and unique cases. However, the framework is promising, and being able to contextualise the ongoing partnership using the framework allows for further evaluation of the existing partnership, while also helping to provide a guide for future successful interactions with corporate industry partners. What follows is a summary of critical insights from the authors about the key attributes of research processes needed to enable deeper and more meaningful relationship with research collaborators.

One of the key determining factors contributing to establishing trust and a meaningful collaboration was in the way the first case study applied principles of dialogic engagement and reversal of expertise to fully embrace a partner-led approach to determining the research focus. This crucial step led to a clear understanding between the youth counselling service provider and researchers of the primary, secondary and tertiary beneficiaries of the research. The main purpose of this was to ensure a shared frame of reference that young people's priorities and help-seeking needs were the primary drivers of every aspect of the partnership. The impact of this step in establishing a foundation of trust and shared value defined not only the research design, but the tone and quality of the research partnership for the remainder of the project, and subsequent initiatives underway.

A key insight from the second case study was the critical importance of prioritising the initial stages of the research partnership with corporate industry by focussing on establishing trust (Canhoto, et.al., 2016; Ulaga, 2003) as a necessary foundation to longevity. In this case, once the initial three steps were achieved, academic buy-in, face-to-face interaction and helping the partner to achieve wins first were crucial before any other successes were achieved. The impact of this dialogic approach has led to establishment of a high-profile international collaboration spanning undergraduate teaching, commercial research and R&D.

The main insight from the final case study is in acknowledging that, where multiple stakeholders with diverse needs and self-perceptions of alignment to a given project are involved in constant and complex communication, researchers should remain open to ongoing and iterative evaluation of research priorities so as to create space for co-design to emerge *in situ*, as a result of open dialogue. In contrast to the first two case studies, the impact of such an approach allows for the project to be agile and for Phase 1 (co-design of research priorities) to remain an ongoing and iterative process. It demonstrates how such an openness can lead to deeper trust within a partnership through a tested commitment to the initial project vision.

Across all cases studies, several key underlying principles of applying co-design as an enabler of meaningful partnership emerged, including that co-design as a framework enables:

- A relational dynamic between researchers and collaborators
- Mutual benefit, shared value and reciprocity
- Application of co-design principles in varying modes and with varying scalability
- Collaborators to feel valued as experts of their reality

The principles extrapolated from the case studies and summarised here illustrate that co-design has the potential to enable research impact, whether it is embedded across all research phases, implemented in the early planning stages or allowed to emerge over time. A common principle to all three case studies was inverting the notion of expertise by foregrounding stakeholders' local and operational knowledge (in case of industry or community partners) or lived experiences (in case of individuals engaged as co-designers), and, by this, signalling commitment to a partner-led approach and establishing trust. While use of formal strategies to build connections, e.g., project development workshops, is important, the use of informal strategies to build connections should not be underestimated as it builds rapport and creates a chance for more open discussions about perceived mutual benefit, with opportunity to form a value-based alignment. Lastly, the case studies show that identifying a primary beneficiary of the research enables researchers to take a partner-led approach to research design.

13.5 CONCLUSION

This chapter has examined a relationship between research impact and research process, and demonstrated how design-led methodologies like co-design might be applied as a framework for creating meaningful research partnerships. The chapter has also provided examples of three contexts where co-design was used to plan research impact at three levels: (1) in the context of a specific project; (2) to build a foundation for partnership spanning multiple initiatives; and (3) as a framework for deepening engagement even as research priorities crystallise or emerge over time. The case studies demonstrate that application of co-design to research design creates non-linear, causal interdependencies, framed as "a dialogic process, an intense (and perhaps endless) 'conversation' between researchers and collaborators" (Nowotny et al., 2001, in Greenhalgh et al., 2016). More empirical data, case studies and rigorous documentation of how research co-design is conceptualised, implemented and evaluated is needed. In conclusion, this chapter aims to show that, in the context of research impact and creating meaningful research partnerships, designers contribute a necessary depth of expertise in design research methods and a critical skillset – specifically in co-design, participatory design, collaborative design, consensus design – to what is an underrepresented area of research co-design.

ACKNOWLEDGEMENTS

The researchers would like to thank QUT Design Lab, research collaborators and funders, including young people involved in *niggle* design, yourtown (KidsHelpline), Australia Research Council and QUT Cube; the BMW Group; and the Queensland College of Teachers for their long-running support of the TeachConnect project.

REFERENCES

Australian Research Council. (2019). Research impact principles and framework. Retrieved from www.arc.gov.au/policies-strategies/strategy/research-impact-principles-framework

Barnes, T., Pashby, I., & Gibbons, A. (2002). Effective university–industry interaction: A multi-case evaluation of collaborative R&D projects. *European Management Journal*, *20*, 272–285.

Canhoto, A.I., Quinton, S., Jackson, P., & Dibb, S. (2016). The co-production of value in digital, university–industry R&D collaborative projects. *Industrial Marketing Management*, *56*, 86–96.

Centre for Social Impact. (2018). The compass: Your guide to social impact measurement. www.csi.edu.au/research/tools-and-guides/compass-your-guide-social-impact-measurement/ accessed March 30, 2020.

Chisholm, J. (n.d.). Q&A what is co-design? Design for Europe. http://designforeurope.eu/what-co-design

Collins, A., Joseph, D., & Bielaczyc, K. (2004). Design research: Theoretical and methodological issues. *The Journal of the Learning Sciences*, *13*(1), 15–42.

Dawes, L., Loughlin, W., Whitehouse, H., Adams, P., Diezmann, C., Nugent, M., … Mascadri, J. (2017). Transforming the nature and delivery of mathematics and science secondary teacher education in Queensland. Final report.

Edmondson, G., Valigra, L., Kenward, M., Hudson, R.L., & Belfield, H. (2012). *Making Industry-University Partnerships Work - Lessons from Successful Collaborations* (SciencelBusiness Ed.): SciencelBusiness Innovation Board AISBL, *94*(2), 392–429.

Greenhalgh, T., Jackson, C., Shaw, S.., & Janamian, T. (2016). Achieving research impact through co-creation in community-based health services: Literature review and case study. *The Milbank Quarterly.*

Hacker, K. (2013). *Community-Based Participatory Research.* Sage Publications.

Kelly, N., & Antonio, A. (2016). Teacher peer support in social network sites. *Teaching and Teacher Education*, *56*, 138–149.

Kelly, N., Clará, M., Kehrwald, B., & Danaher, P. (2016). *Online Learning Networks for Pre-service and Early Career Teachers*. UK: Palgrave Pivot.

Kelly, N., Clará, M., & Kickbusch, S. (2015). *How to Develop an Online Community for Pre-service and Early Career Teachers*. Paper presented at the ASCILITE 2015, Fremantle, Western Australia.

Kelly, N., Kickbusch, S., & Dawes, L.A. (2019). TeachConnect: An online platform for teachers.

Kelly, N., Kickbusch, S., Hadley, F., Andrews, R., Wade-Leeuwen, B., & O'Brien, M. (2018). Raising the quality of praxis in online mentoring. In *Educating Future Teachers: Innovative Perspectives in Professional Experience* (pp. 123–134). Springer.

Khazragui, H., & Hudson, J. (2015). Measuring the benefits of university research: Impact and the REF in the UK. *Research Evaluation*, *24*(1), 51–62.

Lantz-Andersson, A., Lundin, M., & Selwyn, N. (2018). Twenty years of online teacher communities: A systematic review of formally-organized and informally-developed professional learning groups. *Teaching and Teacher Education*, *75*, 302–315. doi: 10.1016/j.tate.2018.07.008

Productivity Commission. (2010). *Contribution of the Not-for-Profit Sector*, Research Report, Canberra.

QUT Creative Industries Faculty. (2019). Research impact symposium, chaired by Oksana Zelenko and held at QUT Institute of Health and Biomedical Innovation (IHBI), 8 April 2019, Brisbane, Australia.

Rappaport, N., Alegria, M., Mulvaney-Day, N., & Boyle, B. (2008). Staying at the table: Building sustainable community-research partnerships. *Journal of Community Psychology*, *36*(6), 693–701.

Rybnicek, R., & Königsgruber, R. (2019). What makes industry–university collaboration succeed? A systematic review of the literature. *Journal of Business Economics*, *89*, 221–250.

Sanoff, H. (1990). *Participatory Design: Theory & Techniques*. Raleigh, NC.

Slattery, P., Saeri, A., & Bragge, P. (2020). Research co-design in health: A rapid overview of reviews. *Health Research Policy and Systems 18*(1), 17.

Thorgren, S., & Wincent, J. (2010). Interorganisational trust: Origins, dysfunctions and regulation of rigidities. *British Journal of Management*, *22*(1), 21–41.

Times Higher Education. (2019). University impact rankings 2019. *THE World University Rankings*. Retrieved from www.timeshighereducation.com/rankings/impact/2019/

Times Higher Education. (2020). World university rankings 2020. *THE World University Rankings*. Retrieved from www.timeshighereducation.com/world-university-rankings/2020/world-ranking

Ulaga, W. (2003). Capturing value creation in business relationships: A customer perspective. *Industrial Marketing Management*, *32*(8), 677–693.

Werker, C., & Ooms, W. (2019). Substituting face-to-face contacts in academics' collaborations: Modern communication tools, proximity, and brokerage. *Studies in Higher Education, 45*(7), 1431–1147. doi: 10.1080/03075079.2019.1655723

White-Cooper, S., Dawkins, N.U., & Kamin, S.L. (2009). Community-institutional partnerships: Understanding trust among partners. *Journal of Health Education & Behaviour*, *36*(2), 334–347.

Wrigley, C. (2016). Design innovation catalysts: Education and impact. *She Ji: The Journal of Design, Economics, and Innovation*, *2*(2), 148–165.

Zelenko, O. (2012). Design and ethics in digital mental health promotion. In Felton, E., Zelenko, O., & Vaughan, S. (Eds.), *Design and Ethics: Reflections on Practice* (pp. 97–110). London: Routledge (Taylor & Francis Group).

Zelenko, O., & Bridgstock, R. (2013). Design: A meta-level discipline. In *School of Design Symposium: Design Thinking in Higher Education*, 1 October 2013, Brisbane, QLD.

Zelenko, O., Stoyanov, S., Dalgleish, J., Hides, L., Sade, G., Staneva, A., & Cheers, J. (2018). Co-designing mobile help-seeking toolkit with young people: An Australian case study. *Design 4 Health*, 4–6 September 2018, Sheffield, UK.

14 Design Curricula
Navigating Process and People

Deanna Meth, Sheona Thomson and Dean Brough

CONTENTS

14.1 INTRODUCTION

> ***Senior Manager:*** *I sense a fair bit of school-wide grumbling about our new course. We've worked so hard to get through the paperwork, taking on board QUT's strategic vision, responding to suggested changes from the school committee, reworking our proposals following faculty and institutional comments. But now the resistance seems greater than ever. How on earth have we come to this point?*
>
> ***Program Director:****Yep, I just don't get it. So many of our students have told us they wanted something different. Just yesterday when I met with them, one said again how important it was for 'Design disciplines to be able to interact and work alongside each other' and his friend added: 'I think a double degree would be the best way to have industry ready graduates'. They really want change. So we're giving students what they want, and our industry friends as well have repeatedly told us things like 'design needs connection with business, law, science, etc., which all use design at their core' and graduates need to have 'respect (for) differing ways of working and thinking, being open to a changing world'. All these things – connectivity, collaboration, risk-taking, and becoming a responsible designer who does good in the world – they're all there in the new degree, and yet the grumblings persist. I'm perplexed and quite frustrated by it all.* Conversation between Senior Manager and Academic Program Director c.2017.

The conversation above reflects a fairly typical one at times of large-scale curriculum development and change in education. It's all there – the raft of institutional approval processes to be churned through, the tough job of leadership, demands and pulls from a range of stakeholders and disciplines and, finally, people feeling disenfranchised from it all. Whilst some would view this as academics resisting change and seeking to protect themselves, others see it as integral to defending an education they feel to be right, possibly threatened by the implications this change may have for them, their practice and, importantly, their professions.

What, after all, is a Design School for if not to educate future Design professionals and further knowledge and research in Design as a discipline? Somewhere, buried amongst the weighty demands of policy, process and people, lies the education – the Design curriculum and students' learning experiences and achievements. When developing curricula, whether at a large or small scale, staying on track in the midst of this complexity can be overwhelming. It is important to remember that the end goal is a curriculum that does what it needs to do for the specific context within which it is situated.

14.1.1 CHAPTER OVERVIEW

Cross-referencing sector-wide literature and theoretically underpinned approaches related to developing higher education curricula, pedagogies and design education in particular, this chapter provides authentic examples from our own experiences of developing a new Design curriculum at Queensland University of Technology (QUT). Acknowledging the breadth of curriculum design literature already available on the aspects identified above, in addition to introducing the most commonly referenced literature and approaches, examples from our own recent curriculum re-design exercise (*in italics*) and simple prompt questions aid as practical tips in each section, guiding you through these multiple considerations. This combination of theoretical and practical perspectives offers you, the reader, deeper insights into the complex environment within which such work is situated and the ways in which it might be navigated.

Underpinned by a future-focus for curricular development, the opening section looks at developing a vision for the course. "What should our graduates be like? What do they need to be able to do?" ensuring we define a set of desired knowledge, skills and attributes or capabilities as graduate outcomes and consider how we might seek to ensure a balance of such outcomes in our curriculum. Note at this point that, at QUT, the term "course" denotes the entire course or program of study; for example, towards a degree or diploma qualification. These entities are subdivided into "units" of study, sometimes called "modules" at other institutions. In the midst of these questions, where are the pressures coming from for answers? Evans, Kerr and Kelly (2021) highlight that the curriculum reflects the nature of key stakeholders' design practices and processes, and that this will vary by context. As the opening scenario implies, these include disciplinary drivers, forces from within and without specific disciplines; institutional nature and strategic context, which will require and lead to different educational offerings and, importantly, the student cohort – their backgrounds, learning needs and aspirations.

This is the key point from which to proceed incrementally. Progressing beyond the vision, start to develop your curriculum, including intended learning outcomes, how these will be tested as being achieved, i.e., the assessments, and the knowledge and skills content necessary to ensure students are able to accomplish this. The chapter moves to explore a constructivist approach to aligning these curricular elements for such intended outcomes (Biggs and Tang, 2007, p. 52). That is, considering how individual units, their intended learning outcomes and assessments combine to make a coherent program or learning journey with an overall set of intended outcomes for students. For design disciplines, this is made substantially more complex by notions that design curricula are inherently "sticky" (Orr and Shreeve, 2018) and ambiguous in their enactment due to the space for individual students' interpretation and creativity (p. 73).

"But this just feels like a process list" you might say, and to some extent you would be right. This is the mechanics of curriculum design, but for this seemingly straightforward set of steps to be meaningful, it cannot simply be a "tick box" exercise. It is underpinned by the need to adequately and fully explore the disciplinary context, all stakeholder requirements and, importantly, ensure an evidence base that substantiates both what students will learn and the best way to go about achieving this. Recognising this complexity and the specific needs of design education is a fundamental step in being able to successfully develop and deliver it.

After examining how best to engage and partner with students during the curriculum development process, the chapter considers a set of areas so often forgotten in the whirlwind of course development: the elements that "wrap" around the curriculum that can make or break new courses, such as the physical spaces and digital platforms where learning takes place, the industry-relevant software, the budgets and staff; and, finally, and most critically, the people involved, their expertise, buy-in and ownership. Such local "cultures" and contexts can add additional layers of complexity and challenge to this task (Trowler and Cooper, 2002, p. 231) and are crucial to consider.

Ultimately, curriculum development will always be a site of negotiation and compromise involving all elements, and the result will be about balancing in the best possible way, the range of tensions, requirements and innovative ideas on the table. The inevitable progression of any new curriculum involves establishing how best to teach it, and Evans et al. (2021), Scott and Thomson (2021) and Thomson and Scott (2021) showcase a range of examples and opportunities for innovative teaching in design.

14.2 DEVELOPING A COURSE VISION AND CURRICULAR ARCHITECTURE

14.2.1 A Future-Focused Framework

> *Our institution offers a 'Future focused curriculum design framework' which 'provides support and resources for curriculum leaders and teams to conceptualise and realise future focused learning and to assist with accreditation and reaccreditation of courses and programs. The approach is organised around design stages. These stages are also linked to key feedback and approval milestones in the course accreditation/reaccreditation pathway'* (QUT intranet resources, retrieved 29 March, 2020).

In developing a course vision, it is important to also incorporate mechanisms that will ensure the vision informs all stages, and is realised in an holistic course implementation and delivery. Unfortunately, vision and aims are too easily lost once the process and "paperwork" begins and course elements are distributed across staff teams. If work is divided at this stage, it is difficult to reconstitute the holistic perspective, which often leads to staff teaching in disconnected siloes. With a diminished sense of the whole, the disconnected delivery of parts translates at best to students experiencing a lack of coherence across units, and, at worst, unnecessary duplication of content as well as curricular gaps. In this case, the whole is most definitely greater than the sum of its individual parts. The future-focused curriculum design framework mentioned above was used in developing the new Bachelor of Design degree at the university. This framework takes the course team through four key stages, the first of which, "Visualise," has a focus on future needs. In this framework, the processes and "paperwork" sides of the course are knit together with the developmental considerations. Meth, Russell, Fitzgerald and Huijser (2020) outline this framework in full, drawing out the multiple ways in which curriculum development when referenced to relevant scholarly evidence and practice can be much more meaningful than a tick-box process, no matter what your context.

14.2.2 Anticipating Future Needs

> *In 2019, our School of Design rolled out a new degree encompassing seven major study areas, 19 double degree pathways, and an international pathway. Whilst graduates are expected to emerge as specialists in their disciplines, the degree was developed*

in response to stakeholder needs, including industry and students, as well as institutional strategic context which prioritises transdisciplinary 'real world learning' experiences encompassing work-integrated learning, authentic learning and assessment and recognising and fostering deep connections with community organisations and industry – the future employers of our graduates.

It is not unusual for a degree program or course to be subject to numerous and diverse requirements. In our case, the new curriculum had to encompass a range of components to satisfy the breadth of current and anticipated future needs. From the outset, to anticipate and best serve future needs it is important to understand who the various course stakeholders are, including students (and their parents), academics and employers. Barnett and Coate (2005) use the term "zones of influence" to characterise these multiple claims on courses (p. 71). A summary of key influences and the ways in which such influences might play out are presented for consideration in Table 14.1. Recognise the inherent tensions between key zones of influence and the balance to be struck in satisfying related demands, as well as opportunities for compromise.

The new Bachelor of Design degree curriculum is structured to give learners in each of the seven majors a mixture of foundational skills units streamed by broad discipline clusters ("spatial" and "experiential"), discipline-specific units and a common core of transdisciplinary, real-world problem-based units (Figure 14.1). The course model preceding this predominantly focused on disciplinary knowledge and skills, and whilst strong in this area, it was arguably more siloed, potentially undervaluing multidisciplinary and other pathway opportunities both within and beyond the School of Design. The new curriculum structure also enabled the introduction of 19 new double degrees combining design with business, law, engineering and urban and regional planning. Delivering design education and research that is firmly situated within, and responsive to, an increasingly interdisciplinary world was the principal aim of our curriculum transformation (as expounded through the overarching theme of this book).

As a direct response to the interdisciplinary agenda, the course structure and content considers what it means for the future of graduates, disciplines and the design sector overall. It balances depth of disciplinary knowledge with a breadth of understanding – including the broader contexts within which design disciplines sit. Learners are invited to operate in real-world, transdisciplinary settings and appreciate diverse stakeholder perspectives and needs, as well as understanding their motivations and positionality as responsible and ethical designers.

14.2.3 Curricular Domains and Course Learning Outcomes

A transferable model for you to consider in your curricular developments, and articulate what has been described above in terms of educational theory, is that of Barnett and Coate (2005). They conceptualise a curricular framework with three distinctive but overlapping domains of "knowing, acting and being" (p. 70). More recently, Parkin (2017) developed an expanded version of this model that is more nuanced

TABLE 14.1
Key 'Zones of Influence' on Curriculum Development (Concept Adapted From Barnett and Coate, 2005, p. 71)

Zones of Influence	Description
Institutional nature and strategic context	Depending on the institution within which the curriculum sits, the educational focus may vary; for example, highly theoretical and research-heavy, more focused on practical application or somewhere in between. Additionally, every institution will have declared strategic aspirations for their curricula, such as those related to employability skills, delivering programs online, embedded sustainability, industry engagement or future researchers. Whilst academics developing new design courses may not always necessarily warm to these, it is important to engage with such requirements from the outset, to avoid becoming a victim of last-minute, and often less than ideal, "bolt-on" solutions.
Disciplinary drivers and professional and accrediting bodies	Recognising that there might be a wide range of disciplines situated within a "design discipline" package, these are likely to present a diverse set of curricular needs and Evans et al. (2021) have noted some of the differing expectations and requirements such variations bring. Additionally, in some instances such as architecture, related external professional or accrediting body requirements can bound what is necessary or possible within your course. Navigating what is critical, with space to explore what is necessary for your institutional and disciplinary context, as well as possible spaces for beneficial overlaps across disciplines, is essential.
Employers	Universities worldwide are under increasing pressure to produce graduates who are ready for the workplace. In Australia, the education minister, Dan Tehan was recently cited directly linking university performance funding to their ability to produce "job-ready graduates with the skills to succeed in the modern economy" (Bolton, 2019). This necessarily will play out directly in university curricula, and the focus given to specific skills, knowledge and capabilities. For design disciplines, the skills agenda will manifest as a struggle between design-specific skills and what are termed "soft" or generic skills also needed in the workplace, such as communication and leadership. Through such pressures, researchers have warned against the dangers of education tending towards "training" (Parker, 2003, p. 530).
The student cohort	Your intended student cohort will be linked to the rationale for the course, as well as ultimately its marketing, with close connections to skills and employability discussed above – students' expectations of the curriculum and their intended career and/or study pathways. Bring the "future-focus" into these considerations. Where might you see design graduates being employed following their studies? What will their learning support needs be for them to achieve this? Is the course a pathway to another necessary qualification? It may be that yours is only a short course – what are the implications for learners and the curriculum in this instance?
Underpinning assurances of quality	For a course to be considered viable, marketable and aligned to external expectations and institutional strategic intent, there will be institutional assurances of course quality and standards. The way in which such assurances are gained through benchmarking, performance measurements and internal and external evaluations of stakeholders, will also play a role in the nature of the curriculum.

with four curricular domains: knowing, doing, creating, being (p. 127). This "fully engaged curriculum" model draws on key concepts related to learning and development, including Bloom's Taxonomy of cognitive levels (1956), and Meyer, Land and Baillie's (2010) threshold concepts of knowledge, as well as students' identity and transformations (p. 128). These frameworks are also not dissimilar to that of Cross (2011, p. 55), related to "knowing, thinking and acting" dimensions of the design activity. The ultimate aim of all models is to ensure curricular balance and integration across the domains so that, for example, there is not only a focus on knowledge acquisition, but the necessary means for students to apply that knowledge, whilst also understanding and reflecting on their own positions and identities (professional or personal). The latter aspect relates to the domain of "being": an area that the new transdisciplinary Impact Lab units seek to develop more strongly in design graduates.

Figure 14.2 shows how the domains of knowing, acting and being, annotated on an extract from Figure 14.1, are interpreted as elements across the course in varying proportions. Importantly, they too should not be siloed, and different curricular areas are noted as encompassing several dimensions.

In developing curricula, some course teams are more comfortable with using the commonly used classifications of knowledge, skills and application for examining the balance of components in their course. The authors have noted the recent use of classifications such as "knowledge, practice and values, disposition" for a health course, and "knowledge, design and professional practice" for a design course to denote aspects of the curriculum. Importantly, Barnett and Coate (2005, p. 70)

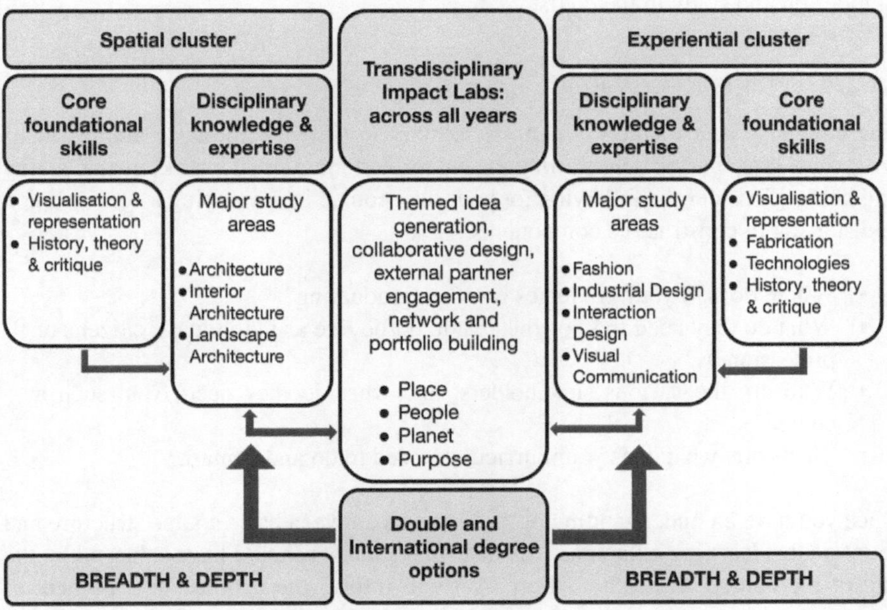

FIGURE 14.1 Diagram showing the curricular architecture of the Bachelor of Design degree

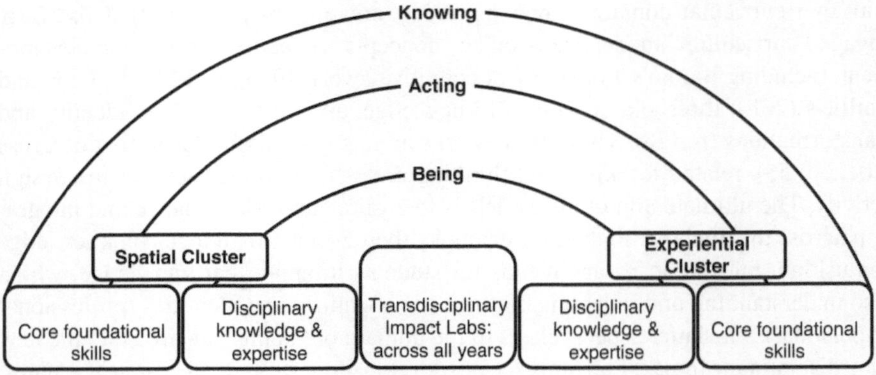

FIGURE 14.2 Sketch interpreting curricular aspects related to 'knowing, acting and being' domains across different curricular elements of the Bachelor of Design degree

evidence how the balance of these domains varies across discipline and knowledge areas – key to the zones of influence outlined above. Depending on the specific focus of your design course, and potentially also the nature of your institutional setting, your balance may look somewhat different. In our curriculum, for example, the architecture major, as part of a nationally accredited professional program, must include compulsory profession-specific skills and knowledge content. On the other hand, study areas such as fashion and visual communication have more flexibility with both the course structure and content, but also in the elements students choose to focus on and study in the course.

14.2.4 Points to Consider

The nature of your institution and disciplinary (or interdisciplinary) area of focus will play a large part in determining the balance of domains such as knowing, acting and being. Consider the following for your own course in developing a course vision and starting to construct its components:

- Where do/will your graduates go when graduating?
- What do they need to know/be able to do/be like as individuals, citizens or professionals?
- Who are the various stakeholders, and what do they need from such a course?
- Therefore, what does your curriculum need to do and contain?

Once you have an understanding of the elements above, a basic course structure and set of course learning outcomes, "the learning students should have achieved by the end of the course" (Mackh, 2018, p. 47) will inform the detailed development of your curriculum. At this point, you should be armed with a full set of desired knowledge, skills and capabilities (or attributes) you have determined through course team discussions. For further guidance on this area, Mackh (2018, p. 48–55) outlines a detailed process for incorporating all aspects discussed above.

Developing responses to the questions above should be a negotiated process amongst your colleagues and stakeholders. A common concern for all course teams will likely relate to the necessary balance of elements in the curriculum: some will feel they are making sacrifices and the final outcome may feel like a compromise to many. What it will be, however, is a collectively owned outcome where the best possible solution to fit the multiple influences has been found. And such is the nature of change. This "people" aspect of curriculum development should never be underestimated and is discussed in more detail in Section 14.5.

14.3 THE CURRICULUM DEVELOPMENT PROCESS

To undertake curriculum design and development, most higher education experts would recommend that team members possess an introductory understanding of educational theory, as well as a shared team philosophy of learning and teaching, or at least an appreciation of the range of learning and teaching philosophies possible. Opinions are likely to vary between individuals, often linked to their own beliefs on the nature of knowledge and disciplinary traditions, as well as their prior educational experiences. Institutional approaches or strategic visions within the zones of influence highlighted earlier, which themselves often have specific philosophical underpinnings, may also need to be incorporated.

Mackh (2018) gives a good overview of the range of theories often drawn on across the areas of student development, learning and teaching (p. 15). The most commonly adopted and widely used theoretical basis for higher education curriculum development today is that of constructive alignment (Biggs and Tang, 2007). Ashwin et al. (2015) note that, while some educators are uncomfortable with the constructivist grounding for this, their discomfort is generally outweighed by its key strength – the notion of complete congruence across a curriculum (p. 162), where all curricular elements are visibly "aligned." In substantiating the notion of congruence, the important point is made that this is not only for the use of the teacher, but needs to be made explicit and visible to the students.

14.3.1 Connecting and Mapping Curricular Elements

Drawing on constructivist theory where "learners use their own activity to construct their knowledge or other outcome" the principles of constructive alignment provide a framework for curriculum design by aligning curricular components to the intended learning outcomes (Biggs and Tang, 2007, p. 52). That is, ensuring coherency between the course learning outcomes and individual unit learning outcomes (extending Mackh's [2018, p. 47] definition of course learning outcomes cited earlier, the learning students should have achieved by the end of the unit), and considering the ways in which individual units (including their assessments) are taught and how students learn, all combining into a coherent program of learning for students.

Key elements to consider within your aligned curriculum should include:

- Course vision and course learning outcomes as a starting point, articulating the knowledge, skills and capabilities you wish your graduates to be able to attain
- Assessments that appropriately test learners' attainment of such knowledge, skills and capabilities, as well as the nature, timing and volume of these assessments
- Curricular content; what needs to be taught to ensure that students are able to acquire the knowledge, skills and capabilities, and, within this
- Consideration of how best learners might learn and develop through scaffolding their learning and knowledge and skills acquisition across the course
- Pedagogic (learning and teaching) models for best enabling this, including plans for the support and development of students, feedback mechanisms for students to further develop, and formative assessments

14.3.2 Assessment

A framework for assessment which sits alongside all curricular elements will ensure that students are both adequately assessed and not over-assessed (a frequent issue in courses with many small units), and that there is a range of assessment modes appropriate not only to the content being taught and outcomes expected, but also to inclusively accommodate a range of student learning needs and preferences. For a useful step-by-step consideration of assessment in and across your course, including standards, grading and feedback, see Ashwin et al. (2015, pp. 250–273).

Across our Design degree, types of assessment include: drawings, models, design portfolios, professional portfolios, workbooks, reports, essays, posters, pitches, reflections, and even an exam. These might be individual and/or group, disciplinary and/or interdisciplinary, and with or without external partner involvement. Each type of assessment is designed to suit the knowledge, skill or attribute being developed by the students. Predictably, across all design disciplines, there is a dominance of assessments that test students' design skills and knowledge, and this is necessarily so as students emerge as professionals from their studies. Furthermore, university policy requires that a proportion of all assessments are 'authentic', that is, assessments need to represent that 'performance or product' (Ashford-Rowe, Herrington and Brown, 2014, p. 208) where such skills, knowledge and capabilities play out in the real world, i.e., through the diverse pathways and workplaces of professional designers.

As with other elements of your course, the authenticity of your design assessments will vary depending on your course and unit learning outcomes, and influenced by institutional context and intended or actual graduate destinations. Designing assessments that are aligned to outcomes while being meaningful and rewarding for students to undertake is challenging. Students will question the point of assessments if they do not understand their purpose (Ashford-Rowe, Herrington and Brown (2014, p. 208). Authentic assessment that is relevant with a clear sense of what it is building towards can help to alleviate this.

Although authentic to real-world practice, group work as assessment can be problematic from the student perspective. While the benefits for students of group work are manifold (Race, 2007, p. 125), learners are often reluctant and unhappy to engage in it, because of the perceived, and sometimes real, risk to their individual academic record. This risk lies in situations where group members each receive the same mark for a group output when the work has not been evenly distributed during the process. Carefully consider any implementation of assessable group work, and prepare to mitigate for student tensions, including considering how you might establish groups and create space for students to adequately evidence their individual contributions. As the specifics of group work and group assessment are beyond the scope of this chapter, for a comprehensive discussion of group teaching and assessments, see Race (2007); and for practical examples, see Scott and Thomson (2021).

It is important to set and communicate clear and consistent expectations of standards, and how students' performance in assessment will be measured. At our institution, criterion-referenced assessment (CRA) and a standards-based approach help students understand how their assessments will be marked, detailing the specific knowledge, skills and capabilities they are expected to demonstrate. A criterion-referenced approach also supports consistency in grading. A rubric of expected performance standards against a set of criteria (the CRA sheet) is provided to students and markers for each assessment. There is an inherent tension here, however, and Ardington and Drury discuss the constraints such criteria might place on creativity and freedom of expression (2017, p. 158). We urge all who develop CRAs to do so with this in mind. Perceived subjectivity in design assessment, related in particular to its creative nature and the judgements exercised, for example, in relation to studio-based teaching (Ardington and Drury, 2017), can be highly problematic for students and for ensuring consistency of levels of expectations and achievement. This is part of the inherent "stickiness" of art and design curricula examined in detail by Orr and Shreeve (2018).

Norm-referenced assessment, where students are compared to one another, with grades distributed "on a curve" and students succeed or fail based on the average of the cohort is arguably less appropriate for design education, not least due to the aforementioned perceptions of subjectivity in grading. Mackh (2018) notes that norm-referenced assessment also creates a highly competitive environment amongst students. The "bell-curve" of norm-referenced assessment means that there will always be instances where students must fail because their performance is lower than that of other students (p. 64). This can foil the creation of an exploratory and collaborative studio learning culture.

Since it is important that students receive timely feedback on the path to a final grade, not all assessment needs to be for marks. Consider which assessments form part of students' "summative" grades, and other points where assessments as "formative" learning experiences, with opportunities for students to receive feedback on such learning, might work well.

Our assessment policy notes that all assessment should be standards-based and criterion-referenced. To avoid over-assessment of students, it stipulates a maximum of three assessment pieces in each unit of learning. In an effort to balance assessment of students' individual performance with collaborative group contributions, it also stipulates that the percentage of group work across a set of assessments cannot be higher than 50%. To create risk-free spaces where students have more freedom to experiment, either in their designs or collaborations, there is also the opportunity to remove numerical grades entirely, and have the grading classified as simply 'Satisfactory or Unsatisfactory', with an accompanying rubric stating the expected criteria to meet the satisfactory bar. This grading system has been introduced across all transdisciplinary Impact Labs (noted in Figure 14.1) and has been very well received by students, with no decline in the level of students' outputs and achievements noted.

Your institution is likely to have policies in place for the processes by which you manage assessment, as well as what elements they feel constitute good assessment practice.

14.3.3 Scaffolding Students' Learning and Feedback That Feeds Forward

Integral to an aligned curriculum is the notion that students cannot be expected to display comprehension of advanced concepts or achievement of advanced skills if they have not yet been given the building blocks on which to grow this knowledge. For most design disciplines, this developmental pathway likely commences with a suite of foundational skills units that include representation and visualisation skills (both manual and digital), fabrication (making) skills and foundational theory related to design and design histories, often interwoven with more conceptual and creative disciplinary units. Mackh (2018) notes that these need to be "systematically and deliberately" taught (p. 107) before expecting students to perform at higher levels of "mastery" The concept of scaffolding introduces when, where and how students may then build upon such knowledge, skills and capabilities across their degree, expanding and developing their capabilities in these areas – multiple opportunities to practise and ultimately demonstrate a particular level of achievement.

A fundamental example of this occurs in our design curriculum where foundational skills are taught in the first year of study alongside, but feeding into students' main design disciplines through their studio units. These visualisation, fabrication and history/theory skills are then constantly applied and developed across their second and third years of study, gradually becoming more complex and advanced. A more nuanced example is seen in the suite of Impact Labs which gradually grow students' understandings and portrayals of themselves as designers, as well as their skills of transdisciplinary collaborative problem-solving and working with and communicating to external community and industry partners (see Figure 14.1).

To ensure your curriculum is effectively developmental and scaffolded, it is useful to map where specific skills and knowledge are acquired as students progress through the course. At our institution, we map the level of content and assessment across the course lifecycle as "introductory," "developing" and "mastery" with the assumption that, by final year, students will predominantly be working at "mastery" level. A range of other terms describing this "novice-to-expert" continuum may be used.

Linked to the more general concept of mastery of learning, Gusky (2007) builds on Bloom's (1971) concept of the mastery of learning, noting the critical role of ongoing formative feedback for students. Ensuring students receive feedback on their formative learning as well as their summative assessments is key to their development, and, as part of this, ensuring they are able to feed-forward such feedback into their subsequent learning is important (Race, 2007, p. 74). Build in points where students can give and receive peer feedback, reflect on their own achievements and how they can use past feedback to improve future performance. Returning to the complexity of assessment in design disciplines, and particularly the interpretation of criteria and standards between educators and students, you should be aware of how such issues, often related to subjectivity around the creative endeavour, might translate into the feedback discourse. Ardington and Drury (2017) advocate for greater transparency of feedback discourse in design education to bring both consistency and clarity to the experience of learners (p. 164).

14.3.4 Mapping for Balance and Congruence

The institution 'Blueprint' and accompanying Real World Learning 2020 vision promises a course that encompasses real world learning, employs authentic learning and assessment methods, develops broad student perspectives, appreciation for and skills of engaging in diverse cultural perspectives, and supports the development of professional profiles, networks and employability. Echoes of this are therefore threaded across our Bachelor of Design course.

With reference to the need to balance curricular domains, it would be nonsensical for every unit or module to include, for example, work-integrated learning components, or specific content on employability. Similarly, it would be illogical for each piece of assessment to be conducted in the same mode, particularly in a design degree where there is such a wide range of skills and knowledge to be learned. Linked to the discussion on scaffolding above, it is also not good practice to include core elements only once in the curriculum, such that students are not able to revisit and build upon such core elements. Be alert to the common problems of modularised and unitised curriculum, that of the duplication and isolation of content. Subdivision inevitably introduces the potential for siloed, disconnected or minimised learning experiences that prevent students from attaining a deep and coherent education.

While the features of our institution's Blueprint noted above represent many aspects valued in contemporary higher education today, you will likely also have your own institutional vision to inform your curriculum design. In general, mapping of where and how such features play out across the curriculum can be overlaid on course (and unit) learning outcomes and assessments, ensuring necessary coverage and progression across the course. Table 14.2 provides a sample framework for mapping the main elements in your course. Depending on detail required by your institution and the purposes for which you are creating your course documentation, you may wish to include columns with additional details, e.g., the level of the mastery, the number of credit points for each unit or module, volume of learning in hours (e.g.,

TABLE 14.2

Example Adaptable Framework for Mapping Course Learning Outcomes to Unit Learning Outcomes and Assessments, Including Assessment Timings, Volume and Appropriateness (to Strategic/Contextual/Course Needs)

Unit Or Module Name/Level	Unit Learning Outcomes (ULOs) (List)	Corresponding Course Learning Outcomes	Assessments List/ Types	Mapped to Which ULO Number/s?	% of Total Mark	Timing / Due Date	Other Key Features (Add Columns As Required)
Add and cluster lines as required by level or theme							

contact time versus independent study), or some of the features listed in Figure 14.3, which in our course require clear mapping.

Other key features to map will depend on your context and discipline. These may need to be benchmarked against national qualification frameworks or professional standards. In architecture, for example, courses are required to be mapped to professional accrediting body criteria and standards. Some universities have course management software that is able to cross-reference all course documentation to create and represent this mapping and benchmarking. Regardless of how this mapping is produced, the value of the exercise is to visualise how the parts create a meaningful whole for the benefit of your course team, students and stakeholders.

Ashwin et al. (2015) note that, whilst curriculum development ideally starts with a blank slate and is "top-down" and driven by vision and outcomes, it is commonly experienced as "bottom-up" working from a collection of sometimes disconnected units (p. 166). The principle of congruence and cross-mapping applies equally well, as do notions of considering the fuller context of the degree and revisiting whether the degree is leading students towards intended outcomes and achievements.

14.3.5 Points to Consider

- What are the threads or themes that hold your course together?
- What pathways might learners follow through the course? Map some of these for different student personas'
- Have you considered where students will gradually build skills, knowledge and competencies listed in course and unit learning outcomes across the course from novice-to-expert levels?
- Have you considered how all learning outcomes will be achieved through teaching and learning activities?
- Do all learning outcomes map to specific assessments and is there a balance of appropriate assessment types across the course?
- How and where do you build in opportunities for students to receive feedback upon which they have the chance to act, e.g., as feed-forward?

14.4 STUDENT ENGAGEMENT AND PARTNERSHIPS

In developing the new Design degree, we engaged with our student body in a range of ways to shape our new curriculum. As development got underway, we talked with current students about their course experience and sought their aspirational guidance for the new course. These discussions surfaced opinions and frustrations which tallied with those of external partners and employers. Some of these have been highlighted at the opening of the chapter. Once the curricular shell was in place, we employed seven 'Student Associates for Course Design' – current students representing each design discipline – and partnered them with academics from across the course team to shape the course further. Together the associates undertook a large body of research to define designerly capabilities, and develop a conceptual framework for a new Graduate Professional Portfolio element in the course (Meth, Finger and Brough, 2020). They worked alongside academics to inform the design of units and trial and give feedback on learning and assessment activities. The associate role recognised the expertise and unique perspectives that students have to offer, and academics were unanimously confident about the positive difference this made to their curricula and pedagogies.

In recent years, there has been a sector-wide approach to shift students from being passive recipients of education, to active contributors to governance, curricular and extra-curricular activities at educational institutions. The broad agenda of student engagement goes beyond ensuring that students are engaged in their own learning, to include them in the broader educational endeavours of the school and university (Trowler, 2010). Whilst this area is known for its complexities (Matthews et al., 2019), including implications for the ongoing management of student–staff relationships, Cook-Sather, Bovill and Felten (2014) note three foundational beliefs that value students' specific engagement in learning, which acknowledge:

- The insights students bring that can impact positively on academics' practice
- The additional insights to be gained in student–staff design collaborations
- The way in which student–staff partnerships "change the understandings and capacities of both sets of partners – making us all better teachers and learners" (Cook-Sather, Bovill and Felten, 2014)

Stages of curriculum development detailed in this chapter offer the key points at which student–staff partnerships may bring valuable contributions. Meth, Russell, et al. (2020) provide examples of a range of such partnerships that have brought an important scholarly evidence-base to curriculum development processes. Areas where students may be engaged in discussions and curriculum development across a spectrum of student–staff partnerships include:

14.4.1 Consultation

Either as one-off or a series of discussions, or through active student representation and participation on your learning and teaching committees. Appreciate, however, the tendency for one-off conversations to become tokenistic, and also consider the importance of meaningful consultation, as well as equitable partnerships. For example, consider carefully the power imbalance when a single student sits within a committee of 12 academics, expected to speak to the views of a diverse student body, unless processes and structures are clarified as to how they represent and draw on views of the student body in a regular and formal way.

14.4.2 Feedback From Students

As part of an iterative cycle of course evaluation for ongoing improvement and refinements, build in opportunities across the course to evaluate the effectiveness of your new curriculum in order to hear from your students about their learning experiences, and aspects of the curriculum and teaching that they feel help or hinder their learning. Know that no curriculum is ever perfect when rolled out for the first time, and embrace opportunities to continually improve upon it. It is likely that you have institutional and national student evaluation surveys, which bring with them sector-acknowledged issues related to what is being measured (for example, student satisfaction versus students' engagement in their learning, or academics' performance). You will need to work with such data and use it for ongoing curricular and pedagogic revisions, but we suggest that you consider building in other opportunities to gain such feedback. For example, as part of our new curriculum, we have developed a longitudinal research project that involves conducting focus groups and interviews with students and staff about the new curriculum, and, in particular, new aspects within the curriculum that need careful consideration, such as the Impact Lab suite of units (Meth, Finger et al., 2020; Meth, Russell, et al., 2020, p. 353).

14.4.3 Students as Paid Employees Working With You on Course Design

Recognising when expectations on a student–staff partnership are higher than "student voice," Fitzgerald et al. (2019) provide a rich case study of student–staff partnerships in curriculum development at our institution, including the example cited above from our Design degree, exploring the complexities in working relationships and noting ways in which these might be managed. Furthermore, Meth, Finger, et al. (2020) detail what is possible from such collaborations through work undertaken with the Student Associates to develop the conceptual framework for a Graduate Professional Portfolio in design.

14.4.4 Sustained Partnership Conversations between Students and Staff

To enable this, it's important to sustain open spaces not dissimilar to a community of practice. At our institution, recognising the success of our student engagement experiences in Design, a wider network of student partners has now been created across our faculty to ensure that such ongoing conversations, where any issues related to courses, or teaching, as well as ideas for improvement and/or further development, may be discussed.

14.4.5 Points to Consider

- How, at what point, and to what extent might you seek to engage future, current and past students in developing your course?
- Have you ensured that your discussions and collaborations are representative of the diverse student body and perspectives?
- How might you seek to provide opportunities for all students to contribute equitably to discussions?
- Ensure your students are adequately supported in this role so that they feel comfortable and safe as equal collaborators

14.5 BEYOND THE "CURRICULUM BUBBLE": ESSENTIAL YET OFTEN FORGOTTEN CONSIDERATIONS

Although you might meticulously follow the above steps with sound disciplinary and pedagogical intentions, there remain multiple areas of risk that can make or break your curriculum. Within each course team there exists a unique microculture that brings its own characteristics – teaching and learning regimes and perspectives that cannot be wholly surfaced or integrated in such a standard set of steps. Within such regimes reside a series of "rules, assumptions, practices and relationships" (Trowler and Cooper, 2002, p. 224). This contributes to the "stickiness" or variety and unpredictability, to which Orr and Shreeve (2018, p. 71) refer, bringing additional challenges to those "engaging as learners and teachers." Additionally, there are enabling processes and systems that must be in place and working effectively to see your curriculum through to successful implementation. Whilst these are felt by some to be peripheral, if they are not considered as integral to your developmental journey, the perils of designing in a "curriculum bubble" may be realised, and your curriculum vision rendered unattainable. How you navigate the socio-cultural "terrain" that surrounds, but also intrinsically intertwines with your curriculum, is important.

14.5.1 Curriculum Design as a Site of Negotiation

The curriculum, where beliefs around knowledge, teaching and education play out, is noted as a site of negotiation and "struggle" (Ashwin, 2014, p. 124), a very complex sociological space (Trowler and Cooper, 2002). The woes expressed in the opening conversation are real, and reveal that some academics perceive curriculum change as profoundly threatening. This debate is not unique or new to the School of Design, nor is it unique to our institution or even the wider higher education sector. It is the ongoing educational battleground between traditional and progressive curricula (Muller and Young, 2014, p. 128), the former with a focus on pure knowledge, and the latter a focus on applying that knowledge in experiential, real-world contexts. Concerns stem from the notion that, with this "progressive" shift, curricula have become genericised, in a move which Young (2009) terms the "evacuation" of knowledge (p. 196). Obviously, if you are aiming for a generic course that seeks to build "soft" skills with little focus on disciplinary knowledge, this shift is unproblematic. There is also the perception that these changes will bring about a loss of criticality to tend towards "training" rather than "education." This perception then leads to questioning the purpose of higher education in itself. It is important to be aware of such underlying tensions as possible sources of anxiety for some academics.

14.5.2 Leadership, Team Building and the Collective Endeavour

The success of your curriculum will depend heavily on collective buy-in to, and ownership of, the process and implementation by the course team. Drawing on combined knowledge and areas of expertise, as well as appreciating how the pieces fit together is important for all, having direct impact on how students will experience their learning during the course. Beyond forms and processes, it is people that are key to developing, driving and implementing the curriculum. Helping the course team to be honest about the zones of influence surrounding the course, and agreement on how much these will be allowed to influence the outcomes, is important.

But during times of change, in particular when change is difficult, a close-knit course team does not magically form, there is likely to be resistance, and leadership will play a critical role in ensuring spaces for dialogue, engagement and open decision-making (Gentle and Forman, 2014, p. 76). Qualities of empathy and organisational awareness are noted as critical in higher education (Gentle and Forman, 2014, p. 89) and using these to predict where tensions might arise, as well as support staff and address potential hostile behaviours, is important. In this, relationship-building and developing collaborative behaviours is key. For a comprehensive and accessible guide to the "ins and outs" of leading course design, including curriculum development, and leading and engaging others, readers are referred to Parkin (2017).

To help cohere the new curriculum and openly acknowledge shared ownership of its development across student partners, academic staff, academic leadership and staff from support service areas, a living whiteboard wall, known affectionately as 'The Wall' was used to capture developmental conversations held in our shared course design space (Figure 14.3). Importantly it featured the names of all involved, and showed interrelationships between different areas of the curriculum, as well as information on timings, key elements within individual units, links, and also queries outstanding. This helped to sustain transparency in the development process.

FIGURE 14.3 Photo depicting the course design wall for the Bachelor of Design degree

14.5.3 The Complex Needs of Design Teaching: Other Enabling Systems and Processes

You may have designed an educationally sound curriculum that meets the aspirations of all involved; however, alongside this are the many small elements that, whilst apparently procedural, will need careful planning. These include: access to suitable learning spaces and computing hardware and software; a timetable that works for students' learning; ensuring adequate budgets for equipment and materials; and related to budgets, determining class sizes, student–staff ratios and employing expert staff.

The new design curriculum saw some academics shifting from large traditional lectures to interactive classes, with lectures punctuated by in-class activities, giving students an opportunity to collaborate with one another. Critical to these classes was the acquisition of suitable 'turn-and-learn' lecture spaces – in high demand and short supply. One academic was adamant that based on past experiences, a bid for these was highly unlikely to be successful. The academic program director and curriculum designer worked closely with the timetabling officer who, knowing their limited availability, pointedly asked if these spaces were necessary. Because of good working relationships and a clear vision for the overall course, we impressed upon her those elements within the curriculum (and ensuing learning experiences) that were not negotiable, leading to a win for the academic concerned, and increased faith in the new curriculum being given appropriate space to play out as planned.

Our most important suggestion here is to make friends with your timetabling officer or the student admissions and systems teams, who are likely to find complexities in your curricular plans you never believed could exist. Good working relationships mean that you can weather such complexities together rather than at last minute loggerheads. Recognise that they too face multiple demands; for example, from many academics vying for the best teaching spaces at the best time of day. Include such support areas in developmental conversations from the outset so they too are aware of the educational rationale – so classes are not merely a list on a spreadsheet, but rather a set of carefully designed learning experiences. In developing our curriculum, we also ensured a close working relationship with staff from the library and student support services, linking them to learning design and student support discussions in an ongoing way.

As a designer, you will appreciate the pedagogic need for studio spaces and certain modes of teaching and learning where the curriculum is not simply delivered in a traditional lecture- tutorial-exam style. Furthermore, Kaygan and Aydınoğlu (2018) note the increasing importance of collaborative spaces for interdisciplinary design education, and this has proved to be the case particularly for our suite of transdisciplinary Impact Labs. This need to sustain a healthy studio culture is a strong element within the "sticky design curricula" narrative touched on across the chapter (Orr and Shreeve, 2018), where both sophisticated technologies and equipment combined with frequent access to, and learning from, expert practitioners is key. Sector-wide overreliance on casual or "sessional" staff, often on short-term, hourly paid contracts, as well as pressures for academic staff performance in research and a general perception that research is rewarded over teaching, can hinder course team commitment. Our School of Design staff have worked hard to secure academics' contracts and ensure stability in academic staffing and leadership, resulting in a more coherent course team. Fairly unique to design disciplines, institutions are often not set up to deal with those who straddle the academic–professional practitioner boundary – something so important to a design education and sustaining a healthy studio culture. Lindgren (2021) considers the critical role played by such people in teaching design.

14.5.4 Curriculum Development as Iterative

Alongside this, design thinking as a practice in its own right is now seen as a popular way to approach the task of curriculum development, bringing notions of cycles of ideation, iteration, implementation, reflection and refining. Mackh (2018) provides a comprehensive guide to how one might follow this process, with user-focused and collaborative ideas exchange at its heart. As designers, drawing on designerly ways to approach curriculum development makes complete sense, and Evans et al. (2021), referencing key literature, note how the concomitant mindset, steps and tools offer a clear framework for this. In essence, following our institutional Future-Focused Curriculum Framework (Meth, Russell, et al., 2020) reinforces the importance of bringing evidence to inform developments, and there is an in-built review and refine process after each instance of delivery, leading to continued course improvement. In truth, the processes detailed above regarding relationship-building and collaboration at every stage are not dissimilar to those labelled within design thinking practices (Mackh, 2018), and, as Evans et al. (2021) so rightly note, this is a "repackaging" of a way of broaching design-like problems. With its contemporary spin, many find design thinking approaches to be a more palatable form of achieving the same end via different means, where previously the process was seen as tired, bureaucratic and somewhat institutionalised.

However you choose to label it, it is important to strategically build in "flexible and practical evaluation" opportunities across the course to ensure that your curriculum plays out as intended, being prepared to make changes along the way (Blackmore and Kandiko, 2012, p. 175). Bear in mind the issues associated with evaluations discussed in Section 14.4.2, and maintain awareness of the context and ownership of such work to ensure that you are ready for any implications following evaluation results (Blackmore and Kandiko, 2012, p. 171).

14.5.5 Points to Consider

- Recognise that the diversity of opinions and expertise in your course team will enrich your course. What strategies can you use to create a safe and productive space for the course team to debate and share ideas, and give ownership of the direction and delivery of the curriculum?
- Brainstorm early with your team to surface the myriad possible aspects that are important to the successful enactment of the curriculum so as to reduce the possibility of surprise hurdles late in the process

- Acknowledge any constraints within which the course operates and get to know all related administrative and operational elements upon which the course depends, working with appropriate staff and systems from early on to know what's possible
- Consider how and where you have built in opportunities to evaluate and refine your curriculum

14.6 CONCLUDING THOUGHTS

Negotiating the multiple contexts and drivers surrounding curriculum development can, in reality, be messy. Ashwin et al. (2015) acknowledge that beyond the seemingly straightforward alignment of a set of curricular elements, there is much added complexity, and an urge that a holistic, and ultimately realistic, perspective be applied to the process. Within such a holistic approach, consider the aspirations and learning pathways of your students, as well as the wider educational environment, including any extra-curricular opportunities that might be available to enrich students' learning experiences. For design educators developing curricula, there are arguably added complexities in higher education environments, and some of these have been noted across the chapter, including unique studio learning environments and subjectivities in assessing work with creative elements.

Whilst the chapter has spanned a huge topic area, we have attempted to provide relatable examples from our own experiences along the way, with practical points for your consideration, supplemented by a valuable set of specialist reference works for you to consult when undertaking such a curriculum journey. Whilst the chapter by necessity highlights the many tricky corners in this process, the successes have also been numerous.

ACKNOWLEDGEMENTS

The authors would like to acknowledge the many academics, professional support staff and student partners who work tirelessly in bringing their expertise, commitment and passion to ensuring that the Bachelor of Design course achieves the best possible outcomes for students, their future needs, and those of the discipline.

REFERENCES

Ardington, A., & Drury, H. (2017). Design studio discourse in architecture in Australia: The role of formative feedback in assessment. *Art, Design & Communication in Higher Education*, *16*(2), 157–170.

Ashford-Rowe, K., Herrington, J., & Brown, C. (2014). Establishing the critical elements that determine authentic assessment. *Assessment & Evaluation in Higher Education*, *39*(2), 205–222. doi: 10.1080/02602938.2013.819566

Ashwin, P. (2014). Knowledge, curriculum and student understanding in higher education. *Higher Education*, *67*, 123–126. doi: 10.1007/s10734-014-9715-3

Ashwin, P., Boud, D., Coate, K., Hallett, F., Keane, E., Krause, K., et al. (2015). *Reflective teaching in higher education*. London: Bloomsbury Academic.

Barnett, R., & Coate, K. (2005). *Engaging the curriculum in higher education*. Maidenhead: Open University Press.
Biggs, J., & Tang, C. (2007). *Teaching for quality learning at university*, 3rd ed. Maidenhead: Open University Press and Society for Research into Higher Education.
Blackmore, P., & Kandiko, C.B. (2012). *Strategic curriculum change: Global trends in universities.* Society for Research into Higher Education Series. London: Routledge.
Bolton, R. (2019). University funding to be tied more to producing students who get jobs. *Financial Review*, 2 October, 2019. Retrieved 25 May 2020 from: https://www.afr.com/policy/health-and-education/university-funding-to-be-tied-more-to-producing-students-who-get-jobs-20191001-p52wkz
Cook-Sather, A., Bovill, C., & Felten, P. (2014). *Engaging students as partners in learning and teaching.* San Francisco, CA: Jossey-Bass.
Cross, N. (2011). *Design thinking: Understanding how designers think and work.* Oxford: Berg.
Evans, R., Kelly, N., & Kerr, J. (2021). Being a Design Academic: Design Process, Practice, Thinking. In A. Blackler & E. Miller (Eds.), *How to be a design academic*. Boca Raton, FL: CRC Press, 13–28.
Fitzgerald, R., Huijser, H., Meth, D., & Neilan, D. (2019). Student-staff partnerships in academic development: The course design studio as a model for sustainable course-wide impact. *International Journal for Academic Development*, *25*(2), 134–146. doi: 10.1080/1360144X.2019.1631170
Gentle, P., & Forman, D. (2014). *Engaging leaders: The challenge of inspiring collective commitment in universities*. Abingdon: Routledge.
Gusky, T.R. (2007). Closing achievement Gaps: Revisiting Benjamin S. Bloom's "Learning for Mastery". *Journal of Advanced Academics*, *19*(1), 8–31.
Kaygan, P., & Aydınoğlu, A.U. (2018). The role of space in interdisciplinary collaboration in design education. *International Journal of Technology and Design Education*, *28*, 803–817. doi: 10.1007/s10798-017-9407-2
Lindgren, T. (2021). Industry to academia: The enhanced academic. In A. Blackler & E. Miller (Eds.), *How to be a design academic*. Boca Raton, FL: CRC Press, 29–44.
Mackh, B.N. (2018). *Higher education by deesign: Best practices for curricular planning and instruction*. New York: Routledge.
Matthews, K.E., Dwyer, A., Russell, S., & Enright, E. (2019). It is a complicated thing: Leaders' conceptions of students as partners in the neoliberal university. *Studies in Higher Education*, *44*(12), 2196–2207. doi: 10.1080/03075079.2018.1482268
Meth, D., Finger, M., & Brough, D. (2020). The graduate professional portfolio as "synergy tool": Navigating the complex role of portfolios in future-focused design education. In S. Boess, M. Cheung, & R. Cain (Eds.), *Proceedings of DRS2020: Synergy*, Volume 4 Education (pp. 1803–1816). London: Design Research Society. doi: 10.21606/drs.2020.317
Meth, D., Russell, H.R., Fitzgerald, R., & Huijser, H. (2020). Enabling scholarship of teaching and learning activities across a curriculum design framework: A lever for faculty engagement. In R. Plews, & M. Amos (Eds.), *Evidence-based faculty development through the scholarship of teaching and learning (SoTL)* (pp. 347–364). Hershey, PA: IGI Global. doi: 10.4018/978-1-7998-2212-7.ch018
Meyer, J., Land, R., & Baillie, C. (Eds.) (2010). *Threshold concepts and transformational learning*. Rotterdam, The Netherlands: Sense Publishers.
Muller, J., & Young, M. (2014). Disciplines, skills and the university. *Higher Education*, *67*, 127–140. doi: 10.1007/s10734-013-9646-4
Orr, S., & Shreeve, A. (2018). *Art and design pedagogy in higher education: Knowledge, values and ambiguity in the creative curriculum.* Abingdon: Routledge.

Parker, J. (2003). Reconceptualising the curriculum: From commodification to transformation. *Teaching in Higher Education*, *8*(4), 529–543. doi: 10.1080/1356251032000117616

Parkin, D. (2017). *Leading learning and teaching in higher education: The key guide to designing and delivering courses.* Abingdon: Routledge.

Race, P. (2007). *The lecturer's toolkit: A practical guide to assessment, learning and teaching*, 4th ed. Abingdon: Routledge.

Scott, A., & Thomson, S. (2021). Escaping the studio: A guide to design intensives outside the studio. In A. Blackler & E. Miller (Eds.), *How to be a design academic.* Boca Raton, FL: CRC Press, 115–134.

Thomson, S., & Scott, A. (2021). Watch this!: Using video and blended learning. In A. Blackler & E. Miller (Eds.), *How to be a design academic.* Boca Raton, FL: CRC Press, 135–152.

Trowler, V. (2010). *Student engagement literature review.* York: Higher Education Academy.

Trowler, P., & Cooper, A. (2002). Teaching and learning regimes: Implicit theories and recurrent practices in the enhancement of teaching and learning through educational development programmes. *Higher Education Research & Development*, *21*(3), 221–240. doi: 10.1080/0729436022000020742

Young, M. (2009). Education, globalisation and the 'voice of knowledge'. *Journal of Education and Work*, *22*(3), 193–204. doi: 10.1080/13639080902957848

15 Student Engagement through Mentoring

Shayne Beaver

CONTENTS

15.1 INTRODUCTION: WHY WE STARTED A PEER MENTORING GROUP

With the changing profile of students and the university, my colleagues and I had noticed that the design community culture and relationships between our students was diminishing within our school and our discipline of Industrial Design. Technologically adept students, with very busy study, work and social lives, had reduced the duration of time spent on campus, especially outside of scheduled classes. Bigger university communal spaces were sometimes viewed as daunting, and there was no guarantee anyone from their design discipline would be there, never mind someone from their classes that they recognised. Where once lecturers might have seen students helping one another in their cohorts or mixed-year groups, chatting and supporting one another with sketching, physical or computer modelling, design projects or just having general design discussions – now these spaces are filled with scheduled classes and, only rarely, students working on group projects outside of class. First-year students in particular struggle to find their place within large cohorts of students.

For these reasons, in 2015, I began a Students as Partners (SaP) peer mentoring group. I wanted to include students in the process of engaging their colleagues to build a supportive design community culture that builds peer (and industry) relationships, fostering their knowledge and passion for design, through educational and social activities – both virtually and on campus. We define SaP as students and staff working together to ensure positive outcomes, by sharing perspectives and jointly making decisions with regard to learning, assessment and engagement (Shaw, Rueckert, Smith, Tredinnick and Lee, 2017). For the Industrial Design peer mentoring group, I focussed on the engagement and learning that happens outside of scheduled classes, with this chapter reflecting on how SaP positively transformed our design community culture.

15.2 FORMING THE PEER MENTORING GROUP IN INDUSTRIAL DESIGN

As Meth, Thomson and Brough (2021) discuss earlier, our university runs a formal "Students as Partners" (SaP) initiative. I drew extensively on this knowledge of best practice when it came to starting, running, evaluating and evolving our peer mentoring group – which we called ID Pilots. I started a recruitment drive via email and our design Facebook group, and also personally encouraged students I knew to volunteer as peer mentors; volunteer students who help other students via an institutionally endorsed program. We started with just over 15 volunteers, which has increased to over 35 active members. We worked together to define the purpose of the group with these peer mentors. The university's Peer Leader Support has provided valuable training to build the students' leadership skills, including online modules about peer learning facilitation, collaborative learning, cultural awareness and occupational health and safety. A face-to-face leadership training event (LeaderCon) happens twice a year, including modules such as "listening deeply," "peer mentoring facilitation" and "I've done some cool stuff. How do I talk about it?" as well as keynote speakers. These courses also provides the mentors with formal university certificates for their participation, recording both leadership skill development and volunteered hours.

The mentors and I organised a variety of learning and social activities for Industrial Design students, which the group facilitated throughout each semester. As well as information sessions, specialist skills training and assistance, social gatherings with design-related themes and one-on-one mentoring, the mentors provided design and academic skills advice, and referrals to university and design resources to their peers. The majority of the students that are helped by the peer mentors are first-year students, but students at any stage of the course and from other design disciplines have been welcomed and supported too. Industrial Design alumni have been involved in the program as mentors and mentees as they enjoyed the sense of community – as shown in Figure 15.1.

15.2.1 Our Strategic Focus and How a Peer Mentoring Group Benefits the Community Culture

This ID Pilots peer mentoring group benefited the entire Industrial Design discipline, including staff and alumni, as it fostered a close community culture. Having student mentors to turn to for support has been enormously beneficial to students who are struggling, but also to those students who are just looking to do better or forge networks. More broadly, our group benefits the School of Design as it serves as a model for other peer mentoring groups and helps reduce attrition.

As part of our initial development process, student volunteers brainstormed what they wanted the group to be, and what they wanted to call themselves. The name "ID Pilots" is a reflection of their goal to "guide" or "navigate" fellow students through the Industrial Design undergraduate degree. The strategic focus that we wrote together in early 2016 was as follows:

> *The program facilitates learner connectedness and fosters a community of confident and resilient Industrial Designers, in line with QUT's 2020 vision. It will support the ID Pilot peer leaders in developing their professional profiles, networks and employability through training and participation. ID Pilots will demonstrate key graduate capabilities such as collaboration, communication and leadership. The volunteer peer mentors will be engaged in producing visionary, adaptive, responsible and*

FIGURE 15.1 Contextual Analysis of Participants in the ID Pilots

valued contributions to the discipline and its community as a whole, which is part of the Industrial Design discipline's strategic plan.

The program's purpose statement, aim and objectives were originally to provide a variety of extracurricular peer-to-peer activities that our Industrial Design students have identified as being beneficial in their studies and the building of a distinct Industrial Design culture, with the aim of developing a responsive program that supports students through peer to peer mentoring. Key objectives ranged from helping peer mentors develop graduate skills (such as leadership, communication, organisation and negotiation) and social responsibility by contributing to a community of practice, to building a sense of community within the ID discipline, better engaging students with the course and helping reduce their stress and anxiety.

15.2.2 Design of the Program

There are many decisions that need to be made when forming and running a peer mentoring group, and this perhaps becomes more complex when you consider all members to be of equal importance. The question and corresponding decisions that we came up with when forming the group are shown in Table 15.1.

It was decided at the very beginning that we would be as adaptive and responsive as we could to students' changing needs. This means that, before each semester, we discuss the current issues that students are having and re-evaluate what we have previously done. We consider the time needed to implement an activity, skill sets available to run it, peer mentors' passion for the topic, university protocols, school or discipline strategies, finance and many more. Our most successful and ongoing activities include sketching evenings/afternoons where students gathered together to sketch and render. Different mentors have tried different strategies in these sessions to encourage attendance and engagement, including themed sketching topics; running through tutorials or sketch-what-you-like days; providing pizza or baked goods; various forms of advertising; and trying out various locations, times and frequencies and online participation options for these events. Students felt that the biggest benefit of these was the sense of community and the opportunity for advice and learning feedback, with students valuing the discussions about design and sketching skills in both digital and analogue formats, and just socialising and networking.

First-year information sessions are also one of our most popular ongoing events. This event is hosted annually by staff and the ID Pilots to help first-year students transition into second year. Students in our Bachelor's degree are able to make more informed decisions about customising their learning through the selection of study tours, internships, electives, second majors/minors/specialisation units, exchange and opportunities for entering design competitions. For many students, making these decisions and navigating the university process in order to do so can be quite challenging. This event covers general course progression information, followed by ID Pilots presenting information about the study decisions that they have made and showing examples of the assessments and/or outcomes, and discussing the benefits

TABLE 15.1
Basic Decision Making When Forming a Peer Mentoring Group

Questions That Needed Answering	ID Pilot Solutions
What mode of support would we provide?	Both face-to-face and virtual
What platforms for support would we use?	Mixture of organised events, drop-in collaborative learning opportunities, written, visual and video content delivered through social media
What would be our optimum ratio of Mentors to Peers during the activities?	One peer mentor to a group of eight student peers at maximum when face to face, while online content such as infographics or videos would be widely distributed
What are the timeframes for our deliverables?	These are dependent on the activity within the program – some are pre-semester, others focussed on particular weeks during the semester and others are to be carried out all year, or on just one day of the year
Who is responsible for which activities?	This is volunteer dependent, as each mentor would like to select their own areas of involvement based on their areas of skills, interests and availability
Service role of peer mentors?	This was decided by the coordinator. As there was no immediate and ongoing funding for the group, the role of the peer mentors had to be completely voluntary. All volunteers are required to sign a legal document prepared by the university to acknowledge that no financial remuneration, or remuneration of any kind, will be provided
How many hours would each volunteer need to contribute each year?	Initially we identified 30 hours as the preferable amount, but this has dropped to 10 hours to accommodate the students' workload, particularly those in their final year of study. At a bare minimum, the following activities were identified as being essential for all peer mentors to participate in: • Check in regularly with the ID Pilots Facebook Group • Select at least two activities to participate in each year (e.g., sketching evenings, compiling infographic content, Design Development Record feedback sessions, World Industrial Design Day event, etc.) • Encourage student participation and collaboration • Attend meetings with the coordinator and other ID Pilots to reflect, discuss and develop the program at least twice a year
What resources would be available to the group?	Resourcing included the following: Ongoing funding: NIL Support from the Industrial Design Discipline in managing the program: Shayne Beaver Materials: from Faculty, providing physical spaces and resources such as printing of handouts and advertising fliers if required In-kind support: QUT Peer Leader Support programs provides online and face-to-face training for the mentors Ad hoc funding: there are occasionally funds made available for events to supply catering or prizes for the attendees. This is given on a case-by-case basis by our School

(*Continued*)

TABLE 15.1 (CONTINUED)
Basic Decision Making When Forming a Peer Mentoring Group

Questions That Needed Answering	ID Pilot Solutions
What is the role of the Program Coordinator (Discussed with the Mentors but also a workload decision made by my supervisor)	Responsibilities of the Program Coordinator include : • Recruit ID pilots • Arrange training • Provide ongoing support for ID Pilots • Promote the program to students and staff • Evaluate and report on the program • Record participation and hours of ID Pilots • Rewards and recognition of ID Pilots, especially through the organisation of volunteered hour certificates given by the QUT Peer Leader Support program • Organising regular meetings and managing/overseeing the ID Pilot social media accounts

of this knowledge to their skills as an Industrial Designer. This is followed by a very informal opportunity for students to approach the staff and ID Pilots over pizza and drinks with further questions they would like answered. The result is that many first-year students can more accurately understand the opportunities available and weigh up what they are most interested in based on other students' advice, not just staff, and this is seen as being incredibly valuable.

The other activities that have consistently encouraged engagement from students are the Facebook groups run by the ID Pilots. There are 2 groups. One focusses on visualisation, where staff, alumni and students post anything to do with digital or analogue sketching and rendering. This can include design work from university or personal projects, artwork, tutorials on sketching or rendering, designers to follow on social media and other relevant articles, as well as queries about supplies, techniques or software, for example. October in particular is a very active month on this Facebook group as some members participate in Inktober (Inktober, 2019). The second Facebook group is focussed on 3D design modelling and includes posts related to Computer-Aided Design (CAD) Modelling, as well as physical model prototyping. Again, this group encourages university and personal projects, as well as knowledge from articles, other designers and tutorials, information about technology, programs, suppliers or conferences, amongst other topics. ID Pilots are the administrators of these groups and are regular posters of content, feedback or advice in order to foster a sense of community in this online environment. These Facebook groups are also great for advertising and encouraging participation in relevant face-to-face events that the ID Pilots run, particularly the sketching events, which are photographed and shared to the visualisation group. We have tried other social media platforms and strategies, such as Instagram and Pinterest, but these Facebook groups have by far been the most engaged in and supportive environment.

Other activities or events that happen perhaps once or twice a year, or to suit the needs of specific cohorts, are diverse. We have run many successful World Industrial Design Day events, which have included industry panels, design competitions and social events. Skills workshops on various software for CAD or visualisations, Arduino, portfolios and resumes, time management, prototype modelling with cardboard and foam, laser cutting and 3D printing assistance have also been popular. Online resources including "Quick Tips" infographics posted to social media to help with skills like the design process, getting used to university life, group work, project management and time or stress management have also been regularly implemented. We have curated online resources into PDFs or Pinterest boards to help develop skills and inspiration – these have included book recommendations, sketching videos, model-making tips, famous designs and designers. We have also held "Co-Pilot" weekly drop-in sessions for students to work collaboratively on assessments and get advice when needed.

15.2.3 Setting Limits

For some of these ID Pilot activities, we have developed very particular strategies or resources for the peer mentors and/or the students. For example, based on the online and face-to-face leadership training that the ID Pilots received, they have come up with an outline of what a drop-in session or one-on-one peer mentoring session might look like:

- Set a time limit on the discussion
- Identify the question or problem to be addressed – sometimes it take a few minutes to truly identify what the real question or problem actually is
- Outline habits and lifestyle (what motivates and rewards the student)
- Provide a strategy (perhaps using unit information such as handouts and lecture notes, ID Pilot or university resources, or online resources)
- List personal skills and weaknesses
- Prioritise how to move forward based on personal skills and assessment requirements
- Planning of timing of tasks – allowing for contingency, efficiency and effectiveness
- Get them to reiterate this process (write it down if necessary)

ID Pilots also have rules they developed about what they will and will not do when supporting students in particular settings or situations (Table 15.2).

15.3 POSITIVE OUTCOMES OF ID PILOTS

There have been so many positives that have resulted from the formation and running of the ID Pilots peer mentoring group. Most of these are very closely aligned with the existing literature on the topic, and I will cover these in some detail dividing the information in three key areas: positive outcomes for (1) the student mentees; (2) the peer mentors; and (3) the staff and institution involved.

TABLE 15.2
The Dos and Don'ts of the ID Pilots When Supporting Student Peers

What Do the ID Pilots Do?	What Don't the ID Pilots Do?
Hold scheduled sessions for peers to get help	Proof read assignments
Take any feedback of the program seriously – report to the group	Lend out equipment
Help develop written assessment outlines with peers	Show you old assessments that could be plagiarised
Encourage help-seeking behaviour	Tolerate disrespectful or rude students
Listen deeply and non-judgementally	Solve your design problems for you
Facilitate reflection and problem-solving	Check through entire Design Development Records
Direct to other University support services when necessary	Negatively talk about others – staff or students
Follow the university's Code of Conduct	Provide 24/7 support
Provide anecdotal examples from past experience	Give out our personal contact details
Encourage independent actions and get you to do the work yourself	Meet up in a social setting outside of university for ID Pilot mentoring
Be professional and stay neutral	Be disrespectful to you or about your work
Support you at pre-organised events, e.g., sketching evening	Solve personal problems
Show processes, not answers	Provide unlimited hours of support
Guide you to helpful sources, such as websites	Share our own design ideas
Encourage extracurricular activities for ID	Accept gifts
Encourage collaboration	

15.3.1 Positive Outcomes for the Student Mentees

First-year student experience is a strategic focus for higher education institutions, QUT included, due to the "diversity and changing social and educational needs and aspirations of its entering cohorts" (QUT, 2016). A peer mentoring program supports junior students through the transition and adjustment process of first year through new learning experiences, and can help to develop and create a community of practice and collaboration (Shook and Keup, 2012). This can help with the enhancement of student satisfaction, course experience and retention for all students involved (Zamberlan and Wilson, 2015). Students' inability to succeed academically in their new university is often linked to their peer group (Light, 2001 cited in Shook and Keup, 2012), and it is both the interactions with the peer mentors themselves and the colleagues met through the extracurricular activities of the ID Pilots that provide the opportunities for these essential and influential interpersonal relationships to develop.

Durkin and Main (2002) found that, although many universities provide generic study skills courses, the discipline-based peer mentoring and workshops had distinctive benefits, greater demand and increased engagement in comparison. These discipline-based activities were in addition to normal classes, but were observed to positively impact student mentor and mentee marks in both assessments and exams. Evidence showed that students felt these activities were integral to their studies because they were discipline focussed, whereas the generic university courses were seen as an addition to their curriculum and therefore less important and even unnecessary. Peer mentors can play an important role in providing discipline-specific skills and academic knowledge to junior students (Durkin and Main, 2002; Shook and Keup, 2012). Within the ID Pilots program we have provided many opportunities for student to engage with resources and activities that provide both generic study and/or soft skills such as communication, time-management and group-work dynamics, for example, as well as discipline-specific skills. Both soft skill and discipline skill development can help with the maturation, moral and intellectual development and academic performance of students involved (Shook and Keup, 2012).

ID Pilots are also trained to know when to refer students to other support services within the university, and this type of referral practice has been seen to have a significant impact on students personal development and satisfaction (Shook and Keup, 2012). Greater opportunities for social and academic integration during the first year of university can have a profound effect on the satisfaction of students. Having a group like the ID Pilots who are a rich source of information dedicated to their success can ensure students feel supported and valued within the institution. This can be particularly important for specific populations of students, most notably those who have been historically underrepresented or at risk in higher education such as those from LGBTQ, international and first-generation backgrounds, as well as those with mental or physical disabilities (Shook and Keup, 2012), who are perhaps unaware of the specific resources offered within the university to assist and support them.

Overall, our personal experience shows that students can experience a sense of satisfaction as they feel they are being cared for and listened to by the peer mentors and the Industrial Design discipline as a whole. It is the culture of care and encouragement that has ensured that students who perhaps have felt isolated and unsure of themselves can find their place and recover from any perceived shortfalls through changing their mindsets or working on soft and hard skills in order to better their learning experience.

15.3.1.1 Reflective Practice

Boud, Cohen and Sampson (2002) define peer learning as a reciprocal learning activity in which participants share knowledge, ideas and experiences, including reflections on understanding and skills. Schon's (1987) work on understanding reflective practice in the creative domains, such as the design studio, highlighted the complexity of professional practice and the need to develop expert tacit knowledge through "reflection in and on action." Webster (2008) built on this understanding to develop a more holistic approach to professional practice as situated knowledge through an "action and learning" lens, suggesting that learning takes place in settings other than the design studio. Students' total experience (both inside and outside of academia) informs their development as design professionals, and extra-curricular activities and reflection through discussions with others outside of studio can therefore be beneficial to their development (Webster, 2008). Reflective practice is therefore a skill that ID Pilots can guide other students through in order to increase their professional design abilities. This is a skill that is encouraged through discussions during both group and one-to-one activities.

15.3.1.2 Emotion and Stress

Design pedagogy often focusses on elicitation of emotions as a result of experiences with designed artefacts or spaces (Desmet and Hekkert, 2007); however, emotion is also part of the process itself. Emotional intelligence, interpersonal skills, creativity and stress all contribute to the experience of designing and can have both a negative and positive impact on the studio and the students' design outcomes. Emotions experienced in the learning environment by students can have an impact on academic success and academic adjustment, as well as student health and wellbeing (Saklofske, Austin, Mastoras, Beaton and Osborne, 2012). Stress is experienced by most students, more so in the first year than subsequent years as they are transitioning into new study habits, making new relationships, modifying old relationships and becoming more independent (Parker, Summerfeldt, Hogan and Majeski, 2004). The change from the familiarity of didactic high-school learning to a more self-motivated environment that may favour risk-taking as part of the creative process can be incredibly challenging.

These daunting changes can leave students feeling stressed or overwhelmed, and can lead to students underperforming and even leaving the degree, which has an impact on the individual student, the design discipline and the university. The perception of having low control in academic learning situations can create negative emotional reactions (Stupnisky, Perry, Hall and Guay, 2012). Design teaching and learning researchers have highlighted that emotional intelligence, interpersonal and social skills (Savage, Davis et al., 2010) and stress (Smith and Lilly 2014) impact the design-learning context. Pekrun, Goetz, Titz and Perry (2002) identified three academic settings for emotional experience: class, tests or exams and outside of class, each with their own types of emotions. Few studies have focussed on situations outside of the class (Niculescu, Tempelaar et al., 2016).

This out of class experience can be particularly important in the context of first-year students as they change from high school to an environment that requires more self-reliance and autonomy (Perry, Hladkyj, Pekrun and Pelletier, 2001). The large time expenditure of design students in these out of class situations can cause stress if students are unsure of themselves, or unsure of what is required in the design process and how to make key design decisions. Boud, Cohen and Sampson (2014) define peer learning as a reciprocal learning activity in which participants share knowledge, ideas and experiences, including reflections on understanding and skills. It therefore follows that having peer mentors to turn to in times of stress and uncertainty can have a significant impact on mentees' emotional experience. Apart from course work, discussions between mentor and mentee can and should involve topics such as stress levels, what is considered normal stress, how to lower stress levels, but perhaps also how to embrace stress in order to be creative or just get the work done. The ID Pilots use their own experiences and those of their peers to identify problems regarding engagement, skill development and a sense of community belonging that can be easily rectified though student intervention. The group is agile and adapts each semester to the needs of the students. Continuing and evolving the work of the group to meet current student needs and utilising existing research is essential for its ongoing success in ensuring that students feel supported and in control of their experiences.

15.3.2 Positive Outcomes for the Student Mentors

Astin's (1999, in Shook and Keup, 2012) involvement theory states that "the amount of student learning and personal development associated with any educational program is directly proportional to the quality and quantity of student involvement in that program." Peer leaders are considered to be highly invested due to their dedication to training and ongoing time expenditure. The application of skills learnt in leadership training and their design education through their engagement with mentees can reap true educational and personal benefits. This includes a greater awareness of themselves, their leadership and design abilities, their sense of belonging, meaningful interpersonal relationships within the institutional environment and greater awareness of the university and the design community (Astin, 1999, cited in Shook and Keup, 2012).

Collaboration, leadership, self-direction, oral communication, intercultural skills, civic engagement and critical thinking are all skills that can be developed as part of a mentoring role and these are all highly valued by universities and employers alike (Russel and Skinkle, 1990, and Association of American Colleges and Universities, 2011, cited in Shook and Keup, 2012). Applying these skills to a variety of new settings, with a diversity of people and in solving possibly complex problems, builds the mentors' skills and therefore increases critical thinking, confidence, empathy and emotional intelligence in the mentors (Keup, 2010, cited in Shook and Keup, 2012). It is not just these individual skills but the integration of all of these skills as "learners address real world problems that are unscripted and sufficiently broad to require multiple areas of knowledge and multiple modes of inquiry" (Rhodes, 2010, p. 50, in Shook and Keup, 2012).

Academic skills have previously been reported as being the lowest areas of development for mentors (Keup, 2010) and this is true of our mentoring group as well. In some cases being a mentor perhaps has a slightly negative impact on academic performance due to the time dedicated to the group at particular times during the semester, and the additional stress attributed to participating in the group instead of being entirely focussed on assessments. This negative impact is supported by Shook and Keup's (2012) research and is why we are cautious when allocating or selecting the type and number of activities and the time dedicated to the group by individual mentors.

15.3.3 Positive Outcomes for the University and Staff

As a university and design discipline, we benefit tremendously from this SaP group. This can be in the form of possible reduced attrition and a definite increase in mentor and mentee satisfaction through our extremely cost-efficient initiatives. The ID Pilots group accommodates sometimes high numbers of students for some activities, with very little to no financial commitment from the School. The group also provides valuable feedback from students on the university and school support services, course content, discipline units and knowledge, staff activities and the design community, identifying both the positive and negative aspects. Student leaders from the group have advocated on other committees and panels, such as curriculum changes and graduate exhibitions. Students involved both as mentors and mentees feel a stronger connection to the university, discipline and design community, and are often likely to stay engaged even after graduation. Participating students are likely to be more prepared for the work environment and this of course reflects well on the university's reputation (Shook and Keup, 2012).

15.4 THE CHALLENGES EXPERIENCED BY THE PROGRAM AND OPPORTUNITIES FOR CHANGE

15.4.1 Burnout and Over-Commitment

All of our student mentor volunteers are extremely passionate and compassionate people, and it is therefore often the case that I, as the coordinator of the program, have to step in and curb or question how much we try and tackle as a group and individually each semester. It has been essential to keep our goals and activities realistic and achievable in terms of time commitments by all involved. I have seen a few cases of students over-committing to hours before the semester's activities start and therefore have students either burning out and not having the time or energy to accomplish all that they would have liked and therefore letting their peers and mentees down, or perhaps prioritising the ID Pilots over other aspects of their lives, such as academic achievement, in order to meet the obligations they set for themselves. Neither of these outcomes is positive and it is the coordinator's role to ensure that these cases are avoided. There is no hard and fast rule for how many hours a year each volunteer should or can contribute as this is dependent on the personal situation and activities that they intend to sign up for. We do, as a group, try to run activities at times of the semester when we know that both mentors and mentees are likely to engage when they are not overwhelmed with immediate assessment deadlines. It is for this reason that a close but professional relationship between all members of the ID Pilots, including myself, is essential to ensure we can all be supportive, but also realistic with one another, both in planning for the semester and in re-evaluating or reprioritising throughout the semester.

15.4.2 Demanding Students

As design educators we experience at least a couple of students in every cohort that perhaps demand and/or need more support and guidance than the majority of other students. The reasons for this are varied, but as staff we know how to deal with these students in a way that is both timely and helpful in most cases. For peer mentors, this is a difficult situation to deal with as they do not have the knowledge, experience, training or informed support network of the staff. I have had peer mentors report back cases where mentees have tried to contact them at all hours of the day and night with extremely compelling reasons for reaching out, where mentors have been stuck in conversations for unreasonable lengths of time (sometimes hours) as the student seems incredibly distressed. Mentees have become demanding and even aggressive if mentors are unwilling or unable to give answers to assessment solutions or queries, and mentors have experienced students that are so despondent about their own work and/or capabilities they have felt like no amount of peer assistance is enough. While some of these situations are similar to the scenarios students are exposed to in their leadership training, it is still confronting and challenging when they are dealing with a sometimes very distressed or despairing student.

There is, however, a number of ways that we have addressed these issues in the past and plan to continue to explore solutions in the future. Currently, face-to-face interactions begin with setting a realistic time limit for the exchange to avoid lengthy and perhaps unproductive interactions, and mentors are encouraged to show students our set of "dos and don'ts" if they become uncomfortable with anything being asked of them. All peers are also required to have the contact details of the coordinator and university security in case of more unpleasant interactions, but there has been no need for anyone to resort to this. There are, however, gaps in our strategies that need to be addressed; we have not kept a record of which students access the majority of our activities, what they need help with or how long each session lasts. This is currently being rectified. Coordinators will be able to see if there are students who require an unreasonable amount of support from the mentors, and perhaps step in to offer higher levels of university support when necessary. It is also clear that a few of these demanding students have had underlying problems such as mental health, as I discuss below.

15.4.3 Mental Health Issues in Tertiary Education

Mentors have reported that they struggle with understanding the boundaries and etiquette of what they should (and should not do or say) in their role when it comes to mentees' emotional states when involved in one-on-one discussions, both in person and online, and in moderation and commenting on social media posts. In rare situations, mentees have revealed distressing details about their personal situations and mentors have been unsure how to respond appropriately to this.

There has been a documented increase in the prevalence and severity of mental health issues in tertiary education students (Daddona, 2011) and a noticeable increase in observed and disclosed cases within our own discipline. Mental health issues cover a huge variety of issues including traumatic events, sexual abuse or assault, eating disorders, personality disorders, self-injury, anxiety and depression (Daddona, 2011). As a result, peer mentors should be equipped to deal with a variety of situations should they arise and be confident in their abilities to refer students to university counselling or other appropriate services. There have been a few cases of peer mentor to student mentee discussions becoming very personal and confronting for those involved, and this has been identified as a priority area for training and reporting improvement in the future. Mental Health First Aid training has been undertaken by staff within the School and this could perhaps be made available to the student mentors in the future.

15.5 OUR ONGOING AND EVOLVING STRATEGY

A five-step strategy was implemented in mid-2020 by the mentors in order to further evolve their leadership abilities to better deal with the emotions of their fellow students. From a student mentor's perspective, how can they tell when a student's emotions are a natural result of the design process and university life, and when their emotions are having a detrimental effect on their wellbeing? This is a tough question to answer and has been further compounded by the occurrence of an international pandemic that has disrupted students personal and university lives. There is no single solution, but we must identify strategic ways to increase the skills, benefits and positive outcomes of a design peer mentoring group. Daddona (2011) developed a five-step process to assist peer mentors in dealing with the issues of mental health in mentees, which served as a starting point for an evolved ID Pilot strategy. The five steps are explored in the following sections (15.5.1–15.5.5).

15.5.1 Effective Training

To ensure mentors are familiar with university processes, but also to unlock their potential, effective training is key (Durkin and Main, 2002). Peer mentors participate in a variety of activities with students; some are less personal and involve group activities, while others can involve one-on-one time with students. Mentors should be equipped with excellent listening skills and be competent in their verbal and non-verbal communication skills to deal with all of these interactions. Training needs to encourage mentors to facilitate learning, actively listen and avoid didactic methods of teaching. The mentors have an understanding of their own experiences regarding adapting to university life, forming new friendships and dealing with stress and other emotions from both university and external factors. How integral these issues can be to success or failure at university is covered in some of the training information, but there is room for better integration of this into the ID Pilot strategy. In addition to this, mentors need to be well versed in reflective learning if they are to foster critical thinking in their peers (Durkin and Main, 2002) and so this area of research should also be summarised and distributed, particularly the research done specifically in the field of design. Resources should be developed with mentors to help distinguish the boundaries between emotional issues they are comfortable offering assistance with, and when a mentee should be referred to other support services.

As expected, the generic training for peer mentors in the university environment is currently effective in providing students with the communication, listening and referral skills they need to get started as a peer mentor. It is, however, clear from the mentor feedback and the literature that more discipline-specific training needs to be implemented, particularly in areas regarding reflective practice and the role that emotion can and should play during the design process.

15.5.2 Effective Interactions

Effective communication skills will make mentees more likely to disclose personal information and to ask for advice on issues or problems (Daddona, 2011). Knowing when to ask open or closed questions to get the required details or avoid discussing particular issues is a skill that might be needed. Body language, such as facial expressions, posture and gestures is also an essential tool for effective communication. Knowing when to talk, when to listen and when to be silent also needs to be mastered (Daddona, 2011). These are skills that are already part of the training that the ID Pilots receive, but again this could be reinforced or better actioned in some of our discussions and activities.

15.5.3 Effective Referral Practices

Mentors should also be skilled referrers to campus services. They should know details about the counselling and other support services available on campus and have strategies for broaching the issue of seeking further support, as some students can take offence to the suggestion. The more information mentors have about the services, the more confident they can be in talking about them, hence easing anxiety for both parties (Daddona, 2011). Referrals can be difficult as counselling does still have a stigma associated with it for some (Daddona, 2011), so mentors should be prepared with facts that ease the student's mind about how normal it is to seek out this type of help. If mentors are not confident discussing this with the student then their staff supervisor (myself) should be informed and should become responsible for referrals. Whilst maintaining the integrity of the mentor–student relationship is important, the mental wellbeing of the student should take priority.

15.5.4 Continuing Peer Support

If a student is referred to another service they should not feel like they are being offloaded. Referrals should leave students feeling like they now have more options for support. Referring students to counsellors or other support services, should leave the mentor and the student with a clear goal of focussing on the discipline-specific issues that the student might have. It should be clear to both parties where the boundaries are for these discussions, though, and this may require mentors to receive some training on how to articulate this.

15.5.5 Support for Peer Mentors

Mentors may themselves need support following interactions with particular students (Daddona, 2011), and this should be made available to them via their staff supervisor (me) and through professional counselling. Ensuring that mentors are debriefed regularly can help to remind them of their roles and discuss situations that they have faced in order to reflect on their practice. Group sharing can help all involved mentors deal with past situations and better plan for future situations. These group sessions should not involve revealing too much detail about personal information (Daddona, 2011). This is the part of the process that needs the most work in order to document the success or failure of any changes to existing training and procedures. It is also important to ensure that these debriefs are regular, structured and documented enough to cover all the issues that may arise during semester.

15.6 THE FUTURE OF THE ID PILOTS

Investigation into existing ways that peer programs are being evaluated is underway, in order to leverage best practices and inform development of a possible universal evaluation framework in the future. The aspiration of this framework is that a versatile evaluation process can be implemented in order to allow coordinators and the university to better understand the value and impact of peer programs, across a diverse group of students and activities. This will allow individual programs to better identify the needs of the students that they are helping and evolve accordingly. We know that peer-to-peer support provides some positive impact and outcomes for student participants (particularly first-year students), and for peer leaders who deliver support, from our own experiences, as well as existing research such as Shook and Keup (2012). However, more robust data sources are needed to better understand, measure and evidence our program's impact, provide accountability and contribute to ongoing development and improvement.

15.6.1 Program Expansion

In late 2019, our School decided that the peer mentoring program would expand to include other design disciplines. Visual Communication, Interaction Design and Fashion Design had identified similar problems with community as those experienced by Industrial Design in 2015 and decided they too would like a mentoring program. As the ID Pilots already had a strong following and reputation within the School, it made sense to expand the program. It was decided that the group will now be called the Design Pilots. Students already participating in the ID Pilots are proud of the work they have done to create such an enviable program, and took on more of a leadership role in order to help with the launch of Design Pilots. New student mentors were recruited and planning for the first semester of 2020 took place early in the year. The Design Pilots program became instrumental in helping students navigate the transition to online learning that occurred from March 2020. We are looking forward to helping more mentees in the future and diversifying the activities that we do in order to cater to many more students.

15.7 CONCLUSION

The personal satisfaction I have gained from being a part of this group has alone been worth the effort of coordinating it. Seeing the changes in all the ID Pilot mentors who have participated in the program and the fantastic opportunities it has provided them in terms of personal development and resulting employment has been very rewarding, both for them and myself. The true impact the program has already had on students will likely never be measurable, but even the small number of students that have acknowledged their appreciation for the program indicates that it has all been worth the time and effort of all those involved. We will continue to ensure that we are responsive to the changing needs of students and endeavour to continue to provide activities that engage and encourage community within our School of Design and design community at large.

REFERENCES

Boud, D., Cohen, R., & Sampson, J. (2002). *What is peer learning and why is it important.* Retrieved December, 20, 2014.

Daddona, M.F. (2011). Peer educators responding to students with mental health issues. *New Directions for Student Services, 2011*(133), 29–39. doi: 10.1002/ss.382

Desmet, P., & Hekkert, P. (2007). Framework of product experience. *International Journal of Design, 1*(1), 57–66.

Durkin, K., & Main, A. (2002). Discipline-based study skills support for first-year undergraduate students. *Active Learning in Higher Education, 3*(1), 24–39. doi: 10.1177/1469787402003001003

Inktober. (2019). *Inktober.* https://inktober.com

Keup, J.R. (2010). National context & institutional practice: Findings from a national survey of peer leadership experiences & outcomes. *National Resource Center for The First-Year Experience and Students in Transition's Institute on Peer Educators*, Indianapolis, October, 17–19.

Keup, J.R. (Ed.). (2012). *Peer leadership in higher education: new directions for higher education*, Number 157 (Vol. 110). John Wiley & Sons.

Meth, D., Thompson, S. and Brough, D., (2021). Design curriculua: Navigating process and people. In A. Blackler and E. Miller (Eds). Boca Raton: CRC Press, 245–268.

Parker, J.D., Summerfeldt, L.J., Hogan, M.J., & Majeski, S.A. (2004). Emotional intelligence and academic success: Examining the transition from high school to university. *Personality and Individual Differences, 36*(1), 163–172.

Niculescu, A.C., Tempelaar, D.T., Dailey-Hebert, A., Segers, M., & Gijselaers, W.H. (2016). Extending the change–change model of achievement emotions: The inclusion of negative learning emotions. *Learning and Individual Differences, 47*, 289–297.

Pekrun, R., Goetz, T., Titz, W., & Perry, R.P. (2002). Academic emotions in students' self-regulated learning and achievement: A program of qualitative and quantitative research. *Educational Psychologist, 37*(2), 91–105.

Perry, R.P., Hladkyj, S., Pekrun, R.H., & Pelletier, S.T. (2001). Academic control and action control in the achievement of college students: A longitudinal field study. *Journal of Educational Psychology, 93*(4), 776.

QUT. (2016). QUT – MOPP – C/6.2 first year experience and retention. Retrieved from www.mopp.qut.edu.au/C/C_06_02.jsp

Saklofske, D.H., Austin, E.J., Mastoras, S.M., Beaton, L., & Osborne, S.E. (2012). Relationships of personality, affect, emotional intelligence and coping with student stress and academic success: Different patterns of association for stress and success. *Learning and Individual Differences, 22*(2), 251–257.

Savage, S.M., Davis, R.M., & Miller, E. (2010). *Professional education in built environment and design.*

Schön, D.A. (1987). *Educating the reflective practitioner: Toward a new design for teaching and learning in the professions.* Jossey-Bass.

Shaw, N., Rueckert, C., Smith, J., Tredinnick, J., & Lee, M. (2017). Students as partners in the real world – A whole-institution approach. *International Journal for Students as Partners, 1*(1), 1–8. doi: 10.15173/ijsap.v1i1.3079

Shook, J.L., & Keup, J.R. (2012). The benefits of peer leader programs: An overview from the literature. *New Directions for Higher Education, 2012*(157), 5–16.

Smith, D., & Lilly, L. (2014). Navigating student stresses in the interface between creative and technological competence: A case study in Interior Architecture.

Stupnisky, R.H., Perry, R.P., Hall, N.C., & Guay, F. (2012). Examining perceived control level and instability as predictors of first-year college students' academic achievement. *Contemporary Educational Psychology, 37*(2), 81–90.

Webster, H. (2008). Architectural education after Schön: Cracks, blurs, boundaries and beyond. *Journal for Education in the Built Environment, 3*(2), 63–74.

Zamberlan, L., & Wilson, S. (2015). Developing an embedded peer tutor program in design studio to support first year design students. *Journal of Peer Learning, 8*(3), 5–17.

Section 6

Leading a Group or School

Section 6 focuses on leadership, which requires a group of or a team and something that brings people together. Both chapters in this section use an interview format to reflect on and share the experience of leading as a design academic. Chapter 16 focusses on how to build and lead a research group, reflecting on the importance of developing a brand, securing physical space, and developing positive research culture through regular weekly meetings online communication channels and an annual retreat. Chapter 17 documents the experience of leading a design school, both in Australia and the US, when change is constant. As well as advice on how to lead in a crisis, focussed on the challenges of COVID-19, this chapter discuses different leadership styles and philosophies, as well as how to communicate and facilitate change. Combined, both chapters provide critical insight into leading, in different contexts, as a design academic.

16 Building a Research Group

Urban Informatics 2006 to Now

Marcus Foth and Markus Rittenbruch

CONTENTS

16.1 INTRODUCTION

The QUT Design Lab at Queensland University of Technology was established in 2016. It was not built from scratch, but has been fortunate to be able to build on and leverage the previous and current research strengths, long-standing track record of achievements, as well as local, national and international recognition of various research groups within the QUT School of Design, including the Urban Informatics Research Lab that Professor Marcus Foth founded in 2006. Ahead of its time and before the term "smart cities" became popular, the Urban Informatics Research Lab pioneered a new field of research, study and practice that examined people creating, applying and using information and communication technology and data in cities and urban environments. The QUT Design Lab aims to be the nurturing ecosystem that incubates new research groups and fosters their growth until they reach critical mass and produce impact. In this chapter, Professor Marcus Foth is in conversation with QUT Design Lab Associate Director Associate Professor Markus Rittenbruch, who himself is a long-standing member of the Urban Informatics Research Lab. They reflect on their experiences, perspectives, share lessons learnt and provide advice to help you build and grow your own research group.

16.2 GROUP FORMATION

MR: *You started the Urban Informatics Research Lab in 2006. How did it all start?*

MF: The Urban Informatics Research Lab emerged quite organically and more or less unplanned when I finished my PhD and started my first PostDoc appointment in 2006.

My undergraduate studies in Germany and Australia brought together two broad areas: **people** (media and communication studies, art and design) and **technology** (computer science and multimedia). During my PhD studies from 2002 to 2005 I added a third area of great interest to me: **place**. At a time when the Internet was heralded as the information superhighway causing the death of distance, and policy makers and commercial entities were promoting telework, e-commerce and distant education, I went quite the opposite way by suggesting that "place still matters." My PhD thesis "Towards a Design Methodology to Support Social Networks of Residents in Inner-City Apartment Buildings" (2006) looked at the way that web-enabled technologies could be useful for local communication and interaction within community networks.

The way my PhD studies brought together people, place and technology formed the cornerstone for the Urban Informatics Research Lab. Right after my PhD graduation I was part of a great team of academics across media and communication studies, urban sociology and architecture, and we won a three-year Discovery grant from the Australian Research Council, including an Australian Postdoctoral Fellowship that supported my research between 2006 and 2008. We were successful with an additional two ARC Linkage grants the following year, and I spent some time at the Oxford Internet Institute at the University of Oxford as a Visiting Fellow in 2007 supported by an ARC Linkage International award. So this early success with funding applications meant we had income to hire and appoint research staff and PhD students, and this in turn organically created a growing team of researchers and research students. So first there was momentum that created the group, and then we eventually started to call ourselves the Urban Informatics Research Lab. So the group very much originated from a bottom-up approach. We were not formally recognised as a research group for a long time, and did not bother applying for recognition until 2012. In hindsight, us "flying under the radar" of the official university administration and its performance assessment criteria proved to be a good thing. It allowed us to be more agile and focus on our research, rather than on reporting and bureaucracy.

16.3 BRANDING

MR: *You have been recognised as having coined the term urban informatics, which has now been adopted by universities and industry worldwide. What is urban informatics? And what role does branding play in forming a research group?*

MF: After over a decade of being part of the Urban Informatics group, we describe ourselves as applying diverse, both established and novel methods of inquiry to identify challenges and opportunities in urban environments, across socio-cultural, economic, ecological and technological spheres. We design innovative technologies, interventions and services to respond to such challenges and opportunities through meaningful engagement with individuals, communities and organisations across the public, private, not-for-profit and education sectors. Our aim is to research and co-create urban futures that are more liveable and equitable. Thus we aim to ask difficult questions and go an extra mile to ensure we deliver the most insightful, impactful outcomes. However, giving the group a name predated this level of clarity of purpose unfolding.

Before we settled on the name Urban Informatics, it was clear that the group's common denominators were people, place and technology. I did not like existing terms such as urban technology, urban infrastructure or urban computing, because I felt they were too technocentric, and I wanted to emphasise the human and social elements of our research.

During my PhD, I was part of the *Community Informatics Researchers Network* (CIRN), which was founded in 2003 primarily by colleagues at Monash University on the basis of Michael Gurstein's work (Gurstein, 2000). Informatics, with its implied reference to information systems and information studies, slightly shifts the attention away from the hardware and more towards the socio-cultural aspects of information exchange, communication and interaction, social networks and human knowledge. Gurstein described community informatics as being primarily concerned with improving the wellbeing of people and their communities through more effective use of information and communication technology. Likewise, Urban Informatics – in my original conception – is concerned with the impact of technology, systems and infrastructure on people in urban environments.

While I believe our work made "urban informatics" popular, and the term has now been adopted by universities and industry worldwide, the invention of the term is not mine. The earliest use of the term I found in Mark E. Hepworth's 1987 article *The Information City* (Hepworth, 1987), which mentions "urban informatics" on page 261. However, Hepworth's overall discussion is more concerned with the broader notion of "informatics planning." Considering the article pre-dates the advent of ubiquitous computing and urban computing, it does contain some visionary thoughts about major changes on the horizon brought about by information and communication technology and the impact on cities. Later, in September 2003, Howard Rheingold, author of *Smart Mobs* (2002), wrote an article for the now discontinued TheFeature.com entitled *Cities, Swarms, Cell Phones: The Birth of Urban Informatics* in which he introduced his interviewee Anthony Townsend, who later became the author of *Smart cities: Big data, civic hackers, and the quest for a new utopia* (2013), as an "urban informatician and wireless activist." I was honoured that Townsend accepted my invitation to write the foreword for my first edited book: *The Handbook of Research on Urban Informatics: The Practice and Promise of the Real-Time City* (Foth, 2009).

At our first retreat in 2010 we came up with a definition based on two key characteristics: First, the new possibilities (including real-time data) for both citizens and

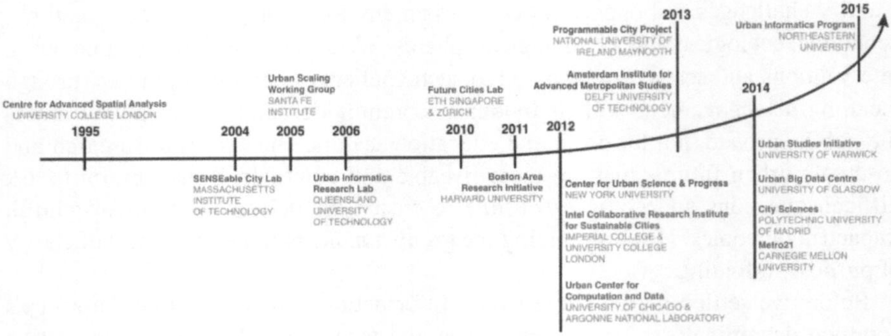

FIGURE 16.1 Timeline of research centres that focus on urban informatics. Source: Townsend (2015, p. 12)

city administrations afforded by digital technology, big data and ubiquitous computing, and; (2) the convergence of physical and digital aspects of the city.

> *Urban informatics is the study, design, and practice of urban experiences across different urban contexts that are created by new opportunities of real-time, ubiquitous technology and the augmentation that mediates the physical and digital layers of people networks and urban infrastructures* (Foth, Choi, et al., 2011).

Starting in 2012 and fuelled by the increasing popularity of commercial opportunities under the label of smart city and big data (Townsend, 2013), subsequent definitions of urban informatics emphasised big data analytics for better planning outcomes in the city contexts. This direction of urban informatics scholarship has been referred to as "data-driven, networked urbanism" (Kitchin, 2015) or urban science (Batty, 2013). As a result, a range of new research centres focussing on urban informatics have been established since our group started at QUT in 2006 (Figure 16.1).

What is noteworthy with regard to our group's name is the consistency of our identity, which helped us to establish a brand and brand recognition beyond the university campus. Urban informatics was used as our "departmental" affiliation in addition to mentions of our faculty and university. It was also a default keyword on papers, theses and funding proposals. This in turn contributed to our visibility and national and international standing over the long term.

16.4 TRANSDISCIPLINARITY

MR: *You mentioned that a crucial aspect that sets the Urban Informatics group apart from the more narrowly defined urban science research centres is its broad, interdisciplinary scope. Can you please explain this further?*

MF: What I very much appreciated throughout my education and research training is being academically polyglot, a bit of a polymath and being able to learn and use theories, methods and insights from multiple disciplines and combining and applying them in new ways to different problems and "opportunity spaces" (Hornecker et al.,

2006). I started my education with an undergraduate degree in Medieninformatik – a computer science program at Furtwangen University in Germany with a major in media studies, but already back then in 1997 the curriculum was designed to create generalists working across multidisciplinary knowledge and skill sets with broad lateral thinking and capabilities. What followed was a Bachelor at Griffith University and a Masters at QUT, and those experiences can be characterised as being *cross-disciplinary* and *multidisciplinary*. It was not until my PhD that I started integrating and synthesising theories and methodologies from different disciplines. This was the first time I produced *interdisciplinary* approaches and findings.

The Urban Informatics Research Lab became a testbed for honing the skill of conducting interdisciplinary research studies, as it requires a broad perspective across the triad of people, place and technology (Figure 16.2), and the associated disciplines of the social sciences (humanities and arts); computer sciences (and software engineering, ICT); and spatial sciences (architecture, planning, and design). Its modus operandi thus became *transdisciplinarity*, which Rheingold describes as follows:

> *Transdisciplinarity goes beyond bringing together researchers from different disciplines to work in multidisciplinary teams. It means educating researchers who can*

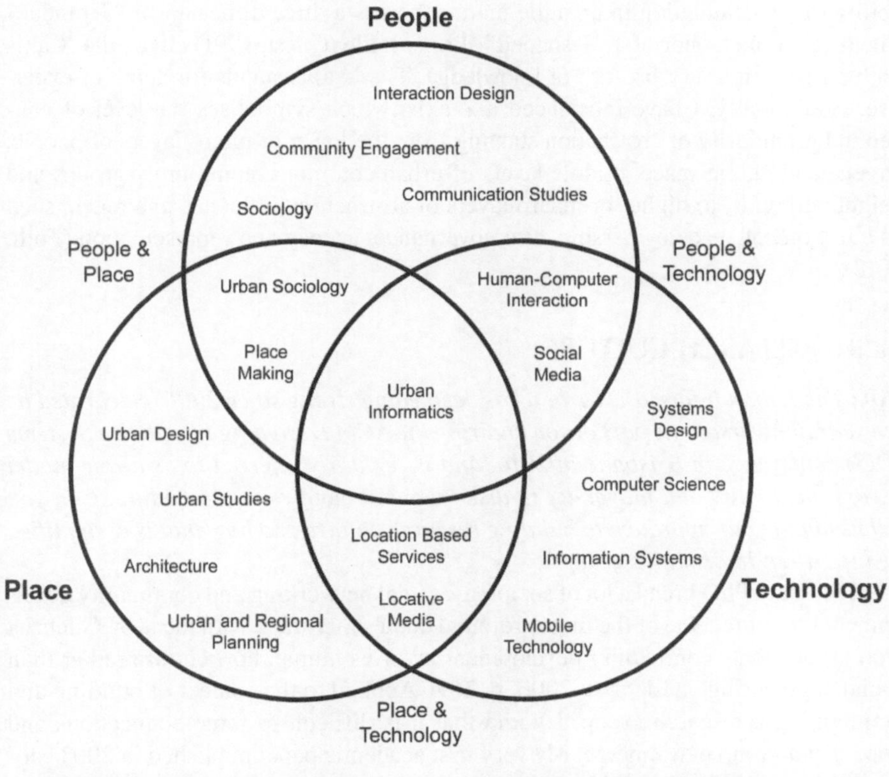

FIGURE 16.2 Transdisciplinary research across people, place and technology. Source: Foth (2006, p. 32), CC-BY-4.0 (https://doi.org/10.6084/m9.figshare.12269504.v1)

> *speak languages of multiple disciplines – biologists who have understanding of mathematics, mathematicians who understand biology* (Institute for the Future, 2006).

The Urban Informatics Research Lab has thus not only been conducting *inter*disciplinary research studies, it has also been producing *trans*disciplinary PhD graduates. Over the years, we have had urban planners and architects studying the impact of digital technology on placemaking; multimedia designers researching young people's use of social networking platforms; software engineers looking at situated community engagement strategies for better urban planning outcomes; and business graduates examining sustainability concerns in domestic environments. The group regularly comprises and collaborates with architects with degrees in media studies, software engineers with expertise in urban sociology, human-computer interaction designers with a grounding in cultural studies and urban planners with an interest in digital media and human geography.

The group's transdisciplinarity reflects the field's increasing demand for transdisciplinarity, which requires dissolving the rigid boundaries between disciplinary silos. Nomadic, polyglot researchers are better able to traverse seamlessly between academic schools and bring disparate fields together. This is a crucial skill in order to be innovative and to keep the bigger picture in mind. This in turn requires an ability to move attention in an agile manner across a three dimensional "T model." The original metaphor of a "T-shaped" skills profile (Guest, 1991) aligns the X axis with transdisciplinary breadth of knowledge. The Y axis stands for depth of expertise. Additionally, I have introduced a Z axis, which symbolises the level of conceptual granularity of abstraction starting from the lower or micro layers of people, streets, and public space; middle layers of urban commons, community groups and neighbourhoods; to higher or macro layers of abstraction, influence and reach, such as questions of space ownership, city governance, agency and representation (Foth, 2018).

16.5 RESEARCH CULTURE

MR: *The Urban Informatics group has been going strong since 2006, celebrated its ten year anniversary as part of you chairing the ACM Designing Interactive Systems (DIS) conference in Brisbane in 2016, and is regularly referred to as a role model across the faculty and university in discussions around research culture. Can you tell us about your approach to building research culture and how that is exemplified by the Urban Informatics group?*

MF: In my PhD I read a lot of sociology, social networking and community development literature. One of the most profound quotes that stuck with me is by Gilchrist who argues that "community development involves human horticulture rather than social engineering" (Gilchrist, 2000, p. 269). Applied to the context of building and maintaining a research group, I found that it is difficult to force connections and hope that a group may emerge. My very first academic paper published in 2003 captures this, too: "If you build it, they will not necessarily come" (Foth, 2003). In the case of the Urban Informatics group, we started and grew quite organically based on shared interests and shared purpose. A lot of the initial arrangements that we may now consider evidence of a thriving research culture were originally put in place for more pragmatic and logistical reasons. We started with a weekly "all hands on deck" team meeting. This has always been kept informal without an agenda and minutes in order to encourage group members to talk freely about their work, progress, challenges, seek feedback and peer support, not just from me and other academics but also from the group at large. Unless you are part of a research project, the PhD journey can be quite lonely, and so the weekly team meetings are useful to foster a sense of belonging, create ambient awareness of what everyone else is working on, and this in turn also encourages information to be shared across the group. When someone comes across something that is relevant to someone else's PhD project, it gets passed on or even discussed by the whole group.

Another key reason why being part of a research group is so important is the way it adds vital oxygen to research conversations and scholarly debate. I find that, in the last two decades working at QUT, as well as speaking with colleagues at other universities in Australia and overseas, sadly there seems to be not more but less oxygen for academic conversations of this nature overall. While the pressure is increased by performance measures and reviews, and ever higher KPIs to produce more publications, more PhD completions and more research income, the neoliberal university affords the researcher less and less time devoted to the actual core business of research; that is, reading, writing and thinking (Morrissey, 2015). Research groups thus fulfil a crucial role in offering a space and home for the academic pursuit of knowledge, a safe haven providing refuge and shielding group members from the bureaucratic and disillusioning treadmill of meetings, reporting, management and strategic planning. Early on, there were many moments where the one hour set aside for the weekly team meetings was not enough to satisfy people's appetite for these research conversations. So we set up what has come to be known as our "Friday Rumbles" – a one or two hour time slot reserved every Friday 3 PM, where we schedule ideation sessions, dry-runs of upcoming conference or milestone seminar presentations, invited guest speakers and workshops on theories or methods. The Friday Rumbles can also be of a more social nature, e.g. watching a documentary followed by a group discussion, or playing board games.

In addition to these two regular weekly items in our group's diary, the group has created other forms and ways of expressing and fostering its research culture (M. Bilandzic and Foth, 2014). For example, since 2010 we have been organising annual research retreats that take the whole group to a beach side village near Kingscliff in Northern New South Wales for four days (Figure 16.3). This intense time period is spent advancing research papers, ideating around new project ideas and engaging in theory and methods workshops. It also offers a great opportunity to onboard newly arrived PhD students and visitors, build a team spirit, and at the end we all get to know everyone else much better.

Communication is obviously essential. We maintain two email mailing lists – one is a public interest-based list that anyone across the university can self-subscribe to, and another just for the immediate group members. We use the team communication platform *Slack* and have set up a number of channels, which get used regularly

FIGURE 16.3 10th Urban Informatics research retreat in Kingscliff, NSW, May 2019. Source: author supplied, CC-BY-4.0

to post information about events, jobs, summer schools, research methods, publications and call for papers, or just random and light-hearted posts. Each PhD student also has their own private channel to chat with their supervisory team in between in-person meetings, and research projects also have their own dedicated channels. Not only does this take a lot of informal "chatty" communication traffic off everyone's email inbox, it also supports a culture of care and peer support. If someone is on leave, sick or travelling, they can still maintain a connection with the other group members. When someone comes across a research paper or news article that may be of relevance to someone else's PhD project or the whole group, they post it on Slack and tag relevant users, and often further discussion emerges. We have also created a shared "dashboard" using a Google Drive spreadsheet, which everyone can access and edit. There are a number of tabs where we capture and track the status of research publications submitted for peer review, schedule the Friday Rumble series, list everyone's upcoming travel dates and itineraries, and collect information about relevant local meetup groups, journals and mailing lists.

MF: *What were your first impressions and insights when you joined Urban Informatics?*

MR: I joined QUT and subsequently the Urban Informatics Research Lab in 2010. My previous position was with NICTA (National ICT Australia, now CSIRO Data61) where I led a project on designing joint video conferencing and large-scale multitouch interaction environments. Before that I worked on projects at the University of Queensland, the Distributed Systems Technology Centre CRC (DSTC) in Brisbane, and ProSEC – a CSCW and HCI research group situated at the University of Bonn in Germany. The first thing that struck me about Urban Informatics was the way that PhD and Master students were embedded into the group. I was familiar with project-based workflows where researchers and sometimes project-based PhD students worked together to meet project milestones.

Students would participate in project or PhD meetings, but were rarely involved in the discussion of emerging research opportunities, topics and developments. What surprised me about Urban Informatics was how open and inclusive group discussions were. While students were focussed on their research topics, the group's research culture promoted an open exchange of ideas, methods and approaches, publication plans and practical knowledge amongst academics, postdoctoral and research fellows, and PhD students. "Practical knowledge" involved best practices about things like how to sign up for international summer schools and fellowships, connecting with other research labs, as well as sharing handy tips for how best to deal with internal university procedures and guidelines for travel, funding, etc. I think the critical importance of this latter aspect is often overlooked. Becoming part of a new organisation either as a PhD candidate or academic staff member can be quite disruptive as a lot of procedural knowledge does not translate between workplaces. While many organisations try to compensate by offering handbooks and information sessions, the localised, contextualised and applied knowledge of members of the Urban Informatics group was invaluable for me for hitting the ground running when I started at QUT.

Another aspect that stood out was the strong international outlook of the group right from the start. The PhD cohort was quite international, which is not unusual in the Australian context, yet, on average, the group has more overseas students than comparable groups in our university. In addition to PhD students the group hosts a series of international visitors as well as overseas interns, who in many cases returned to enrol in the PhD program with the group. While I had experienced similar international collaborations while working in Germany and other parts of Europe, this experience was less common in Australia due to the country being more geographically isolated. The strong focus on international collaborations across PhDs and research projects created a culture that saw successive generations of students enrol in international summer schools (for instance, the Oxford Internet Institute's Summer Doctoral Program at the University of Oxford in the UK, and the UBI Summer School at the University of Oulu in Finland) and conduct overseas fieldwork.

One last aspect that for me epitomises the group's impact over the years was my experience that people tended to "stick around." When I started as a Research Fellow the group comprised three postdocs: Dr Christine Satchell and Dr Jaz Choi who were ARC Postdoctoral Research Fellows, and myself funded by your Queensland Government Smart Futures Fellowship and NICTA. Over the years some of our PhD graduates joined as academic staff or research fellows and the group attracted a growing number of academics from across the School of Design and the Creative Industries Faculty. While people's individual levels of engagement has waxed and waned, there remains a strong sense of cohesion and belonging to the group. The Urban Informatics retreat has been a crystallisation point and an excellent opportunity for this wider community to connect.

MR: *What do you see as the key elements in establishing a research culture?*

MF: My experience founding and setting up the QUT Design Lab in 2016 proved again that genuine research culture takes time and is greater than the sum of its parts.

So even if we were to quickly replicate all the aforementioned "tangible" appearances and evidence of our group's research culture, we would not necessarily succeed in kick-starting a group afresh. Whilst referring to place-based communities, some quotes I cited in my PhD thesis describe this well: "Those geographic communities already rich in social capital may become richer thanks to community networks, and those communities poor in social capital may remain poor" (Jankowski et al., 2001, p. 113), or simply put, "connectivity does not ensure community" (Blanchard and Horan, 1998, p. 302). There are a number of other factors that I believe are noteworthy with regards to fostering a strong research culture and engaging in "human horticulture."

First, our group thrives on having a set of **shared values**. While we have never had an explicit, formal strategic plan that spells out vision and mission statements and a set of strategies, we do have shared values. As you already pointed out, I think it is absolutely crucial that PhD students are never treated as "just" students, but as junior researchers and equal team members. Beyond their PhD topic, I believe that a well-rounded research training journey should also give them exposure to the realities of academic life, e.g. management decisions from the top and how to deal with them, finance and budget considerations, funding applications and peer review, industry and government liaison, etc. Another critical value is appreciating and cherishing diversity and – as I explained above – transdisciplinarity. This also means that our definition of scope is quite broad and inclusive, and this in turn makes the group permeable and accessible. I try to ensure the group's membership is diverse and regularly comprises not just designers and people from a technology background but also artists, architects, business graduates, and industry and community practitioners. Furthermore, the group is made up of a mix of different gender, nationalities, age groups, and socio-cultural backgrounds. This creates fertile ground for debate, as innovation thrives on diversity (A. Bilandzic et al., 2018). However, for a research group to be productive a balance has to be struck between homogeneity and diversity, as "birds of a feather flock together." Yet, in my experience the likemindedness that holds a group together is better derived from shared values and – over time – shared experiences and shared memories.

Second, I strongly believe a research group requires **shared purpose**. A tongue-in-cheek saying that has stuck with the group over many years is our aspiration for "world domination." Putting the joke aside, the serious part to this is identifying the group's *raison d'être*. For over a decade, we have consistently subscribed to two interrelated sets of methodologies and research paradigms: participatory action research (Foth and Brynskov, 2016; Hearn et al., 2009) and participatory design (Foth and Axup, 2006; Frauenberger et al., 2018), and in turn sought to contribute to these fields internationally. So in essence, the group shares a purpose in changing the world for the better by demonstrating how rigorous academic research can be translated into societal and environmental impact and benefit. This is reflected in the range of research challenges we sought to tackle over the years, which are all motivated to deliver not just rigorous research but also real world impact. We first started off with a study that significantly shaped the social sustainability strategies of the Kelvin Grove Urban Village, the Queensland Government's flagship urban renewal project in inner-city Brisbane (Foth and Adkins, 2006; Klaebe et al., 2009). The group has also produced a number of technology innovations, such as *CityFlocks* – an early mobile location-based recommender system (M. Bilandzic et al., 2008); *DispoMaps* – an iPhone app to temporarily share a map with your location details with others and then dispose of it safely; *FixVegas* – a mobile app that lets you take a photo of city assets and street furniture that require repair and submit a maintenance request to the local council (Foth, Schroeter, et al., 2011); *CapitalMusic* – a mobile app to visually share what music you are currently listening to with people in your vicinity (Seeburger et al., 2010); and *Discussions in Space* – a hybrid mobile phone and public screen application that allows passersby to contribute content via SMS or tweets (Schroeter et al., 2012; Schroeter and Foth, 2009). Since then, we have studied situated community engagement for better urban planning outcomes; reduced energy consumption in domestic households for improving sustainability; identified digital opportunities to enhance passenger experiences on public transport; re-conceptualised libraries as co-working spaces and living labs for connected learning and innovation; and explored human–food interactions to foster urban agriculture and reduce food waste. While these examples demonstrate diversity, they also evidence a shared purpose and commitment to produce critical research beyond smart cities (Foth, 2019) and towards real-world impact and benefits.

The group's work was recognised twice as the 2012 iAwards Queensland State and National Merit Recipient in the Research & Development category, with a 2012 Vice-Chancellor's Performance Award, and more recently in 2018 as one of 40 highly competitive research areas, where Australia is first in the world according to data analysis company *League of Scholars* (Figure 16.4). Using a granular approach

AT THE PINNACLE

40 fields in which Australian unis top the world

1	Photovoltaics	UNSW	21	Reflective practice	Monash Uni
2	Marine biology	James Cook Uni	22	Brain injury	Monash Uni
3	Information literacy	Qld Uni of Technology	23	Plant ecophysiology	Uni of WA
4	Physical activity	Uni of Sydney	24	Solar thermal	UNSW
5	Optimal control	Curtin Uni	25	Occupational therapy	Uni of Queensland
6	Peptides	Uni of Queensland	26	Reproductive immunology	Uni of Adelaide
7	Business process management	Qld Uni of Technology	27	Ancient DNA	Uni of Adelaide
8	Restoration ecology	Uni of WA	28	Employment relations	Griffith Uni
9	Ecological economics	ANU	29	Scholarly publishing	Monash Uni
10	Electronic materials	Uni of Wollongong	30	Southern Ocean	Uni of Tasmania
11	Accounting education	Macquarie Uni	31	Silicon solar cells	UNSW
12	Social identity	Uni of Queensland	32	Oral epidemiology	Uni of Adelaide
13	Coral	James Cook Uni	33	Coral reef ecology	James Cook Uni
14	Refugees	UNSW	34	Herbicide resistance	Uni of WA
15	Ecological restoration	Uni of WA	35	Hotel management	Griffith Uni
16	Community engagement	Qld Uni of Technology	36	Urban Informatics	Qld Uni of Technology
17	Conservation ecology	Uni of Melbourne	37	Youth mental health	Uni of Melbourne
18	Quantitative genetics	Uni of Queensland	38	Evidence-based healthcare	Uni of Adelaide
19	Environmental remediation	Uni of SA	39	Hypersonics	Uni of Queensland
20	Solar energy materials	UNSW	40	Data linkage	UNSW

Source: League of Scholars

FIGURE 16.4 40 fields of research that Australia is leading the world. Source: League of Scholars as cited by Dodd (2018). With permission.

that zeroes in on specific subject areas, they found that QUT is world-leading in urban informatics. A university is ranked first if it has more active researchers in a particular field than any other anywhere in the world.

Third, a research group really needs a **shared space**. Recently, during the COVID-19 crisis, we found that this shared space can be digital and computer-mediated but, by 2020, our group already benefited from an existing strong research culture and a history of shared, in-person experiences. In general, shared physical space where the group can collocate is imperative, especially when you are just starting out. Our group never occupied lavish or purpose-built spaces, and that is not as important as the ability to collocate itself. Early on, we were assigned an open office space. While that office space was nothing special, it sufficed, and we quickly made it "ours." Placing a prominent sign on the door helped in this endeavour, as colleagues and visitors quickly started to refer to our group as the Urban Informatics group – although we were not officially recognised as a research entity by the university. This also further supported our brand identity as I mentioned earlier.

Certainly, sharing space comes with its own set of challenges and issues, e.g. noise and the governance of the communal kitchen and fridge, but these are outweighed by the benefits of being able to identify a shared space as the group's home. Ideally, the space needs to accommodate the various work practices and research styles of the group members, be they desktop-based research or creative, studio or arts-based practices. It should also include ample room for social, informal conversations without disturbing anyone, for meeting as a group, and allow for placemaking to occur; that is, over time, the group decorates the space with photos, posters and other artefacts and mementoes that in turn strengthen the research culture and the sense of belonging. While the neoliberal university often mistakes the luxuriousness of furnishings as a measure of the quality of a research space, in my experience, factors such as proximity, adaptability, customisation and the group's ability to make the space their own are far more impactful in fostering a thriving research culture. Hot-desking is the antipode to building a strong research culture.

Except for a small, once-off research support grant in 2012, the Urban Informatics group has never been directly funded by QUT. In my opinion, research culture is not so much about how much money is available to invest in infrastructure and facilities, although that certainly helps – but it does not create research culture *per se*. The socio-cultural aspects are far more crucial. A good litmus test of the strength of a group's research culture is when you go away. In 2016, I went on sabbatical to Europe, in 2018 my mother passed away and I had to quickly take leave for a couple of weeks, in 2020 we are dealing with COVID-19. In all these and other instances, I see with delight that the group continues regardless, and that is great proof of its success. Meetings still occur, the Friday Rumble series is scheduled, chatter on Slack continues and things tick over and progress. I think that this continuity and resilience in the absence of the group leader is also evidence of a thriving research culture.

16.6 WHAT'S NEXT?

MR: *The Urban Informatics group can look back on a long and rich history. More recently, the ANZSRC Review Steering Committee* has decided – based on submissions by yourself and other colleagues – to include Urban Informatics as a new Field of Research code (FoR 330412) in the 2020 classification. This is further recognition of the group's work and notability. So, what's next?*

MF: Urban Informatics as a brand has remained consistent, it helped us establish visibility and recognisability, and it has served us well. But times are changing and priorities are changing. Our group benefited from significant interest in the topic of smart cities from government and industry since IBM launched their *Smarter Cities* campaign in 2009, and McKinsey published their 2012 white paper on *Emerging Trends in Urban Informatics* (Bays and Callanan, 2012). However, these developments also meant that urban informatics is increasingly narrowly defined to be just about urban data analytics and often used interchangeably or even equated with the new field of urban science (Batty, 2013), while the wider focus on citizen participation and engagement, civic innovation and design seems to take a backseat.

I thus see a need and an opportunity to pause, reflect, consolidate and refresh our group's identity and purpose. This is further corroborated by the new QUT Blueprint and the broader research performance and innovation policy context, which stipulates a firm commitment to transdisciplinarity, collaboration, openness and agility. My hope is that the larger QUT Design Lab that we are part of since 2016, lays the foundation for an agile and dynamic research ecosystem. It will obviously have to adjust over time according to internal and external factors, such as emerging trends, changes to government and industry research funding priorities and societal challenges, and we are no exception. This ecosystem provides the supportive foundation for a research group lifecycle that starts incubating new research clusters around emerging topics (*ideate*), fosters their growth (*develop*) until they reach critical mass and produce impact (*deliver*).

Against this backdrop, I am currently working with colleagues on setting up a new research group on **More-than-Human Futures**, which is strongly aligned with the QUT Design Lab's research trajectory. Whilst at the helm of the Urban Informatics group, I have consistently focussed on the philosophical and societal issues and challenges of sustainability. For example, in 2008 I co-organised two of the first academic workshops on human-computer interaction and sustainability (Foth et al., 2009; Paulos et al., 2008). My Queensland Government Smart Futures Fellowship (2009–2012) investigated the interplay between energy consumers, technology devices and sustainability behaviours.

More recently, I have critiqued predominant, technocratic concepts of the "smart city" built upon conventional STEM and data science approaches and proposed

* In 2019, the Australian Research Council, the Australian Bureau of Statistics (ABS), Statistics New Zealand (Stats NZ), and the New Zealand Ministry of Business, Innovation and Employment (MBIE) undertook a joint review of the Australian and New Zealand Standard Research Classification (ANZSRC), also referred to as the *Field of Research* (FoR) codes

a post-anthropocentric and more-than-human approach to their design (Clarke et al., 2019; Foth and Caldwell, 2018; Loh et al., 2020; Yigitcanlar, Foth, et al., 2019; Yigitcanlar, Kamruzzaman, et al., 2019). I argue strongly for the importance of a HASS (humanities, arts, social science) perspective grounded in Indigenous epistemologies and planetary health. I have built these perspectives into my contributions to current debates around infrastructuring in participatory design, which extends discourse around concerns of scale, impact and institutioning (Foth et al., 2015; Foth and Turner, 2019; Frauenberger et al., 2018; Teli et al., 2020).

My hope is that this new research group on more-than-human futures will bring together a strongly transdisciplinary and transcultural mix of researchers from across creative practice, communication and design including Indigenous researchers who will guide the group's exploration of Indigenous concepts such as care for and stewardship of Country. The group's vision is to:

> *Forge a transdisciplinary collective of Creative Industries Faculty academics and partners to lead a proactive, creative response to unfolding planetary ecocide. By identifying and co-creating new pathways for all Australians we will embrace more-than-human futures, grounded in Indigenous teachings of caring for Country.*

This group sets out to

> develop innovative research projects, artworks, designs, critiques, and socially engaged actions that are deeply informed by Indigenous perspectives and decolonial theory. Our research will question and ultimately seek to catalyse the urgent social changes required to mitigate and reverse collapsing ecological systems and assure justice for all species.

It is my hope that this new group will take on some of the learnings and successful ways of nurturing a strong research culture in order to eventually grow into a sizable and impactful group of researchers.

ACKNOWLEDGEMENTS

It takes a village to raise a child. We sincerely thank all current and former Urban Informatics Research Lab members, be they research students and graduates, research staff, interns, visitors, examiners, advisors, as well as all our colleagues, study participants, alumni, peer reviewers, supporters, funders and helpers.

REFERENCES

Batty, M. (2013). *The New Science of Cities*. MIT Press, Cambridge, MA.

Bays, J., & Callanan, L. (2012). *Emerging Trends in Urban Informatics*. McKinsey on Society. http://mckinseyonsociety.com/downloads/reports/Social-Innovation/Emerging-trends-in-urban-informatics.pdf

Bilandzic, A., Casadevall, D., Foth, M., & Hearn, G. (2018). Social and spatial precursors to innovation: The diversity advantage of the creative fringe. *Journal of Community Informatics*, *14*(1), 160–182. https://eprints.qut.edu.au/122329/

Bilandzic, M., & Foth, M. (2014). Urban informatics research lab, Queensland University of Technology. *ACM Interactions*, *21*(6), 16–19. doi: 10.1145/2674917

Bilandzic, M., Foth, M., & De Luca, A. (2008). CityFlocks: Designing social navigation for urban mobile information systems. In G. Marsden, I. Ladeira, & P. Kotzé (Eds.), *Proceedings of the 7th ACM Conference on Designing Interactive Systems* (pp. 174–183). Association for Computing Machinery, New York, NY. https://doi.org/10.1145/1394445.1394464

Blanchard, A., & Horan, T. (1998). Virtual communities and social capital. *Social Science Computer Review*, *16*(3), 293–307.

Clarke, R., Heitlinger, S., Light, A., Forlano, L., Foth, M., & DiSalvo, C. (2019). More-than-human participation: Design for sustainable smart city futures. *Interactions*, *26*(3), 60–63. doi: 10.1145/3319075

Dodd, T. (2018, March 13). World leaders in key disciplines. *The Australian*, p. 30.

Foth, M. (2003). *If You Build It, They Will not Necessarily Come: The Need for Online Communication Research into Communities of Place.* Communication Research Forum (CRF), Canberra, ACT.

Foth, M. (2006). *Towards a Design Methodology to Support Social Networks of Residents in Inner-City Apartment Buildings* (G. Hearn & J. Jones, eds.) [PhD, Queensland University of Technology]. https://eprints.qut.edu.au/16655/

Foth, M. (Ed.). (2009). *Handbook of Research on Urban Informatics: The Practice and Promise of the Real-Time City.* Information Science Reference, IGI Global, Hershey, PA. doi: 10.4018/978-1-60566-152-0

Foth, M. (2018). Participatory urban informatics: Towards citizen-ability. *Smart and Sustainable Built Environment*, *7*(1), 4–19. doi: 10.1108/SASBE-10-2017-0051

Foth, M. (2019). Research beyond smart cities. *Technology|Architecture + Design*, *3*(1), 39–41. doi: 10.1080/24751448.2019.1571797

Foth, M., & Adkins, B. (2006). A research design to build effective partnerships between city planners, developers, government and urban neighbourhood communities. *The Journal of Community Informatics*, *2*(2). http://ci-journal.net/index.php/ciej/article/viewArticle/292

Foth, M., & Axup, J. (2006). *Participatory Design and Action Research: Identical Twins or Synergetic Pair?* Participatory Design Conference (PDC), Trento, Italy.

Foth, M., & Brynskov, M. (2016). Participatory action research for civic engagement. In E. Gordon & P. Mihailidis (Eds.), *Civic Media: Technology, Design, Practice* (pp. 563–580). MIT Press, Cambridge, MA.

Foth, M., & Caldwell, G. A. (2018). More-than-human media architecture. *Proceedings of the 4th Media Architecture Biennale (MAB'18)*, 66–75. doi: 10.1145/3284389.3284495

Foth, M., Choi, J.H.-J., & Satchell, C. (2011). Urban informatics. *Proceedings of the Conference on Computer Supported Cooperative Work (CSCW'11)*, 1–8. doi: 10.1145/1958824.1958826

Foth, M., Paulos, E., Satchell, C., & Dourish, P. (2009). Pervasive computing and environmental sustainability: Two conference workshops. *IEEE Pervasive Computing*, *8*(1), 78–81. doi: 10.1109/MPRV.2009.13

Foth, M., Schroeter, R., & Anastasiu, I. (2011). Fixing the city one photo at a time: Mobile logging of maintenance requests. *Proceedings of the 23rd Australian Computer-Human Interaction Conference (OZCHI)*, 126–129. doi: 10.1145/2071536.2071555

Foth, M., Tomitsch, M., Satchell, C., & Haeusler, M.H. (2015). From users to citizens: Some thoughts on designing for polity and civics. *Proceedings of the Australian Computer-Human Interaction Conference (OZCHI)*, 623–633. doi: 10.1145/2838739.2838769

Foth, M., & Turner, T. J. (2019). The premise of institutioning for the proliferation of communities and technologies research. *Proceedings of the 9th International Conference on Communities & Technologies*, 24–28. doi: 10.1145/3328320.3328398

Frauenberger, C., Foth, M., & Fitzpatrick, G. (2018). On scale, dialectics, and affect: Pathways for proliferating participatory design. *Proceedings of the Participatory Design Conference (PDC) 2018*. Participatory Design Conference (PDC), Hasselt, Belgium. doi: 10.1145/3210586.3210591

Gilchrist, A. (2000). The well-connected community: Networking to the "edge of chaos." *Community Development Journal*, *35*(3), 264–275. doi: 10.1093/cdj/35.3.264

Guest, D. (1991). The hunt is on for the Renaissance Man of computing. *The Independent (London)*, *17*.

Gurstein, M. (2000). *Community Informatics: Enabling Communities with Information and Communication Technologies*. Idea Group, Hershey, PA.

Hearn, G., Tacchi, J., Foth, M., & Lennie, J. (2009). *Action Research and New Media: Concepts, Methods and Cases*. Hampton Press, Cresskill, NJ.

Hepworth, M.E. (1987). The information city. *Cities*, *4*(3), 253–262. doi: 10.1016/0264-2751(87)90033-3

Hornecker, E., Halloran, J., Fitzpatrick, G., Weal, M., Millard, D., Michaelides, D., Cruickshank, D., & De Roure, D. (2006). UbiComp in opportunity spaces: Challenges for participatory design. *Proceedings of the Ninth Conference on Participatory Design: Expanding Boundaries in Design - Volume 1*, 47–56. doi: 10.1145/1147261.1147269

Institute for the Future. (2006). *Science and Technology Perspectives* (No. SR-967). Institute for the Future. www.iftf.org/our-work/people-technology/technology-horizons/science-technology-outlook-2005-2055/

Jankowski, N.W., Van Selm, M., & Hollander, E. (2001). On crafting a study of digital community networks: Theoretical and methodological considerations. In L. Keeble & B.D. Loader (Eds.), *Community Informatics: Shaping Computer-Mediated Social Relations* (pp. 101–117). Routledge, New York, NY.

Kitchin, R. (2015). *Data-Driven, Networked Urbanism*. doi: 10.2139/ssrn.2641802

Klaebe, H., Adkins, B., Foth, M., & Hearn, G. (2009). Embedding an ecology notion in the social production of urban space. In M. Foth (Ed.), *Handbook of Research on Urban Informatics: The Practice and Promise of the Real-Time City* (pp. 179–194). IGI Global, Hershey, PA.

Loh, S., Foth, M., Caldwell, G.A., Garcia-Hansen, V., & Thomson, M. (2020). A more-than-human perspective on understanding the performance of the built environment. *Architectural Science Review*, 63(3–4). 372–383 . doi: 10.1080/00038628.2019.1708258

Morrissey, J. (2015). Regimes of performance: Practices of the normalised self in the neoliberal university. *British Journal of Sociology of Education*, *36*(4), 614–634. doi: 10.1080/01425692.2013.838515

Paulos, E., Foth, M., Satchell, C., Kim, Y., Dourish, P., & Choi, J. H.-J. (2008). Ubiquitous sustainability: Citizen science & activism (Workshop). Tenth International Conference on Ubiquitous Computing (UbiComp), Seoul, South Korea. http://eprints.qut.edu.au/14130/

Rheingold, H. (2002). *Smart Mobs: The Next Social Revolution*. Perseus Publishing.

Schroeter, R., & Foth, M. (2009). Discussions in space. *Proceedings of the 21st Annual Conference of the Australian Computer-Human Interaction Special Interest Group (OZCHI)*, 381–384. doi: 10.1145/1738826.1738903

Schroeter, R., Foth, M., & Satchell, C. (2012). People, content, location: Sweet spotting urban screens for situated engagement. *Proceedings of the Designing Interactive Systems Conference*, 146–155. doi: 10.1145/2317956.2317980

Seeburger, J., Foth, M., & Tjondronegoro, D. (2010). Capital music: Personal expression with a public display of song choice. *Proceedings of the 6th Nordic Conference on Human-Computer Interaction (NordiCHI)*, 777–780. doi: 10.1145/1868914.1869027

Teli, M., Foth, M., Sciannamblo, M., Anastasiu Cioaca, I., & Lyle, P. (2020). Tales of institutioning and commoning: Participatory design processes with a strategic and tactical perspective. *Proceedings of the Participatory Design Conference*, Manizales, Colombia. doi: 10.1145/3385010.3385020

Townsend, A. (2013). *Smart Cities: Big Data, Civic Hackers, and the Quest for a New Utopia*. W.W. Norton & Company, New York, NY.

Townsend, A. (2015). *Making Sense of the New Science of Cities*. Data & Society Research Institute. www.starcitygroup.us/wp-content/uploads/2019/12/making-sense-of-the-new-science-of-cities-final-2015.7.7.pdf

Yigitcanlar, T., Foth, M., & Kamruzzaman . (2019). Towards post-anthropocentric cities: Reconceptualizing smart cities to evade urban ecocide. *Journal of Urban Technology*, *26*(2), 147–152. do: 10.1080/10630732.2018.1524249

Yigitcanlar, T., Md. Kamruzzaman, Foth, M., Sabatini-Marques, J., da Costa, E., & Ioppolo, G. (2019). Can cities become smart without being sustainable? A systematic review of the literature. *Sustainable Cities and Society*, *45*, 348–365. doi: 10.1016/j.scs.2018.11.033

17 Leading a Design School
Practices from Australia and the USA

Lisa Scharoun and Gjoko Muratovski

CONTENTS

17.1 INTRODUCTION

Like many industries, universities are in a state of near-constant flux. In addition to shaping and driving learning programs, creating an environment conducive to innovation and cutting-edge research, engaging with industry in strategising new technologies and processes, and streamlining and enhancing internal process, leaders need to have the skills to navigate and communicate changes, and be able to create a culture of positivity that transcends the inherent difficulties of the uncertainty of 21st-century educational systems (Scott et al., 2008: p. 7).

Broadly speaking, the scholarly literature identifies two overlapping categories of university leaders. The first category is associated with institutional and departmental management positions; people who often initiate or respond to institutional changes and strategies. This category of leadership is often examined in terms of "management style." The second category deals with the more traditional work of academics and includes teaching and research. This category of leadership is often examined in terms of "effectiveness." Heads of School sit between these two categories and are expected to lead their academic colleagues to achieve "excellence" in teaching and research in line with the mission and the strategic focus of the university (Juntrasook, 2014).

"Exemplary leaders," explains Kibort (2002), "must promote psychological hardiness in a climate of challenge. They must create a stimulating and pleasant workplace environment, while fostering a more effective workplace." In this chapter, we look at the role of the leader of a design school and the challenges and opportunities that this position presents. Through a discussion with two leaders of design schools at institutions in the USA and Australia, this chapter gives an understanding of how these leaders approach the myriad complexities of working in higher education.

Professor Lisa Scharoun was appointed Head of School – School of Design at the Queensland University of Technology in September of 2018. Lisa received a Master of Arts in Design Studies from the University of Arts London, and subsequently worked in London as a graphic and interior designer before taking on a role as a Senior Lecturer in Visual Communications at Raffles Design Institute in Shanghai. On completion of her PhD in Visual Communications from Griffith University in Brisbane (Australia) she was offered the role of Program Director of the Graphic Design discipline at Deakin University in Melbourne. Subsequently, Scharoun was hired to lead and drive a renewal of the Graphic Design program at the University of Canberra where she eventually took on the role of Program Director of the Design disciplines. Scharoun also co-founded and directed the Cross-Cultural Design Lab whilst at the University of Canberra. A highly awarded teacher and researcher, Scharoun received six Vice Chancellor's Excellence Awards for teaching and research at the University of Canberra along with the Australian Council of University Art and Design School's "Distinguished Teaching Award." Scharoun is also a practising graphic designer and artist who has won many national and international prizes for her work. She was a Top 10 finalist in the Bill and Melinda Gates "Records for Life" competition and was awarded the "Ease of Adding Information" prize for a design solution created to assist mothers in developing countries to remember to vaccinate their children. Her digital photographic design work has won numerous international awards such as the JM Cameron Prize for Women's Photography and the *Prix de la Photographie Paris*.

Gjoko Muratovski is an Endowed Chair and Director of The Myron E. Ullman, Jr. School of Design at the University of Cincinnati – the oldest university-based school in the USA. He was appointed to this role for a five-year term in August of 2016. In this role, he led the school into its 150th anniversary in 2019. The Ullman School of Design is a top-ranked design school in the USA, as recognised by publications such as *Bloomberg Business*, *Business Insider*, *Fast Company*, and *Vogue*. The school is supported with a $10,000,000 (USD) Endowment Fund, established by Myron Ullman, III – the Chairman of Starbucks Corporation. The school regularly engages with Fortune 500 companies such as Procter & Gamble, Johnson & Johnson, General Motors, Fiat Chrysler Automobiles, Ford, Toyota, Macy's, Amazon, Nike, Apple and Facebook on developing new partnerships and research initiatives.

In addition to his appointment at the University of Cincinnati, Muratovski holds appointments as Innovation Consultant at Stanford University (USA), Visiting Professor at the University of Zagreb (Croatia), Guest Professor at Tongji University (China), Fellow of the Designers Institute (New Zealand), Fellow of the Design Research Society (UK), and Inaugural Scholar-in-Residence at Frank Lloyd Wright's

Fallingwater Institute (USA). He also served as a Visiting Professor at Accademia Italiana (Italy), Visiting Professor at the Copenhagen Business School (Denmark), High-End Foreign Expert with the State Administration (China), and Advisor to the White House Presidential Innovations Fellows Program (USA). Prior to these, Muratovski held academic positions in Australia and New Zealand.

Lisa Scharoun (LS) and Gjoko Muratovski (GM) met for a first time in 2010 at the Australian Council of University Art and Design Schools (ACUADS) conference in Canberra. After an initial conversation about design history and theory, they realised they had a shared passion for visual communications and design education, sparking a connection that has seen them continue this conversation over the next decade. Since this first meeting, their careers have evolved to loftier positions and taken them on journeys through many cities and countries. In the first section of this chapter, they provide an overview of their respective design schools and then continue on in a more informal conversational style to discuss their approaches to various subjects such as teaching and learning, change management, diversity and inclusion.

17.2 BACKGROUND OF OUR DESIGN SCHOOLS

LS: Queensland University of Technology (QUT) draws its history back to 1881, starting with the Brisbane School of The Arts. QUT is now home to approximately 50,000 students across 6 faculties. Although Architecture and Design programs have long existed as part of the QUT ecosystem (over 100 years) the current QUT School of Design was created in 2012 as a merger between the subject areas of Industrial Design, Interior Design, Landscape Design and Architecture, formerly hosted in the Faculty of the Built Environment, with the subject areas of Fashion Design and Interactive and Visual Design, housed in the Creative Industries Faculty. Currently the School of Design encompasses seven discipline areas: Architecture, Fashion, Industrial Design, Interaction Design, Interior Design, Landscape Architecture and Visual Communication.

The School of Design is a part of the Creative Industries Faculty (CIF), which has approximately 7,000 students. QUT was the first university to adopt the Creative Industries concept and established the world's first Creative Industries Faculty in 2000. CIF is one of the largest Art, Media and Design entities in the Asia Pacific region, and consists of three schools: School of Creative Practice, School of Communication and the School of Design. Our Media Communication area is ranked #1 in Australia and #16 in the world (QS World Rankings 2019) with Art and Design: #1 in QLD; #5 in Australia; and #41 globally (QS World Rankings 2019). Our Architecture program is Top 51–100 in the world (QS World Rankings 2019).

The QUT School of Design is a global leader in driving and shaping the area of social design. In 2019 our Design Lab secured the highest ranking in the country for "Design Practice and Management" (Excellence in Research Australia, Field of Research Code – 1203). As a school we have a mission to drive and shape an understanding of "responsible design." We believe that to be responsible is to try and create design solutions, teach and research in a way that attempts to influence the world in a positive way. Wherever possible, through our teaching, research and practice, it is our aim to take responsibility in shaping artefacts, values and social norms in order to influence the creation of more sustainable and ethical design solutions. We have a deep understanding of how our disciplines affect people, the planet, community, government and industry in both local and global contexts. Working with principles of mutual respect, we embrace design as a change agent to challenge existing perceptions, practices and approaches to the world's wicked problems. Addressing areas of environmental change, education, health, resilient communities, food security, advanced manufacturing and sustainable urban development, our research and teaching form a nexus that allows us to actively and effectively engage in solving problems associated with Australian national priorities and UN Sustainable Development Goals. We follow a transdisciplinary approach to teaching and research in which all areas intersect and collaborate. In turn, we facilitate and thrive on building connections across diverse discipline areas such as health, business, education, biomedical science and engineering.

GM: The Ullman School of Design, whose heritage can be traced back to 1869 in Cincinnati, Ohio, was one of the first university-based design schools to open its doors in the United States. Its predecessor, The McMicken School of Design, which was established with the mandate to drive innovation in industry, set the foundations for a design program whose idea still lives on even today. This quintessential American design school was established with the sole purpose of advancing industry by using design as a tool for creating a competitive advantage. Since the early days, the school's unique curriculum and its capabilities have been co-created together with significant industry support. This kind of partnership between the university and the industry continues even today. Currently, the Ullman School of Design offers three core study programs: Industrial Design, Communication Design, and Fashion Design, and one specialised study pathway in Transportation Design. Most of the work produced at the school sits on the intersection of these disciplines – ranging from social innovation and healthcare design to the future of mobility and wearable technologies (Muratovski, 2019).

One of the most distinctive features of the school is its Bachelor of Science in Design degree, a five-year program (compared to other design schools where the undergraduate degree takes three to four years to complete). This is because, in order for the design students to graduate, they must complete five industry placements at up to five different companies during their studies. This model, which was developed by the University of Cincinnati in 1906, is the oldest, and one of the largest, cooperative (co-op) education programs in the world.

Unlike some other design schools, this school seldom uses advertising, promotional materials or other marketing strategies to highlight its profile. In a typical American Midwestern fashion, the school quietly relies on the strength of its graduates and their reputation in order to maintain its own. Its graduates have helped shape the American professional design culture by creating some of the most iconic American designs, and their fellow graduates continue in their footsteps by holding design leadership positions at some of the most influential American design firms and corporations.

17.3 LEADERSHIP STYLE

17.3.1 What Is Your Leadership Style and Philosophy?

LS: The best way to describe my approach would be one of authentic leadership; valuing staff input and creating open and honest relationships that are built on an ethical foundation (Gardner et al., 2011). Facing issues with an optimistic and positive outlook, I promote openness and inclusivity – attempting to give everyone a voice in addressing the school's challenges. I am passionate about design education and in developing people into being the best version of themselves. To me, creating solid and trusting teams are the key to a successful organisation. "Cooperation … does not simply descend out of the blue," explains Coyle (2018:113),

> it is a group muscle that is built according to a specific pattern of repeated interaction, and that pattern is always the same; a circle of people engaged in the risky, occasionally painful, ultimately rewarding process of being vulnerable together.

In order to connect with my staff, I often share my personal experiences in academia and life. I am honest about both my successes and my failures and believe that, in many instances, you learn more from failing than from succeeding. The greatest successes in life are often founded on the lessons learnt from our failures.

I cannot discuss my leadership style without addressing the role of gender in leadership roles in academia as this has played a large part in how I approach and shape my career. Historically, women have had a lower representation in upper management roles in academia. Findings from the American Council on Education's (ACE) 2017 report on Institutional Leadership show that women outnumber men in the attainment of higher degrees. However, women do not hold Associate Professor or full Professor positions at the same rate as their male peers (Johnson, 2017). Women are therefore less likely to ascend to higher leadership roles, and "Women of all races and ethnicities are likely to hold lower ranking faculty positions" (Johnson 2017, p. 6). This trend is changing, however, with programs such as the Athena SWAN (Scientific Women's Academic Network) Initiative, which promotes female leadership in academia, specifically in under-represented areas.

Athena SWAN was established by the UK Equality Challenge Unit in 2005 to celebrate, recognise and establish women as leaders in advancing academic women in science, engineering, mathematics and medicine. The main principles of the Athena SWAN chapter are addressing gender inequalities, tackling unequal representation, addressing the high loss of women in science and related disciplines, as well as addressing the large amount of women on short-term contracts and the ability for women to complete PhDs and transition into the workforce (SAGE, 2019).

The charter was expanded in May 2015 to include non-STEM departments, including arts, design, humanities, social sciences, business and law. SAGE (Science in Australia Gender Equity) is the administrator of the Athena SWAN initiative in Australia. QUT was awarded a Bronze Member of the SAGE Athena SWAN initiative in 2019 and this is significant in that only 13 of the 43 Universities in Australia have achieved this. As a female leader in the Faculty of Creative Industries at QUT, I am grateful for the forward-thinking policies of my institution and fully support the implementation of these principles in my school.

I myself have had a difficult road on the pathway to a leadership role due to issues with gender bias and therefore aim to assist all staff that I supervise, especially women, in developing their careers and providing support in what is a very stressful journey. As noted by Shahtalebi et al. (2011): "women have special capabilities and possess 'soft skills' in human sources management which are more needed currently than before in organizations. … [T]he presence of women in organizations results in the development of a healthy and moral culture." It is my utmost aim in my role to create an inclusive and supportive culture in the school that builds on the capabilities of all staff. However, considering my gender and position, I especially hope to be an advocate and constant support for female colleagues.

GM: As a Director, I always place the needs of the school first by making sure that everything we do represents who we are as an institution and is aligned with our original mission. In doing so, I strive to provide a holistic engagement system for all stakeholders. Working with academic staff and students to develop a career they are passionate about, helping them achieve their academic goals and supporting them to reach their full potential is what I enjoy about being in the higher education sector. I find this to be a profoundly rewarding experience.

When discussing leadership of a design school, it almost becomes inevitable to raise the topic of "design leadership" as well (see Turner, 2013). This kind of leadership can best be described as an "empathy-first" approach to leadership, which is similar in some ways to the concept of "servant leadership" (Greenleaf and Spears, 2002). As such, design leadership is far less self-absorbed or self-serving compared to other types of leadership tendencies, which often come to the detriment of organisations, their members and their innovative capacities.

Discussing such vague and varied terms as "design" and "leadership" can at times be confusing, because these two terms can mean different things to different people. However, considered from the perspectives of leadership and creativity research, design leadership is far less ambiguous a concept than it might first appear. Whether solving everyday problems of "how to live" in the physical or in the social world, core processes of problem-solving, meaning-making and vision can often be seen at the heart of both leadership and design. Also, they both tend to succeed when carried out with creativity and empathy.

While some people believe that creativity is a "gift" bestowed on select individuals, the reality is that creativity is a universal, fundamentally human characteristic of discovering solutions and finding the extraordinary in the ordinary, everyday world (Thornhill-Miller and Muratovski, 2016). Furthermore, it can be argued that being creative – deciding to think or look at the world differently – is, in some ways, actually a choice (Sternberg, 2003). In this regard, leadership, like creativity, is also a choice. Both involve putting oneself forward and taking the risk and responsibility of making something happen in our socially shared space. While leadership or creativity comes naturally for some, for others, creativity, leadership and even empathy are things that can be taught and learnt (see Thornhill-Miller and Muratovski, 2016).

Thus, it is this sort of strategy-based and increasingly constructed and coached leadership that might justify the label "leadership by design."

17.3.2 What Leadership Models Do You Subscribe To?

LS: The leadership style often associated with effective female leaders is "transformational" and is a style that best represents how I lead in my organisation. Transformational leadership as defined by MacGregor Burns (1978) is "the ability to inspire and guide the workforce through change and transformation, it also emphasizes the achievement of goals by gaining employees' trust and respect." Transformational leadership, as defined by Rosener (1990) involves "getting subordinates to transform their own self-interest into the interest of the group through concern for a broader goal." Moreover, I believe that the most important aspect of an educational institution is its people and my role as a leader is to try and develop and support the people in my organisation. "Group performance," explains Coyle (2018, p. 15), "depends on behaviour that communicates one powerful overarching idea: *We are safe and connected*."

The most effective women leaders, according to Rosener (1990), "encourage participation, share power and information, enhance other people's self-worth, and get others excited about their work." When I started my current role as Head of School in September of 2018, I set out to establish a culture in the school that would eschew the values of community, collaborative culture, respect and caring. Prior to my arrival, the QUT School of Design had a lengthy period of shifting leadership and so my first prerogative was to change the perception of management in the school from one of being transient and secretive to one of being open, collaborative and supportive. One way that I set out to achieve this was to connect personally with my staff by meeting with them individually, to map out their careers, try to understand their challenges and to work with them to find the best solutions. I also continually run co-design workshops to enable us to work as a group on the main institutional challenges that are identified in individual meetings and on recent staff surveys. Through co-design workshops we are able to work through our issues as a group and find solutions to our greatest challenges together.

GM: In terms of staff management and leadership, I prefer applying a collaborative leadership model and I aspire to establish a flat management structure whenever appropriate. Flatter organisational structures allow decisions to be made where the work happens and by those with expertise in the specific technology, techniques, challenges and skills in the specific area. This model allows me to focus more on strategic issues and supporting the staff. This makes the organisation nimbler and more competitive. I find this model to work very well when it comes to building institutional trust and a strong internal culture.

For example, at the beginning of my current role, I encountered a situation where the Academic Program Coordinators were given very limited authority, and the decision process in relation to their programs was very much top-down. The role of the Coordinators had been narrowed down to performing simple administrative tasks, primarily revolving around the facilitation of the teaching schedules and addressing student queries. I mentored and empowered the Coordinators to see their roles as leadership roles. I have also given them an operational framework and a range of decision-making authorities within that framework. In turn, this has enabled me to develop a supportive management team with a strong sense of trust in my leadership. Together we were successful in facilitating much-needed change-management exercises, such as the internal restructuring of each of the programs, as well as an upgrade of the entire school curriculum. By delegating some of the routine day-to-day operations to the Coordinators, I was able to focus on building new strategic capabilities, fundraising and developing new relationships with industry.

However, when necessary, I can resort to other leadership and management models to manage a variety of situations and circumstances. For example, I have completed an Executive Education program on Crisis Leadership in Higher Education at the Harvard Kennedy School of Government, in partnership with the Harvard Graduate School of Education. This program, which was developed for senior university leaders, gave me critical information about proactive training and planning so that I could implement effective crisis management systems, protocols and policies in advance of critical events that may affect a university. This leadership training was particularly useful to me during the crisis related to the COVD-19 pandemic.

17.4 HOW DO YOU CONNECT WITH YOUR STAFF?

LS: The School of Design is home to seven discipline areas, a design research lab and a full design and fabrication workshop. Facilitating teaching and research in a school of this size means that we have a large staffing profile. Our staff includes over 70 on-going academic staff and up to 200 sessional staff members depending on staffing needs during the semester. In addition, we have a team of professional staff that support our students, research lab, timetabling and budgeting, as well as workshop support and management. Our school is split across two large campuses approximately five kilometres apart. Due to the number of staff and physical distance it is not always possible to connect with colleagues in person. Therefore, to connect with my staff in an authentic way on a regular basis, I created a monthly newsletter to discuss the challenges we face in higher education. In this publication, which I have designed with a layout and identity in the style of a high-quality special interest magazine, I write personal anecdotes on how I have overcome barriers and/or navigated difficult situations. I also invite our School Management team, as well as all staff members in the "voices" section, to write about their journeys. To date, we have discussed topics such as balance, trust, community, mentoring, giving back to the community, continual improvement and empathy. This newsletter has become popular not only in our school but also in the Faculty in general and I have had beautiful comments from staff on how much they appreciate the discussion. One such comment reflects the ability of the newsletter to connect with staff:

> *Your [newsletter] article comes at a particularly difficult time in my life and has made me reflect more on how I look at every day and the future. I now feel inspired from*

reading all the articles and know that I will find the purpose that has been lacking in my life.

One element of academia that is often cited as a challenge is creating an effective work–life balance. This is something that I struggle with myself and it is my aim to create opportunities for staff to engage with and understand better ways to approach this. I have instigated training in Mental Health awareness for all staff and established a Health and Wellness committee to address issues with lifestyle and healthy balance. To date, our Health and Wellness committee has established a series of regular staff luncheons themed around "mind, body, spirit" in which we meet as a group to discuss food, exercise, meditation and other techniques that can help us to establish a better balance in our lives. We also host an annual picnic to celebrate our successes in a natural setting and unwind as a group before we break for Christmas holidays.

Along with wellbeing initiatives as cited above, I believe that creating stability in the workforce is essential to wellbeing. Since taking on my current role, I have made workforce planning and stability a priority. When I started in my role, we had a large number of short-term contract and sessional staff that had been employed on a continual basis. I have worked to create more sustainable and long-term solutions, such as opening on-going positions in all discipline areas. Providing clear pathways for continuing employment enables staff to invest in the school and have a sense of belonging, and I am more able to help them shape their career and continually improve their outcomes. I have also worked to streamline processes and create clearer guidelines around sessional appointments; making what is often viewed as "casual" employment at the university into an EOI process that allows candidates to showcase their industry experience, quality teaching and research outcomes, as well as providing a clearer set of expectations and understanding of current roles in the school. I have also provided additional training support and opportunities to be on school committees; something that has previously been inaccessible to staff on sessional contracts. Sessional staff also frequently contribute to the "voices" section of our newsletter. This has enabled sessional staff to further develop their skills and feel as though they are a part of our educational community.

GM: I believe in investing in people. When I come across talented staff, regardless of status and rank, my objective is to help them be the best they can be. I do believe that an institution is only as strong as its people. As an example, all of the strategic facilities that I have initiated and secured funding for were created around people who have shown potential to lead and inspire others in their respective areas.

Over the last four years in this role, I have developed an eco-system of a dozen interdisciplinary research facilities, ranging from a Futures Foresight Lab to a Color Management Lab. Most of these labs are embedded within our new Fashion Technology Centre, a Makers Hub and Future Mobility Centre. In line with this, I have developed and launched a Design Museum where we celebrate iconic works produced by our alumni. The design collection of the Museum is broad, diverse and includes items such as the first Apple mouse, the NASA brand style guide, the Nike Flyknit shoe that was introduced at the 2012 London Olympics, Michael Graves' tea set collection for Alessi and some of his other works, as well as the original Star Wars toys from 1977 (Muratovski, 2019). The museum, in particular, has become a major source of inspiration and pride for the staff and students and is a must-see for all our visitors. But most importantly, it tells a story of who we are. By finding a common thread that brings us all together, and by sharing a common narrative, I connect with my staff.

17.5 HOW DO YOU FOSTER DIVERSE AND INCLUSIVE ENVIRONMENTS?

LS: Diversity is not only about including people of different backgrounds in the current culture of the school; it is about evolving and shaping a new culture together. In my previous role at the University of Canberra, I set up a Cross-Cultural Design Lab to lead and shape a dialogue on the importance of cross-cultural collaboration in Design. By educating others and, in exchange, being educated in the meaning of images, symbols and behaviours of different cultures through the design process, I believe we are able to create more effective design solutions. Similarly, I feel that a truly diverse staff and student profile will allow us to create a forward-thinking culture that will provide new and innovative solutions to our greatest challenges.

As an institution, QUT is committed to increasing partnerships and research with Aboriginal and Torres Strait Islander communities.

> *We will live up to these responsibilities by being guided by Indigenous Australians in core policy and by embedding Indigenous knowledges and perspectives into curriculum, by instituting strategies of tailored student recruitment and support, and through staff employment and professional development* (QUT BluePrint6 2019).

Indigenous Australians have been cited as the world's oldest culture and, as such, they have made and continue to make an immense contribution to the Creative Industries. Our aim as a Faculty and school is to foster a deeper understanding of Indigenous Australians; to this end, we have a current focus on the recruitment and retention of Indigenous academic staff and students. We are also working to establish a leading national scholarship program and a fellowship program that would see a higher representation of Indigenous Australians in the School of Design.

The School of Design is currently exploring avenues to further increase Indigenous community participation and other under-represented community groups in design education. We are particularly committed to providing all students the opportunity to engage with the unique perspectives of the Turrabal and Yugara peoples on whose land our Brisbane campuses are located. One of our Impact Labs explores the concept of "place" and we are looking to partner with Indigenous Australian elders to co-design and create new experiences of understanding the land and its history. Through this process, we hope to further educate students and provide education that gives direct exposure to First Nations perspectives in Australia "by working with Aboriginal and Torres Strait Islander people QUT plans to address challenges not only faced by Indigenous Australians but all Australians" (QUT BluePrint6 2019).

GM: While living in Australia and New Zealand, I had the opportunity to work at universities with strong policies and protocols for serving diverse societies, such as Indigenous Australian and Torres Strait Islander people, Maori and Pasifika communities and other diverse communities. Having the opportunity to engage and work with people from many different cultures has been a truly rewarding experience. This has also made me very conscious of the need for having a vibrant and inclusive culture in the university.

When I arrived in the USA, I started administering, developing and implementing a range of equity, diversity and inclusion policies and programs. I followed best practices in the field in terms of hiring, stakeholder engagement and community outreach. But most importantly, we started implementing changes in the curriculum so that diverse communities were better represented in the content. As a result, in 2017, *Insight into Diversity* – the oldest magazine devoted to diversity and inclusion in higher education –recognised the Ullman School of Design at the University of Cincinnati as a "Diversity Champion," when compared to other design schools in the country. *Metropolis*, the leading industry publication in the field of design and architecture, confirmed this further by calling the school a "game changer," and by ranking the University of Cincinnati as having the #1 design school in the USA when it comes to implementing equity and diversity initiatives – with the Harvard Graduate School of Design being #2 (Von Koenig, 2017).

Following this, I have also launched several other equity and diversity initiatives. For instance, I worked with Procter and Gamble (P&G) on securing $500,000 (USD) of financial aid in the form of scholarships to support prospective design students from under-privileged backgrounds and diverse communities. In addition to the financial support, designers and executives from P&G will also provide on-going mentorship and possible career pathways in the company for these students.

17.6 HOW DO YOU COMMUNICATE AND FACILITATE CHANGE IN YOUR SCHOOL?

LS: "*Example is not the main thing in influencing others; it is the only thing.*" – Albert Schweitzer (in Lipman, 2012). When dealing with change, I believe that one of the most effective ways is to lead by example. According to Lipman (2012), "it's difficult to resent managers who roll up their sleeves and wade into the trenches when they need to, and who share the same sacrifices their teams do." One of the first tasks that I had to undertake in my current role as Head of School at QUT was to communicate the overwhelmingly negative results from the 2018 Staff Survey. The Staff Survey is conducted every three years and had been administered just prior to my arrival. The results for the School of Design reflected issues with facilities, school and executive management, communication, feeling valued, change management and work–life balance. As staff communicated to me that their most immediate concern was about issues with facilities, I set out a plan to work together with staff and Facilities Management (FM) to understand and engage with the University's Estate Master Plan (created in 2018). We set up a space workshop and co-design session for the entire school and key partners to try and create a more functional and collaborative approach to forward planning and resourcing for the spaces available to the school. This set the ball rolling for an on-going consultation with FM and key parties in the university. In addition, the creation of a staff-led committee on better usage of space allowed me to better understand needs and pain points experienced by staff and students. Ultimately, this process enabled me to advocate for the school and to establish the best immediate use of our allocated spaces. After a year of meetings and negotiations with our Facilities Management team, I was finally able to secure funding and property for a new facility for the school, which is in the early stages of planning.

To make our current facilities function better also required the consolidation of office space where appropriate. This is a touchy subject, as staff feel a level of ownership over their allocated spaces, particularly those who have been in the spaces for a long time. I therefore sectioned my own office into a shared office so that staff would see the benefit of sharing resources. To address storage issues, I undertook a working-bee to clean up studio spaces and store rooms that had become overrun with years of discarded student projects. During the working-bee I laboured together with staff, cleaning rooms, moving projects and packing down dumpsters. Over lunch, staff mentioned that this was the first time a Head of School had engaged in a studio clean-up process.

Another difficult change for the school was a shift in teaching and learning practices. In 2017, The School of Design at QUT underwent a curriculum renewal which saw a complete transformation of previously siloed undergraduate programs in Architecture, Interior Architecture, Landscape Architecture, Fashion, Interaction Design, Visual Communication Design and Industrial Design transform into a single Bachelor of Design course with a core subject offering of transdisciplinary "Impact Labs" running throughout the program. This was a difficult transition for many staff members, especially as some of the units in the old structure had been running for many years and the new structure required cutting out some of these units to enable the Impact Labs to become the core of the new course. I felt that the newly established Impact Labs, being the spine of our new curriculum, presented something that I needed to champion, and so I have assisted directly in the teaching and developing of these units.

In Semester 2, 2019, I taught into and developed content for the "Impact Lab 2: People" unit. In this unit, students work with a range of non-profit organisations in the community that are dedicated to social needs, to create new design solutions to the issues presented. Impact Lab 2 is a required subject for all first-year design students and we had an intake of over 700 students in 2019. Teaching into this unit not only allowed me to understand the needs of on-going, contract and sessional staff, it also enabled me to connect with our current first-year student cohort directly. Students work in groups to create both a digital and oral presentation to the client and are encouraged to use this project to start building their professional identity. As a Visual Communications academic and social designer, my contribution to the unit was largely around helping students to understand how to present their work in a clear, appropriate and visually appealing way, as well as guiding them in the process of engaging with clients and social design concepts. Teaching into this unit resulted in first-year students connecting with their Head of School, providing them with an understanding of the direction and vision of the school as it directly related to the content of this unit. It also gave an indication to staff in the school that these units are essential to the success of the new curriculum. The outcomes of this unit were significant, and it was nominated for and ultimately selected (from all of the programs across QUT) for a 2019 Vice Chancellor's Excellence Award for quality teaching and engagement.

GM: Change is inevitable in any organisation. As a Director, I drive change based on the information and insights that I have. This, however, is never an easy process. It is natural for people to be upset over changes if they do not understand why they are necessary. That is why I approach change management in a transparent way. I share the insights that I have with my team, provide a rationale for why change is necessary, propose a general plan for moving forward and then initiate a robust conversation covering all aspects of the plan. Change management is a process that should never be taken lightly, as things can easily spiral out of control. Therefore, a strong guiding principle behind the change is necessary. There always needs to be a grand overarching idea of why the institution needs to change.

On the leadership side, for example, there can be populist leaders following the polls or pursuing short-term opportunities wherever they serve their immediate interest, often at the expense of ideals and in the absence of vision, but perhaps doing so in the name of "democratization." However, this strategy can only work for a short period of time because the leader's resources are always limited, while the general demands are not. A more forward-looking leader, on the other hand, might deal with the pragmatic realities of everyday situations, while still acting with some increased vision for the long-term or greater good. Finally, visionary leaders may choose to remain detached from the present situation and focus instead on imaging and shaping a new collective future, although the pressing matters of the day may often work against them.

To pursue this analogy further, a leader can succeed to some extent in the world of design (or, for that matter, in a design school) in three parallel ways: by (1) giving people exactly what they say they want; by (2) giving people something that serves the higher, broadened understanding of their own needs; or by (3) developing solutions that change, enlarge or revolutionise, perhaps in some culture-changing way, the people's sense of what they want or what they need (Thornhill-Miller and Muratovski, 2016).

In the context of a school, the stakeholders that a school leader needs to simultaneously consider include students, staff, community, industry partners and the university. The difficulty, however, is that each of these stakeholders will have a very different understanding of what "success" means to them. Finding a common thread that will bring everything together under one idea and vision is not an easy task. Also, it needs to be taken into account that there is a constant stream of shifting priorities. That is why design schools should be in a constant process of change. And this is not necessarily a bad thing. Change should not be a last resort when something is clearly no longer working or no longer necessary. Change should be seen as a natural process in the pursuit of excellence, quality and distinction. This is important because design is not a fixed construct. Design, in many ways, is a reflection of the world around us in terms of culture, society, economy, environment, technology, industry. As the world around us changes, so does the field of design.

Therefore, a good design school should follow these changes in order to remain relevant. A great design school, however, should inspire, instigate and lead these changes. This is why, as a School Director, I am not content with maintaining a status quo; I constantly ask myself and my colleagues questions such as "why are we doing this," "what can we do better" and "what kind of impact do we want to have?" In this sense, I approach my job from the perspective of design leadership. Essentially, what I am trying to do is to design a School of Design, and my approach is no different to that of a service designer working on a large, complex and multi-faceted problem.

Some of the major change-management initiatives that I have led in my current role include a complete school-wide revision of the curricula in order to create a more student-centric approach, streamline our operations and reduce unnecessary overhead expenses. When I started my role, I encountered a school that was in debt and had a curriculum that was disconnected and far too intense for both the students and staff. I quickly realised that both of these things were related – the curriculum was overloaded with unnecessary courses that increased the number of credit hours far above the requirements for our degree.

The problem with over-delivering content is not only that the students are pushed to their limits in terms of time commitment and overassessments (which adds to stress and anxiety), but also the workload of the staff is stretched beyond what is necessary (which also adds to stress and anxiety, and negatively affects the budget). All of the classes that cannot be covered by full-time staff need to be covered by a large number of tutors, or by paying overload rates to the permanent staff because they are asked to teach above their teaching limit. This in turn unnecessarily drains the non-permanent compensation budget (non-salaried expenses) while, at the same time, the overall quality of the content delivery tends to suffer.

In principle, student tuition is calculated to cover the expenses for an undergraduate program that will deliver 120 credit hours (which is the standard in the USA). Every credit hour taught above this comes at the expense of the school – which is something that was never explicitly explained to the staff when they were making the last curriculum revisions more than a decade ago. Every single program in the school was delivering between 10–18 credit hours more than the prescribed amount. Over time, the cost of doing so started to add up and the school started to incur debt, even though student enrolments were strong and always as projected. This meant that other funding streams needed to be redirected to cover the debt, such as cuts in travel budgets, research funding, staff development, sabbatical provisions, teaching assistants, etc.

By acting transparently and discussing in detail the budget model of the school with the staff, I managed to establish a better sense of understanding why this change is needed, and how they will benefit from this in the long turn. This was not an easy thing to do, given that academics tend to have strong feelings of ownership about courses that they have been teaching for years, but it was a necessary step towards ensuring that the school is future-proofed, solvent and relevant. In order to relieve the school from its debt faster, I have introduced more focused fundraising and industry-engagement strategies that have helped us address immediate needs to invest in facilities, new technology and equipment, as these things were neglected for years due to the budget deficit.

Now, after implementing a range of new strategies over the last three years, the school is completely out of debt, well equipped and all of its discretionary budgets are in surplus. Every decision that was made during this process was examined through the lenses of whether this represents "who we are" as a school, by looking at the core ideals of our founding mission. Every decision that we made was supposed to take us a step closer to our ideal state of who we were always meant to be. This was the overarching grand idea that helped us drive the change forward.

17.7 HOW DO YOU LEAD IN A CRISIS?

LS: COVID-19 has had a global impact on higher education and this impact will be on-going. At QUT, we foresaw the crisis impacting our school as early as February 2020 with the realisation that our Chinese students would be unable to return to study in Australia due to an Australian travel ban imposed on China. Although we had international students from China dispersed throughout our programs, the largest number of international students from China were in our Architecture and Master of Architecture programs. This caused us as a school to immediately come up with a plan for online delivery of units that were impacted. As the scope and scale of the virus escalated, we soon realised that we needed to transition this plan to encompass the entire school. We made an immediate decision to plan for all units to have an online delivery mode and to transition staff to a remote working scenario.

An official university-wide decision was made to enact a teaching pause and transition to online in our fourth week of Semester 1. We had already anticipated this and therefore staff felt supported because a process was already in place. Comments during this period, addressed to me personally, expressed that my method of providing emails that clearly informed all staff of the process and what we were facing helped to make this a manageable transition. In all of my communication to staff I tried to keep the message positive, while making sure to emphasise that this was an unprecedented and stressful situation but that we would be able to do this if we worked together and stayed calm.

During a crisis there is a need for clear and direct communication. Throughout the crisis, my weekly emails proceeded the Vice Chancellor's messages, gave clear direction and translated her statements in relation to the school. In these communications, I made sure to always acknowledge the difficulty of what we were facing and how grateful I was to the staff and their ability to work with their peers and students during this period. If I wasn't sure about something, I also made sure to address this and to note that I would explore it and bring the information back to the school as soon as I had it. The emails were also accompanied by fortnightly all-school video conferences on a Friday morning. These half-hour sessions were a means for the leadership team to address the weekly outcomes and processes going forward and enabled staff to have a virtual "face-to-face" and Q&A with school leaders.

It is also essential to ensure that staff feel connected and supported during a crisis. Staff meetings that included the entire school, as well encouraging regular meetings within individual study areas, were essential. As the online and work-from-home transition evolved, staff began to face video conference fatigue. Therefore, sticking to a half hour in the all-school meetings and giving staff a clear communication hierarchy (communicated by an all-school email) were essential. Staff have reached out to me to stay that they appreciate these biweekly "face-to-face" catch ups as it makes them feel connected during a time of social isolation and that the communication mapping helped them to understand how to get in touch. I worked closely with my school management team to create an integrated mentoring and support program that we have now rolled out across the school. This program includes both formal and casual mentoring in teaching, research and leadership through peer-to-peer, small-group and all-school options. It was important to acknowledge the amazing teamwork that my staff showed under pressure and their ability to adapt to a rapid change in the way they performed their role. To do this, I personally designed and sent out "Thank You" cards to each staff member in the school, as well as support staff in the faculty and our QUT teaching and learning units, with a handwritten note expressing my gratitude for all of their efforts during this semester.

As the crisis and impending economic fall-out is still evolving, it has been crucial to keep in touch and be on the same page with both senior management in my own institution as well as others around the world. It has been a great support to be in touch throughout this crisis with Gjoko as we were able to compare approaches and work through challenges together, as well as to bounce ideas and solutions off each other. I have also benefitted from a great working relationship with the other Heads of Schools in my faculty, as well as the Heads of Schools of institutions in Australia and New Zealand. These connections have enabled me to feel supported and connected to a broader global network – thus bolstering us with the confidence to go forward; to understand that we are all in this together and to know that we will get through this.

GM: Universities are constantly in some kind of state of crisis – for one reason or another. The main difference with the COVID-19 pandemic was that this was a large-scale crisis that affected the entire university sector in the same way, all at once. At the Ullman School of Design, however, we were not severely affected. Like everyone else, we also had to move to remote teaching at the drop of a hat, and everyone had to forcefully adapt to the new reality, but we managed to do that surprisingly fast. I believe that the collective spirit prevailed, and after the initial shock, everyone stepped up to the challenge.

Fortunately for us, we have just completed the school curriculum changes, alongside other strategies to streamline our operations and reduce our overhead expenses. Under my leadership, we also relieved the school of its financial debt and all our budgets were in surplus. This made us well prepared to weather out the "storm," with minimal changes to our original plan of operation. In many ways, we were better prepared than the university as a whole, and, at times, it was the inherent problems of the university that were weighing on us, rather than any of the internal challenges that we had to address in the school. For example, despite the fact that we managed not to lose any student enrolments and managed to put in place an emergency operations plan that further reduced our operational expenses, we still had to use our budget reserves (made up entirely of external funding generated by the school itself) to help pay for the financial shortfall of the central university administration.

Otherwise, from my experience, decisive leadership works best in a state of crisis. In uncertain times, people are looking for guidance and direction. The problem is that not all leaders are well prepared to handle crisis. Some become overwhelmed and lose control of the situation. Others hope that things will go away or somehow fix themselves. I tend to start putting in place contingency plans at the earliest indications that a crisis might be upon us. In this particular case, discussing the effects of COVID-19 with Lisa early on gave me some tactical advantage. Because of the differences on when semesters start, and the speed of government responses to the crisis, universities in Australia were affected slightly before their American counterparts. By observing how the situation was unfolding in Australia, I started anticipating some of the similar challenges that we too may start facing. My prior training in Crisis Leadership in Higher Education also helped me to have a better grasp of the situation. While some of the other leaders in my university were hoping for the best-case scenario and were delaying their emergency responses until the situation presented itself better, I started planning for the worst-case scenarios immediately. I gathered my team and we started analysing various "What if?" scenarios and discussing strategies and tools that we will need to put in place in each case. Of course, we were hoping for the best as well, but, unfortunately, we had to activate all our scenarios as the situation started unfolding. At least we knew what to expect and we had a plan. Many of our pre-emptive proposals that were at first deemed drastic and unnecessary, were quickly adopted by the university soon after. These recommendations included proposals for us to shift the grading system to Pass/Fail, reduce the assignment expectations and prepare all courses for online delivery (not just some), as well as ordering loan laptops for academic staff, adjunct instructors and students that may not have access to computers.

In addition to this, we even managed to provide service to our community and local hospitals. The state-of-the-art technology facilities that we just launched the year before became critical assets for developing Personal Protective Equipment (PPE). For example, our Fashion Technology Center, which had industry-grade equipment, was converted into a micro-factory with the capability of producing 5,000 face masks a week, as well as other medical garments. A number of our academic stuff and instructors volunteered and worked around the clock to provide PPE supplies for those in need.

While we are still in a COVID-19 crisis mode while this chapter is being finalised, I am confident that all of the preparations we have put in place now will make the school stronger and that we will emerge from this crisis better prepared for anything else that may come our way in the future. And I am thankful to Lisa for her support and for sharing her experience with me as things were unfolding.

17.8 CONCLUSIONS

LS: Although we are leading design schools on different continents, Gjoko and I have many similar challenges. Through our conversations over the past year we have discovered that, in many ways, we have taken similar approaches to these challenges. We share a commitment to a collaborative leadership style and the need for having a vibrant and inclusive culture. I agree with Gjoko that empathy is key to being an effective leader and that creativity, leadership and even empathy are not skills that one is always naturally endowed with; they are skills that can be taught and learnt. Mentoring and supporting staff to become the best that they can be is core to our common approach.

If you asked me what one of the most challenging aspects of my role was at the moment, I would say communicating change. Gjoko aptly stated that change is a natural process and "*design is not a fixed construct.*" The Higher Education sector will continue to shift and change along with the expectations of staff and students; keeping a balanced and grounded perspective is what is most necessary in dealing with this constant flux. In both my work and my personal life, I continue to embrace change and to cope with the complexities this brings. In the words of the ancient philosopher Lao-Tzu (2011): "In dwelling, live close to the ground. In thinking, keep to the simple. In conflict, be fair and generous. In governing, don't try to control. In work, do what you enjoy. In family life, be completely present." Keeping balanced in life, particularly when having a demanding role such as Head of School, is not always easily achieved, but I will continue to strive for this outcome.

Leading a design school is a moving feast of opportunities and challenges. During my time as Head of School at QUT I have seen a vast amount of change in a relatively short period. Throughout these changes, I see my role as providing the tools and guidance to create an effective road-map; working with staff to find new routes and enabling them the freedom to forge new paths. I see every roadblock as a learning opportunity and I know that I will never stop learning on this journey.

GM: Design schools can be influenced by many factors, and some of them may not be immediately obvious. From local history, geography, economics and politics, to access to natural resources and industry can all have an impact on the strategy that a particular school will need to develop in a particular city or country. This is why design schools sometimes vary markedly from one another in focus and scope. This is what makes our field so exciting. It is also what makes the leadership of a design school so challenging. Even though every place has its unique set of challenges, problems and opportunities, as Lisa points out, the work of the school leader seems to be similar across the board. A constant barrage of conflicting expectations by a range of stakeholders is the norm. The ability to make critical decisions, often under tough circumstances, is critical. That is why a great deal of stress resistance is necessary when holding such a position. This is something that a person can build over time, which is why it is important to reach this kind of appointment by regular progression through the ranks. Otherwise, such a role can feel far too overwhelming at times. There are many other potentially helpful or interesting connections that one could make concerning the various topics and approaches briefly mentioned here.

Many of these, however, would also only serve to reinforce the idea that leadership of a school not only needs creativity and strategic acumen, but empathy and resilience as well.

REFERENCES

Coyle, D. (2018). *The Culture Code: The Secrets of Highly Successful Groups*. Random House Business Books: New York, NY.

Freyne, C. (2010). The school of the arts movement. Available at: https://dictionaryofsydney.org/entry/the_school_of_arts_movement [Accessed 11 October 2019].

Gardner, W.L., Cogliser, C.C., Davis, K.M., & Dickens, M.P. (2011). Authentic leadership: A review of the literature and research agenda. *Leadership Quarterly*, *22*, 1120–1145.

Greenleaf, R.K., & Spears, L.C. (2002). *Servant Leadership: A Journey into the Nature of Legitimate Power and Greatness*. Mahwah, NJ: Paulist Press.

Johnson, H. (2017). *Pipelines, Pathways and Institutional Leadership: An Update on the Status of Women in Higher Education*. American Council on Higher Education. Available at: www.acenet.edu/Documents/HES-Pipelines-Pathways-and-Institutional-Leadership-2017.pdf [Accessed 11 October 2019].

Juntrasook, A. (2014). You do not have to be a boss to be a leader: Contested meanings of leadership in higher education. *Higher Education Research & Development*, *33*, 1, 19–31.

Kibort, P. (2002). *The Written Word: Literature and Leadership*, *76*, 10–35.

Lao-Tzu. (2011). *Tao Te Ching: The Book of the Way (New Edition)*. Octopus Publishing Group.

Lipman, V. (2012). The best managers always lead by example. *Forbes*. Available at: www.forbes.com/sites/victorlipman/2016/02/04/the-best-managers-always-lead-by-example/#4e0e331f279d [Accessed 11 October 2019].

Macgregor Burns, J. (1978). *Leadership*. Harper Collins.

Muratovski, G. (2019). *The Making of an American Design School*. Cincinnati: The Myron E. Ullman, Jr. School of Design.

QUT BluePrint. (2019). *BluePrint*. Available at: www2.qut.edu.au/jobs/working-at-qut/quts-vision-and-goals.jsp [Accessed 25 September 2019].

QUT History. (2019). Available at: www.qut.edu.au/about/our-university/history [Accessed 25 September 2019].

QUT History. (n.d.) Available at: https://www.qut.edu.au/about/our-university/history [Accessed 7 June 2019].

QS World Rankings. (2019). Available at: https://www.topuniversities.com/university-rankings/world-university-rankings/2019 [Accessed 6 November 2019]

Rosener, J. (1990). Ways women lead. *Harvard Business Review*. Available at: https://hbr.org/1990/11/ways-women-lead [Accessed 25 September 2019].

SAGE (Science and Gender Equity). (2019). The athena SWAN accreditation framework. Available at: www.sciencegenderequity.org.au [Accessed 11 October 2019].

Scott, G., Coates, H., & Anderson, M. (2008). *Learning Leaders in Times of Change: Academic Leadership Capabilities for Australian Higher Education*. Available at: https://research.acer.edu.au/cgi/viewcontent.cgi?article=1001&context=higher_education [Accessed 15 September 2019].

Shahtalebi, S., Yarmohammadian, M.H., & Ajami, S. (2011). Women's success factors from leadership in higher education. *Procedia-Social and Behavioral Sciences*, *15*, 3644–3647.

Sternberg, R. (2003). President's column--Creativity is a decision. American Psychological Association. Available at: www.apa.org/monitor/nov03/pc.aspx [Accessed 13 April 2015].

Thornhill-Miller, B.J., & Muratovski, G. (2016). Design leadership and leadership by design. *Journal of Design, Business & Society*, 2(1): 3-7.

Turner, R. (2013). *Design Leadership: Securing the Strategic Value of Design (New Edition)*. Burlington: Gower Publishing Company.

Von Koenig, G. (2017). Diversity champions: 8 schools that aren't just paying lip service to diversity. *Metropolis*. Available at: www.metropolismag.com/design/design-education/diversity-champions-8-schools-that-arent-just-paying-lip-service-to-diversity/ [Accessed 7 June 2020].

Index